I0483703

LASTING SOLUTION TO OIL-AND-FOOD CRISES

By

Dr. Steve N. ESOMBA

CONTENTS:

Chapter 3:

RENEWABLE ENERGIES & FOOD PRODUCTION

Chapter 4:

FOOD MANIPULATION, BIOTECHNOLOGISATION

LASTING SOLUTION TOOIL-AND-FOOD CRISES LAST

By Dr. Steve N. Esomba

I. Preface

THE WORLD was full of apocalytic happenings in 2008, the eighth year of the 21st Century! They included *Climate Change, Global Warming,* and skyrocketing *Energy-* and *Commodities' Prices.* These doomsday happenings are predicated on Oil Futures, and affect every echelon of living and non-living things. The weak-and-poor bear the brunt of suffering; the strong-and-wealthy suffer the brunt of losses.

Fortunately, there is, and has always been, an *alternative* to prevent these doomsday phenomena and their attendant hardships from happening. The source of these apocalyptic happenings, we all know and admit, is the production and use of *fossil fuels* or *mineral fuels.* We also know we have the capacity to incrementally or speedily *free* ourselves and this civilisation from the fossil fuels *self-imposed burdens.* Nature has provided us with abundant, inexhaustible sources of energy. This is a great incentive for us to put a permanent end to our self-imposed fossil fuel energy use.

There have been, and always will be, Renewable Energies and their limitless, regenerative sources, to meet all of our energy needs! All we need do is to accept these free gifts of Nature to demonstrate: (i) a *common political will; (ii) economic competence; (iii) conservation competence; (iv) energy efficiency competence; and (v) the necessary technologies' competence!* All these qualities are within our reach and realm of available possibilities.

After I finished polishing my *"Renewable Energies and Technologies: Twenty-first Century's Fuel Sufficiency Roadmap"* manuscript, I decided to take on the spiralling, and unpalatable *tsunami* of *surging oil-energy and food prices* problem. This problem took *visible* form as early as 2006 or even earlier. In writing on this problem, my theory is that these oil-and-food prices occurrencies are unwarranted; that they could be *prevented and permanently* eliminated; that this is achievable through fully engaging the production, use and proliferation of *Renewable Energies, technologies and their sources and resources.* I believe that when this theory is put in practice, it could be a *win-win-undertaking* for humanity in the short- and long-terms.

II.   Introduction: Apocalyptic Happenings

THE OIL-and-Food-Crisis of 2008 has shocked the hell out of us all. That it happened, was predictable. Its happening is encrypted on the energy and monetary policies on which the world's economies run. The following summary from Wikipedia on the 2007-2008 world food prices and energy crisis, is eye-opening:

• 2007-2008 World Prices: *"2007-2008 occurred dramatic world food price rises that brought in a global crisis that caused political and economic instability and social unrest in poor as well as rich nations. Initial causes of late 2006 price hikes were unseasonable droughts in grain-producing countries and rising oil prices. Rising oil prices further increased the costs of fertilizers, food transport, industrial agriculture, increased use of biofuels in developed countries, increasing demand for a more varied diet in meat in middle-class populations in Asia, and falling world food stockpiles. Long-term causes include: changes in agricultural production, agricultural price supports, and subsidies in developed nations, diversion of food commodities to high-input food fuels, commodity market speculation, climate change and Greenhouse gas emissions. Since 2006: world price for rice has risen 21.7%, wheat by 136%, maize by 125%, soy beans by 107%, April 2006 rice prices by 24 cents/pound - twice as it was 7 months earlier; world population growth: As of July 5, 2008, the world's population was estimated t o be 6,708,083,942, and continues to grow at rates not unprecedented before the 20th century at 2.2%/year. World population is expected to reach nearly 9 billion by 2042: Africa 2,308 billion; Asia 5,561 billion; Europe 517 million; Latin America/Caribbean 912 million; North America 398 million; Oceania 51 million. Over the past 20 years, the middle class population in Asia has grown in India 9.7, China 8.6 – at a growth rate of 30%-70%, respectively, and at the same time, the vast majority of Asian population remains rural and poor. .."*

In a 20th May 2008 communication that sets out potential policy responses to mitigate the effects of rising global food prices, the European Commission[1] stated that: after 30 years of falling farm prices, recent months have seen a sudden and steep rise in world agricultural commodity prices. In Europe, prices for wheat and dairy products increased by 90% and 30% respectively, between September 2006 and February 2008. Agricultural prices were beginning to fall

---

[1] European Commission: *"Mitigating the effects of rising global food prices."* 18/07/2008 – Global Food Price Rise: Commission Proposes Special financing facility worth €1 billion to help developing country farmers.   EC Agriculture & Rural Development/Food Prices. To implement its mitigation programme, the EC will (i) Prepare a "Health Check Monitoring" of its *Common Agricultural Policy* (CAP), not so much as to re-invent or re-reform it, but to assess if it is working as well as it would in a larger EU and in a shifting international context. The health check is an effort to streamline and to modernise the CAP since its reform in 2003. Where there's need to adjust, the CAP should be adapted to better meet new challenges, such as climate change, water management and bioenergies; (ii) Change the Food Security Programme for the most deprived persons – the elderly, homeless, disabled, families in difficulty, and asylum seekers – as proposed on 11 September 2008: improve the current food distribution in the Eu by increasing the budget by two-thirds to around €500 million as from 2009, and extending the range of products which can be provided, as well as investigate how the food supply is operated; (iii) Take action to improve global food supply by: making biofuels more sustainable, boosting the Renewable & Development (R+D) in agriculture, staying open to, but vigilant on genetically modified organisms (GMO); (iv) Promote open trade and work towards an early conclusion of the Doha Round trade negotiations or Doha Development Round in the World Trade Organisation (WTO). The Doha Development Round was first held in Doha, Qatar, in November 2001 with the objective to *lower trade barriers* around the world, and allow countries to increase trade globally. [Current Doha talks held 23-29 July 2008 in Switzerland, broke down after failing to reach a compromise on agricultural import rules. Trade barriers that restrict international trade include: Import Duties; Import Licenses; Export/Import Licenses; Import Quotas; Tariffs – imposed tax on goods when they are moved from country to country, also known as Protectionism; Subsidies, Non-tariff Barriers to trade (i.e., those that restrict imports but are not in the usual format of tariffs); Voluntary Export Restraints; Local Content Requirements; Embargo; Services; and Trade Remedies. The Doha Trade negotiations are between the developed countries led by the EU, U.S., and Japan, and the developing nations headed by India, Brazil, China and South Africa. The EU and U.S., also have separate tariff conflicts against each other in the maintenance of agricultural subsidies; (iv) Maintain EU humanitarian commitments and scale-up the EU contribution to the global effort to tackle the effects of the crisis on poor populations.

from last Autumn this year [2008], as milk prices dropped by around 30% and wheat prices fell by almost 20% from March to April. The reason for this?

- Increasing global demand for staple commodities and higher value-added food – especially from emerging economies such as China, Brazil and India.

- High energy prices are also affecting agricultural prices, as they push up the price of inputs such as fertilizers, pesticides and diesel, as well as transport costs.

- The development of alternative market outlets like biofuels has also had an impact. While current EU biofuel production[2] has little impact (involving less than 1% of EU cereal production), that of the proactive policy pursued by the U.S., has had a noticeable impact on the maize market, but even so, it has remained a relatively moderate contributor to high food prices in general.

- Slower growth in food yields in Europe and years of under-investment in agriculture in developing countries. Adverse weather and poor harvests in major producing and exporting countries, like Australia (3 droughts in 6 seasons and a 50% drop in production in 2006), North America and Europe. The weather-related cereal supply shortfall of North America, Europe and Australia in 2006 was more than 60 million tonnes, four times more than the increase in cereal use for the ethanol in those countries. High farm prices look set to stay in the medium-term, though probably lower than now, provided weather conditions permit a return to average harvest.

- The impact of high food prices has been felt differently across the EU member states and social strata within each country. Those with smaller household budgets will feel the greatest impact on their purchasing power. While arable farmers have benefited, livestock producers have been hit by higher feed prices. Worldwide, the implications of rising food prices for developing countries offer in the short- and long-term, (although impacts vary from country to country, and within countries), the following:

- Recent violent protests and food riots in Latin America, Africa and Asia, demonstrate the immediate and dramatic impact on the world's poorest populations. Few dispute that the net welfare effect on the global poor is negative, particularly in the short-term.

---

[2] EU Biofuel Directive 2003: The EU establsihed a Biofuel Directive aimed at replacing 2% of vehicle fuel by 2005, and 5.75% by 2010. The 2005 goal was not met, and despite uncertainty that the EU could even reach the 2010 targets, an ambitious goal of total fuel usage by 2020 was put in place in 2007. Now the EU isn't sure that, it is such a good idea, and recommended the target to be *suspended* based on the concerns raised by the European Environment Agency (EEA) Scientifiic Committee study until a new comprehensive study on the environmental risks and benefits can be completed. The EEA study "EUROPE'S EPA Advises Suspending Biofuel Targets" by Clayton B. Cornell, April 10, 2005, (in: Environment, International Issues – Archive for the 'Environment' Category), recommended as follows: (i) Biofuel production based on first generation technologies does not optimally use biomass resources with regard to fossil energy saving and greenhouse gas reduction. Technologies for direct heat and electricity generation should be preferred because they are more economically competitive and more environmentally effective than biofuel production for vehicles; (ii) Biomass utilisation implies combustion of very valuable and finite resources from our living environment. These resources ought to be preserved wherever possible. Therefore, biomass utilisation must necessarily go hand-in-hand with energy efficiency improvements. This is not yet the case for the majority of applications in the automative and residential sectors; (iii) The EEA has estimated the amount of available arable land for bioenergy production without harming the environment in the EU (EEA Report No. 7/2006). In the view of the EEA Scientific Committee, the land required to meet the 10% target exceeds this available land area even if a considerable contribution by second generation fuel is assumed. The consequences of the intensification of biofuel production are thus increasing pressures on soil, water, and biodiversity; (iv) The 10% target will require large amounts of additional imports of biofuels. The accelerated destruction of rain forests due to increasing biofuel production can already be witnessed in some developing countries. Sustainable production outside Europe is difficult to achieve and to monitor. [The overambitions of 10% biofuel target is an experiment, whose unintended effects are difficult to predict and difficult to control. Therefore, the Scientific Committee recommends suspending the 10% goal; carrying out a new comprehensive scientific study on the environmental risks of biofuel; and setting a new and more moderate long-term target, if sustainability cannot be guaranteed; Prof. Laszlo Somlyody, Chairman; and Prof. Manfred Kleemann, Oscar Romero (Press Officer); oscar.romero@eea.europa.eu; Marion.Hannerup@eea.europa.eu.]

- In the medium- to the long-term, rising prices offer new opportunities for farmers to generate income. Given the incentives, they could enhance the contribution of agriculture to economic growth; in the longer-term, rising prices could help rural commodities in some developing countries out of poverty.

These are tangible, *life-and-death* reasons *why* we need to, and *must abort,* the use of fossil fuels *temporarily* and *permanently*, and restore our energy consumption from renewable, unlimited, inexhaustible natural sources. One *very compelling* reason for doing this is, the *advanced, irreversible* stage into *Climate Change* and *Global Warming* that we find ourselves in. Another *most compelling* reason is, the bad shape in which the *global economy* finds itself. These apocalyptic happenings are accomplice to the *unstoppable, virulent, violent surging* of *oil and food prices.* These horror scenarios have increased, exacerbated and accelerated the *hunger* and *state of poverty* of the *poorest of our world,* and the losses of the well-to-do. The *unhindered acceptance* and *unsullied* use of renewable energies for electricity generation and other industrial applications must happen. It would provide a *permanent solution* to a string and chain-reaction of many or all of our world's apocalyptic happenings which include: *Climate Change; Global Warming; High Energy Prices; High Industrial Production Costs; High Medical and Insurance Costs; Disappearing Biological Diversity and Systems; Poisoning of Waterways by Oil and Chemical Spills; Greehouse Gas Emissions; Nitrogen Oxides, Carbon Dioxide and Particulates' Emissions; Destructive and Disruptive Floods and Storms; Aggressive Desertification,* and *High Cost of Living of Everything.* The Intergovernmental Panel on Climate Change (IPCC) explains how climate change, one of the most devastating apocaylyptic happenings of all times, consolidates itself:

- *"Changes in the atmospheric concentrations of Greenhouse Gases (GHGs) are aerosols... landcover and solar radiation alter the energy balance of the climate system and are drivers of climate change. They affect the absorption, scattering and emission of radiation within the atmosphere and at the Earth's surface. The resulting positive or negative changes in energy balance due to these factors are expressed as radiative forcing,...used to compare warming or cooling influences on global climate. Human activities result in emissions of four long-lived GHGs: carbon dioxide (CO_2), methane (CH4), nitrous oxide (N_2O) and halocarbons – a group of gases containing fluorine, chlorine or bromide. Atmospheric concentrations of GHGs increase when emissions are larger than removal processes. Global atmospheric concentrations of CO_2, CH4 and N_2O have increased markedly as a result of human activities since 1750 and now far exceed pre-industrial values determined from ice cores spanning many thousands of years. The atmospheric concentrations of CO_2 and CH4 in 2005 increased by far the national range over the 650,000 years. Global increase in CO_2 concentrations are due primarily to fossil fuel use, with land use change providing another significant but smaller contribution...very likely that the observed increase in CH4 concentration is predominantly due to agriculture and fossil fuel use...increase in N_2O concentration is primarily due to agriculture. Many halocarbons (incl. hydrocarbons) have increased from near-zero pre-industrial background concentration, primarily due to human activities."*

The very welfare and economy of the world and civilisation are at stake of collapse and disintegration. Our modern economy and advanced technological status quo seems suddenly unable and unreliable to *satisfactorily* feed the world's teeming *6.7 billion* people and billions of *pet and domesticated animals!* The supply and demand chain for food and energy is cranking and spitting decay and rot. The global *Food Crisis* is threatening the security of nations and future generations, *because:* Giant multinationals are believed squeezing out millions of small farmers and food producers out of business; *because:* Commodity and oil futures speculators are believed to be manipulating the prices as evidenced by mouth-watering balance sheet profits; *because:* Aggressive biofuel production profit ventures have virtually pitted farmer against farmer, poor against poor, land-grabbing against land-grabbing; *because:* Genetically-manipulated crops and food producers have cashed in weirdly; *because:* The world food production protectors, security and energy sufficiency, and humanitarian organisations dive haplessly to introduce reformative measures and stabilise the precarious situation. Adds the International Assessment of Agricultural Science and Technology for Development (IAASTD):

- *"…Today's world is a place of uneven development, unsustainable use of national resources, worsening impact of climate change, and continued poverty and malnutrition. Poor quality diets are partly responsible for the increase of chronic diseases like obesity and heart disease. Agriculture is closely linked to these concerns, including the loss of biological diversity, global warming and water availability…Small farmers and rural communities in developing countries have often not benefited from opportunities that agricultural trade can offer. Opening farm markets prematuredly to international competition further weakens the agricultural sector of a developing country, causing poverty, hunger and harm to the environment in the long term. Trade reforms make relations more equitable. "* [3]

- *"Giant multinationals are undermining economic development in commodity reliant developing countries by influencing the terms of trade in their favor, edging out small-scale producers and shrinking their income and savings,"* according to ActionAid International.[4] *"The world must shift the focus of trade from being driven solely by profits to serve people-centred development,"* Executive Director of South Centre, Yash Tandon, said. *"With market power concentrated in a few hands internationally, African farmers find it hard to build up enough cash to diversify their crops,"* Dede Amanor-Wuks, ActionAid's *West & Central Africa Director said* citing trends in a global coffee trade report. The report's key findings indict developed countries for failing to eliminate *trade distorting subsidies* and reducing opportunities for developing countries access their markets, thus:

- *In 1994, the 12 top companies that comprise "The Global Giants Seed and Agrochemicals" accounted for 80% of the global pesticides market. By 2002, this number halved and the top six companies – BASF, Bayer, Dow, DuPont, Monsanto and Syngenta, now control 80% of the market between them. Only two companies – DuPont and Monsanto, controlled 65% of the world's maize seed market in 2002. Bulk Commodity Trading: In 1998, the top six coffee trading companies held 50% of the world market. By 2002, the top three – Nemann, ED&F Man and Esteve – held 45% of the market. In 2002, just two companies – Cargill and Archer Daniels Midland, controlled three-quarters of the global grain trade. Two companies – Chiquita and Dole, controlled nearly 50% of world banana trade in 2002. Food Manufacturing*

[3] International Assessment of Agricultural Science & Technology for Development (IAASTD): *"IAASTD Agriculture Questions & Answers."* The IAASTD focuses on agriculture as the provider of food, nutrition, health, environmental services, and economic growth that is both sustainable and socially equitable. This assessment recognises the diversity of agricultural ecosystems and local social and cultural conditions. IAASTD http://www.agassessment.org/

[4] Martin Adhola: *"Giant multinationals squeezing out millions of small producers: The market is just too narrow and we need to open it up more fairly to small-scale producers."* ActionAid/South Centre Accra, Ghana. ActionAid formed in 1972 as an anti-poverty international agency, based its article on a joint Report it presented tat the UNCTAD XII meeting.

and Processing: Only two companies – Nestle and Philip Norris/Altria, controlled nearly 60% of the world market for roasted and instant coffee in 2002.

There is great confusion, tumult and hustle and bustle around the world's communities to the unbalanced State of the World Economy and Social Instability. To regain balance and *restore* normality to *properly, sufficiently and efficiently* feed its citizens - the world can no longer *ignore, playdown* or *suppress* the production and uses of *Renewable Energies*: these natural energy sources would *mitigate, reduce* and *neutralise* escalating oil and food prices! This is everyday *irrefutably* being proved. In countries like Germany and several hundreds of *greening industries, companies, communities* and *organisations* this is happening. There are *temporal,* slightly higher prices for renewable energies' and organic food products. These temporal initial high prices are *superficially* buttressed by fossil energy's gravitational inconvenience and death throes. They will level off significantly as we move from *less* fossil and atomic energy consumption toward *full* Renewable Energies' consumption.

III. Author

DR STEPHEN Nangoh Esomba was born in Mbonge, Cameroon, West Africa in 1946. He attended Sasse Secondary School, Buea, took home study courses in journalism, and worked with the Cameroon Times, Limbe, and Government Information Ministry in Buea. He holds Ph.D Environmental Sciences (Opeinde Open University, Holland, 1992), and Ph.D Political Science (Hamburg University, Germany, 1996). He has authored two books: *Grand Wizard Konkolo,* and *Restore Environmental Order* (PublishAmerica.com, Baltimore, Maryland, U.S.A.). His latest manuscript is: *Renewable Energies: Twenty-first Century's Energy Sufficiency Roadmap.*

Chapter 1

RENEWABLE ENERGIES AND FOOD PRICES

1.1 Introduction I: Fossil Fuels

FOSSIL ENERGIES' fuels or *Mineral* fuels or *Liquid* fuels have lived with the world since plants and vegetation began to decay within rocks and underground waterbeds, but much economic use of them has been made since 1750. Renewable Energies' fuels or *Natural Energy* fuels have also lived with the world since Earth's creation, but little economic use of them has been made. Fossil energies have dominated the world economy and industrial production structures *plus* pumping *billions of greenhouse gases* into the atmosphere since 1750. Renewable energies have been made to remain *dormant* and virtually *untapped* and unused commercially.

Fossil fuels originate from hardened remains or traces of prehistoric animals or plants in strata of earth. Fossil, mineral, and liquid fuels are *hydrocarbons.*[5] The majority of hydrocarbons found naturally in Nature occur in *crude oil* or oil in the natural or raw state, where decomposed organic matter provides an abundance of carbon and hydrogen. When carbon and hydrogen are bonded, they can *catenate* or connect like links of a chain, to form seemingly unlimited chains of hydrocarbon organic compounds.

Organic Compounds are any member of a large class of *chemical compounds* composed of *organic* substances and their combinations. This composition changes under various conditions as they exist in Nature, or synthetically manufactured by man and their molecules contain carbon. A few types of chemical compounds such as *carbonates,* simple *oxides* of carbon *cyanides,* and *allotropes* of carbon - are considered to be *inorganic* or chemical elements other than carbon. Chemical compounds consist of two or more different *elements* chemically-bonded together in a fixed proportion by mass. In essence, a chemical element is a type of *atom* that is distinguished by its *atomic number* according to the number of *protons* in its nucleus. There are currently 118[6] known elements

---

[5] Hydrocarbons: Are organic compounds consisting entirely of hydrogen and carbon classed as *aromatic hydrocarbons* (arenes), *alkanes,* and *alkyne*-based compounds composed entirely of carbon or hydrogen referred to as "pure" hydrocarbons. Other hydrocarbons with bonded compounds or impurities of sulphur or nitrogen are referred to as "impure" – and indeed are not hydrocarbons. Hydrocarbons consist of a "backbone" or "skeleton" composed entirely of carbon and hydrogen and other bonded compounds and lack a *functional group* that generally facilitates combustion. Soot or elemental carbon is impure carbon particles from incomplete combustion of a hydrocarbon. Most soot is produced as pigments such as lampback and carbon black – materials produced by the incomplete combustion of heavy petroleum products such as FCC tar, coal tar, ethylene cracking tar, and a small amount of vegetable oil.

[6] Known Existing 118 Elements in Nature, their Atomic Numbers, Symbols and Atomic Weights: 1 Hydrogen (**H**) **1.0080**; 2 Helium (**He**) **4.0022604**; 3 Lithium (**Li**) **6.941**; 4 Beryllium (**Be**) **9.01218**; 5 Boron (**B**) **10.811**; 6 Carbon (**C**) **12.011**; 7 Nitrogen (**N**) **14.00671**; 8 Oxygen (**O**) **15.9994**; 9 Fluorine (**F**) **18.9984**; 10 Neon (**Ne**) **20.179**; 11 Sodium (**Na**) **22.9898**; 12 Magnesium (**Mg**) **24.305**; 13 Aluminium (**Al**) **26.9815**; 14 Silicon (**Si**) **28.086**; 15 Phosphorus (**P**) **30.9738**; 16 Sulphur (**S**) **32.06**; 17 Chlorine

existing in Nature. These elements are either *metals* or *non-metals*. *Metal Elements* are 48[7] in number classified as *Alkali, Alkaline Earth, Inner Transition Elements* (lanthanides, actinides), *Transition Elements, Metaloids,* and *Non-Metals*[8]. The majority of hydrocarbonate compounds in the elements occur in crude oil.

Crude Oil, raw oil or oil in its natural state found in hydrocarbons in Nature, is the fundamental basis of all fossil fuels. Synthetic crude oil and petroleum products are made from crude oil. Various petroleum products form part of our *household items, clothing* and *food*. Generally, fossil fuels occur or are found contained or embedded in consumer products as diverse as *natural gas, coal,* and *petrol,* through a combination of fossil fuels and chemical additives. This produces more versions of fossil fuels including their imitations. Fossil fuels or liquid fuels are in the form of *gasoline* or *petrol, diesel, biodiesel, alcohols, methanol, ethanol, butanol* and *hydrogen*. *Petrol* is crude oil found in light, yellow, dark brown or black colors. *Extra Heavy Oil* is any crude oil which does not flow easily. *Liquefied Petroleum Gases* (LPGs), gasoline, gasoil, gas, kerosene, fuel-oil, and coke, all have high densities, high viscosities and high impurities content, and need specific solutions to produce, transport and refine. *Liquefied Refinery Gases* (LRGs) are LPGs produced at refineries. Refineries use large quantities of *Natural*

(Cl) **35.453**; 18 Argon (**Ar**) **39.948**; 19 Potassium (**K**) **39.102**; 20 Calcium (**Ca**); 21 Scandium (**Sc**) **44.9559**; 22 Titanium (**Ti**) **47.90**; 23 Vanadium (**V**) **50.941**; 24 Chromium (**Cr**) **51.996**; 25 Manganese (**Mn**) **54.9380**; 26 Iron (**Fe**) **55.847**; 27 Cobalt (**Co**) **58.9332**; 28 Nickel (Ni) 58.71; 29 Copper (**Cu**) **63.546**; 30 Zinc (**Zn**) 65.37); 31 Gallium (**Ga**) **69.72**; 32 Germanium (Ge) 72.59; 33 Arsenic (**As**) **74.9216**; 34 Selenium (**Se**) **78.96**; 35 Bromine (**Br**) **79.904**; 36 Krypton (**Kr**) **83.80**; 37 Rubidium (**Rb**) **85.678**; 38 Strontium (**Sr**); 39 Yttrium (**Y**) **88.9059**; 40 Zirconium (**Zr**) **91.22**; 41 Niobium (**Nb**) **92.004**; 42 Molybdenum (**Mo**) **95.94**; 43 Technetium (**Te**) **98.9062**; 44 Ruthenium (**Ru**) **101.07**; 45 Rhodium (**Rh**) **102.9055**; 46 Palladium (**Pd**) **106.4**; 47 Silver (**Ag**) **107.868**; 48 Cadmium (**Cd**) **112.40**; 49 Indium (**In**) **114.82**; 50 Tin (**Sn**) **118.69**; 51 Antimony (**Sb**) **121.75**; 52 Tellenium (**Te**) **127.60**; 53 Iodine (**I**) **126.9045**; 54 Xenon (**Xe**) **132.30**; 55 Caesium (**Cs**) **132.9054**; 56 Barium (**Ba**) **137.34**; 57 Lanthanum (**La**) **138.9055**; 58 Cerium (**Ce**) **140.12**; 59 Praseodymium (**Pr**) **140.9077**; 60 Neodymium (**Nd**) **144.24**; 61 Promethium (**Pm**) 147 principal isotope; 62 Samarium (**Sm**) 150.4; 63 Europium (**Eu**) **151.96**; 64 Gadolinium (**Gd**) **157.25**; 65 Terbium (**Tb**) **158.9254**; 66 Dysprotium (**Dy**) **162.50**; 67 Holmium (**Ho**) **164.9340**; 68 Erbium (**Er**) **167.26**; 69 Thulium (**Tm**) **168.9342**; 70 Ytterbium (**Yb**) **173.04**; 71 Lutetium (**Lu**) **174.97**; 72 Hafnium (**Hf**) **178.49**; 73 Tantalum (**Ta**) **180.9479**; 74 Tungsten (**W**) **183.85**; 75 Rhenium (**Re**) **186.2**; 76 Osmium (**Os**) **190.2**; 77 Iridium (**Ir**) **192.2**; 78 Platinum (**Pt**) **195.09**; 79 Gold (**Au**) **196.9665**; 80 Mercury (**Hg**) **200.59**; 81 Thallium (**Tl**) **204.37**; 82 Lead (**Pb**) **207.12**; 83 Bismuth (**Bi**) **208.9806**; 84 Polonium (**Po**) 210 principal isotope; 85 Astatine (**At**) 211 principal isotope; 86 Radon (**Rn**) 222 principal isotope; 87 Francium (**Fr**) 223 principal isotope; 88 Radium (**Ra**) **226.0254**; 89 Actinium (**Ac**) 227 principal isotope; 90 Thorium (**Th**) **232.0381**; 91 Protactinium (**Pa**) **231.0359**; 92 Uranium (**U**) **238.029**; 93 Neptunium (**Np**) **237.0482**; 94 Plutonium (**Pu**) 239 principal isotope; 95 Americum (**Am**) 241 principal isotope; 96 Curium (**Cm**) 242 principal isotope; 97 Berkelium (**Bk**) 243 principal isotope; 98 Californium (**Cf**) 244 principal isotope; 99 Einsteinium (**E**) 253 principal isotope; 100 Fermium (**Fm**) 256 principal isotope; 101 Mendelevium (**Md**) a metallic, radioactive, transuranic synthetic element of the actinides; 102 Nobelium (**No**), a synthetic element; 103 Lawrencium (**Lr**), a radioactive synthetic element; 104 Rutherfordium (**Rf**); 105 Dubnium (**Db**); 106 Seaborgium (**Sg**); 107 Bohrium (**Bh**); 108 (**Bh**); 108 Hassium (**Hs**); 109 Meitnerium (**Mt**); 110 Darmstadtium (**Ds**); 111 Roentgenium (**Rg**); 112 Ununbium (**Uub**); 113 Ununtrium (**Uut**), a synthetic in the periodic table; 114 Ununquadium (**Uuq**); 115 Ununpentium (**Uup**), a synthetic superheavy element in the periodic table; 116 Ununhexium(**Uuh**), a synthetic superheavy element in the periodic table; 117 Ununseptium (**Uus**), an undiscovered element in the periodic table; 118 Ununoctium (**Uuo**) in the periodic table.

[7] *The 48 Metal Elements and their Symbols are:* Lithium (Li); Beryllium (Be); Sodium (Na); Magnesium (Mg); Aluminium (Al); Potassium (P); Calcium (Ca); Scandium (Sc); Titanium (Ti); Vanadium (V); Chromium (Cr), Manganese (Mn); Iron (Fe); Cobalt (Co); Nickel (Ni); Copper (Cu); Zinc (Zn); Gallium (Ga); Rubidium (Rb); Strontium (Sr); Yttrium (Y); Zirconium (Zr); Niobium (Nb); Molybdenum (Mo); Technetium (Te); Ruthenium (Ru); Rhodium (Rh); Palladium (Pd); Silver (Ag); Cadmium (Cd); Indium (In); Tin (Sn); Antimony (Sb); Caesium (Cs); Barium (Ba); Lanthanum (La); Hafnium (Hf); Tantalum (Ta); Tungsten (W); Rhenium (Re); Osmium (Os); Iridium (Ir); Platinum (Pt); Gold (Au); Mercury (Hg); Thalium (Tl); Lead (Pb); Bismuth (Bi) .

[8] *Non-Metals:* Are elements on the periodic table (see 118 elements) that can be termed either as metals or non-metals, including a few elements with intermediate properties known as metaloids, and include Hydrogen (H), Carbon (C), Nitrogen (N), Phosphorus (P), Oxygen (O), Sulphur (S), Selenium (Se), the Halogens, the Noble Gases that occur naturally like Helium (He), Neon (Ne), Argon (Ar), Krypton (Kr), Xenon (Xe), and Radioactive Radon (Rn). Non-metals make up most of the earth's crust. *Halogens* are a series of non-metals that comprise Fluorine (F), Chlorine (Ck), Bromine (Br), Iodine (I), Astatine (At), and the undiscovered element 117 temporarily named *Ununseptium*. *Alkali Metals:* Are a series of elements that includes Lithium (Li), Sodium (Na), Potassium (K), Rubium (Rb), Caesium (Cs) and Francium (Fr). Hydrogen also belongs to this group, although it is rarely highly reactive and rarely found in elemental form in Nature; *Alkaline Earth Metals* include Beryllium (Be), Magnesium (Mg), Calcium (Ca), Stronthium (Sr), Barium (Ba) and Radium (Ra) – and are all silver colored, soft and react readily with halogens to form ionic salts, and with water to form strongly alkaline (basic) hydroxides; *Inner Transition Elements:* Are those elements with the highest energy electrons in the atomic ground state divided into the *lanthanoid* and *actanoid* series, thus: Lanthanoids comprise the 15 elements with atomic numbers 57 through 71 (i.e., lanthanum to Luthetium); Actanoid/actanide series comprise of the 15 elements that lie between atomic numbers 89 through 103 (i.e., Actinium and Lawrencium and exhibit a wider range of oxidation states; *Transition Elements:* Are about 40 in number and include Zinc (Zn), Cadmium (Cd), Mercury (Hg);

Gas plant Liquids (NGLs) from fractionators as feedstocks and blendstocks. Smaller amounts of LRGs flow from refineries to fractionators for processing. *Light* crude oil is distinguished from *Heavy* crude oil from the latter's low wax content. There are *three* major forms of fossil fuels: *Coal, Crude Oil,* and *Natural Gas.*

Natural Gas is a mixture of hydrocarbon compounds existing in gaseous state or in solution with oil in the natural underground reservoirs. Hydrocarbon liquids condensed from "wet" natural gas from wells are known as *Natural Gas Liquids* (NGLs) and include *liquid ethane, propane, butane* and mixtures of these, as well as the heavier *petanes,* plus *isopetane, natural gasoline* and *plant condensate. Compressed Natural Gas (CNG)* is used in spark ignition engines for both light- and heavy-duty vehicle applications. *Liquefied Petroleum Gases* (LPGs), propane, butane, jet fuel, kerosene, naphtha-based jet fuel and petroleum coke – are petroleum products applied in various consumer goods, transportation and industrial usages. *Petroleum Products* fall into three major categories: *Fuels* such as motor gasoline and distillate fuel or diesel fuel; *Fuel non-fuel products* such as solvents and lubricating oils; and *Feedstocks* for petrochemical industry such as naphtha and various refinery gases. Petroleum products like gasoline, distillate (diesel) fuel, and jet fuel – provided virtually all the energy consumed in the transportation sector at 67% of all petroleum consumed in 2004. *Petrochemical feedstocks* are used in the production of petrochemicals since the 1920s. They are converted to basic chemical building blocs and intermediates which are used to produce *plastics, synthetic rubber, synthetic fibres, drugs* and *detergents.* Petrochemical feedstocks include *ethane, propane, butane, ethylene, propylene,* normal *iso-butylenes, butane* and aromatics such as *benzene, toluene* and *xylene.*

Coal has three main types: *Anthracite, Bituminous* or Bitumen, and *Lignite* or Brown coal. Anthracite is hardest and has more carbon and low percentage of volatile matter; Lignite is the softest, low in carbon, but high in hydrogen; Bituminous is a mineral pitch, asphalt or any form of various kinds of native oxygenated hydrocarbons such as naphtha, petroleum and charcoal. *Petroleum* is produced from the rock *oil shale; Tar sands, oil sands* and bituminous sands – are a combination of clay, sand, water and bitumen. *Charcoal* is a black substance made by burning wood slowly in an oven, or it is a black residue of partly burnt wood, bones and coal. *Black Charcoals* are usually bones that still contain organic impurities which may impart undesired odors or color to treated waters. *White bonechars* are over-charred bones that present low fluoride removal capacity; *Grey-brownish bone chars* are the best quality chars for absorption applications; Bone chars usually have lower surface area than activated carbons, but present high absorptive capacities for copper, zinc, and cadmium; bone chars are used to remove flouride (a reduced form of chlorine) from water and to filter aquarium water, often used in sugar refining industry for decolorizing, and as a decolorizing agent.

Fossil fuels, petroleum products and their synthetic imitations, dominate the world economy in transportation, consumption, industrial applications and agriculture. Agriculture depends heavily on petroleum products like chemical *fertilizers, pesticides, herbicides* and *hydrocarbons* – many of them developed from fossil fuels, natural gas, oil, *saturated* and *unsaturated cyloalkanes*, and *aromatic arenes*. Gases like *methane* and *propane;* liquids like *hexane* and *benzene* and waxes; and low-melting solids like *paraffin wax* and *naphthalane;* polymers like *polyethylene, polypropylene* and *polystyrene* - are fossil fuels-based. Petrochemical-derived pesticides and fertilizers use machines run on fossil fuels in farming.

1.2 Introduction II: Renewable Fuels

THE PRODUCTION, consumption, marketing and proliferation of Renewable Energies has received *low level priority* in the conduct of business affairs worldwide. But the production, consumption, marketing and proliferation of fossil fuels and their attributes, dominate the world economic activiies. Nuclear energy enjoys the same prerogatives. The production of food enjoys similar prerogatives but on a *lower par* to fossil and nuclear energies. When the oil-and-food-crisis broke record highest prices in mid-2008, fossil oil producers, marketers and commofity speculators came under great suspicion for orchestrating and profiting from it. The suspicion lingers on as evidenced by the record-high profits posted by oil and petroleum companies: record-high bankcruptcies reported by transportation companies and industries that rely heavily on oil; record soaring inflation in economies and consumer goods around the world; and record-high hunger-protests that erupted around the globe especially in developing and underdeveloped nations.

During the tortuous course of this unprecedented modern food-and-oil-crisis, the U.S., known to be the largest producer and user of fossil oil and petroleum products, introduced and intensely debated the production and use of renewable energies on a permanent basis in its 2008 Presidential Election Campaigns: The Democratic Candidate Nominee Senator Barack Obama, and the Republican Candidate Nominee Senator John McCain, both presented their views. Senator Obama's energy programme was specifically for Renewable Energies; and Senator McCain's energy programme was specifically for maintaining fossil fuels and nuclear energies with minimal Renewable Energies. Both candidates aimed at freeing the U.S., from the *burden of dependence* on foreign oil and eventually preventing the repeat occurrence of an energy or food crisis, high energy prices and high food prices.

Renewable Energies are produced direct from Nature and natural phenomena the same way fossil fuels are. But there are *significant differences* between renewable energy sources, and non-renewable energy or fossil energy sources. Renewable Energies are spread around the world and universe, while fossil energy sources are not; Renewable Energies can use the same or slightly adjusted technologies that are used for fossil fuels; Renewable Energies are reusable, recyclable and non-polluting or neglibly polluting compared to fossil energies that heavily and fatally pollute the environment; Renewable Energies pose little or no health risks to Nature as do fossil fuels; Renewable Energies need little investments to exploit compared to fossil and atomic energies. A few Renewable Energy sources like biofuels have been accused of being a source of land-grabbing turf wars and land degradation, and even contributing to the *Food Crisis* and *Rising Food Prices.* But this perceived accusation is due mainly to the unscrpulous pursuits of the owners of biofuel enterprises rather than to the biofuels themselves. Biofuels are only a tiny part of *Biomass,* one of several vastly available renewable energy sources.

Biomass is the general name applied to plant matter renewable energies' sources, which includes: *Biofuels* and *biogas*; *Vegetable Oil Fuels* – cooking and edible oils; *Inedible* oils; oil from *Nuts*; *Vegetable* and *Animal* fats; *Straight* vegetable waste fat and *Drying* oils; oil from *Grass* and *Grass-like* plants and straws; *Woodfuel*; *Essential Oils*; *Essential Fatty Acids*; *Liquid Fuels* from alcohols like *ethanol* and *melanol*; and *Oils* from *Fish*. Various food crops like maize or soybeans are *unnecessarily* being used to produce ethanol fuel, whereas other non-food crops like *Sugarcane* and *Grasses*, can and are suitably being used to produce fuels. Other renewable, inexhaustible natural energy sources include:

Solar, Photovoltaic, Thermal Energy: Sunlight[9] which is in abundance everywhere, is converted by these technologies into heat and electricity. Great sunlight belts of the world have the greatest advantage to harvest solar energy.

Wind[10] Energy from prevailing wind patterns: Wind is always in motion for its energy power to be captured and converted into heat and electricity. Strong and powerful wind areas of the world have the greatest advantage using these technologies.

Geothermal Energy from underground rocks: The vast energy stored under the rocks is converted into heat and electricity. These underground reservoirs of energy are manifested when volcanoes erupt, in hot springs, and in water geysers.

Piezoelectricity or *Free Energy Scavenging* from objects in motion: Free energy surrounds us wherever we find ourselves which could be converted into heat and electricity. Examples of such energies around us include: *noises, sounds, movement* and *activities, raindrops, wind flappings, vehicles in motion,*

---

[9] Sunlight: Sunlight is a self-propagating electromagnetic (EM) radiation spectrum in space or through matter. Sunlight energy emanates from the Sun; the Sun is a massive star or *ionized gas* or state of matter consisting of a collection of particles of *molecules, atoms, ions, electrons* – at the centre of the Solar System; the Solar System consists of the Sun and celestial or *astronomical* objects scientifically confirmed to exist in space. The daylight experienced on Earth is a combination of the radiance of the particles of molecules, atoms, ions, and electrons or *multiple atomic particles.* [Molecules, atoms, ions, electrons are *groups of* particles of matter held together chemically in stable, positive, negative or neutral form by electrical charges.]

[10] Wind: Wind is the flow of air or gases through the Universe. Winds or gases are states of matter consisting of collections of particles (i.e., molecules, atoms, ions, electrons) that are more or less in *constant* motion. Wind is invisible, tasteless and odorless. The gases that make up wind are *oxygen, nitrogen,* some *carbon dioxide* and traces of other gases. [*Particulates* or *Particulate Matter* (PM) or *Fine Particulates* (FM), which are man-made tiny particles of solid or liquid and the gas together suspended in a gas, are dangerous environmetal pollutants. An example of a PM/FM is *Aerosol spray.* Aerosol spray is a suspension of fine solid or liquid droplets in a gas. Concentrated aerosols from substances such as silica, asbestos and *diesel particulate matter* – are sometimes found in workplaces. Some PM/FM occur naturally and originate from volcanoes, dust storms, forest and grassland fires, living vegetation and sea spray; sea spray forms when ocean waves crash and contains a high concentration of mineral salts like salt spray, which is largely responsible for corrosion of metallic objects near coastlines.]

walking, dancing, celebrating, marching, protest-marchings, sporting events, including *mountaineering, and wars and combats.*

Nano, Hydrogen, Fuel Cell, Hybridisation Technologies: Nano engineering is the design and manufacturing of products and other objects based on the manipulation of atoms and molecules and building machines atom-by-atom. The use of nano, fuel cell, and hybridisation technologies side-steps the production and use of energy with carbon dioxide (CO^2), nitrogen oxides (Nox), particulate matter or fine matter (PM/FM) that emit or greatly enhance the emission of GHGs into the atmosphere.

Hydropower, Micro Hydro Technologies: These convert water masses' great power energy into heat and electricity. Great and small-water-endowed world areas have a great advantage of these technologies.

1.3 Rising New Rich, Oil Prices-Profits Up

THE WEALTH of Nations and economies is precariously built on the shifting sands of fossil energy convolutions. The Oil-and-Food Crisis 2008 saw the rise of new millionaires as oil and food prices rose to record mind-boggling levels. Commodity futures speculators, oil-consuming countries' companies and oil-producing countries' companies – all stood to benefit and did benefit from the crisis, as well as lose, as records and events showed.

1.3.1 Commodity Speculators

COMMODITY SPECULATORS are normal human beings pursuing their normal business activities for maximum profit. They are investors who move vast amounts of wealth around the world stacked in commodity markets. Energy traders are believed to have engaged in illegal activities which may have helped drive up the price of oil and food prices, according to independent experts:

- *Akira Yanagisawa*, Senior Economist at Japan's Energy Data and Modelling Center:[11] *"In the most recent terms* (the third and fourth quarter in 2007), *the fundamental prices* [of oil] *are estimated around 50 to 60 dollars. ... it is estimated the premium has risen up to around 40 dollars at maximum."*

- *Michelle Foss*, Center of Energy Economics at University of Texas:[12] *"[G]iven that inventories of crude and products are healthy in many locations and that the very real risk of supply disruption has so far been avoided, the remaining factor left to balance*

---

[11] Akira Yanagisawa: *"Institute of Energy Economics,"* 3/08, http://eneken.ieej.or.jp/en/data/pdf/421.pdf , p. 13.
[12] Michelle Foss: *"Oxford Institute for Energy Studies Working Paper,"* 2/07, http://www.oxfordenergy.org/pdfs/NG18.pdf, p. 34

the price per barrel is speculation. The role of speculation in oil markets has been widely debated but could add upwards of $20 to the price per barrel."

- *Harry Chorn, Chief Economist of Platts:[13] "The actual costs incurred in producing the most expensive oil is only around $70 or $80 a barrel, meaning that about $50 of the current price represents 'the market's risk premium plus speculation.' "*

- *The U.S., Senate Permanent Subcommittees on Investigations:[14] "Several analysts have estimated that speculative purchases of oil futures have added as much as $20 - $25 per barrel to the current price of crude oil, thereby pushing up the price of oil from $50 to approximately $70 per barrel."*

- *Clarence Cazalot Jr., CBO, Marathon Oil:[15] "$100 oil isn't justified by physical demand in the market – it has to be speculation on the futures market that is fuelling this: "Stephen Simon, Executive Vice President, ExxonMobil[16]: "The price of oil should be $50 to $55 per barrel based on supply and demand fundamentals."*

- *John Hofmeister, Press, Shell Oil Co.,:[17] "The proper range of oil is somewhere between $35 and $66/bll."*

- *Gerry Ramm, Senior Executive, Inland Oil Company:[18] "Excessive speculation on energy trading facilities is the fuel that is driving this runaway train crude oil prices."*

- *Prof. Michel Chossudovsky, Centre for Research on Globalisation (CRG):[19] "As detailed in an earlier article, a conservative calculation is, that at least 60% of today's $128 per barrel price of crude oil comes from unregulated futures speculation by hedge funds, banks and financial groups using the London ICE Futures and New York NYMEX futures exchanges and uncontrolled inter-bank or Over-The-Courier trading to avoid scrutiny. U.S., margin rules of the government's Commodity Futures Trading Commission (CFTC)[20] allow speculators to buy a crude oil futures*

---

[13] Larry Chorn, Business Week, 5/13/08, http://www.businessweek.com/bwdaily/duflash/content/may2008/ db20080 513_734146_page_2.htm

[14] U.S., Senate Permanent Subcommittees on Investigations: Staff Report, 6/06, http://levin.senate.gov/newsroom/supporting/ 2006/PSI.gasandoilspec.062606.pdf.

[15] Clarence Cazalot, Marathon Oil: CNNMoney, 11/12/07, http://money.cnn.com/2007/11/12/markets/oil_hundred/index.htm? Postversion=2007111216.

[16] Stephen Simon: House Testimony, 4/1/08, http://globalwarming.house.gov/tools/assets/files/0453.pdf

[17] John Hofmeister: Financial Post, 5/22/08, http://www.financialpost.com/reports/oil-watch/story.html?=532747.

[18] Gerry Ramm: Senate Testimony, http://commerce.senate.gov/ public/_files/RammSenateCommerce 06308Testimony.pdf.

[19] Centre for Research on Globalisation (CRG): This an independent research/media group of writers, scholars, and journalists based in Montreal, Canada. CRG publishes books, organises public conferences and lectures and acts as a thin-tankon crucial international and geopolitical issues. Editor: Prof. Michel Chossudovsky. GlobalResearch.ca; crgeditor@yahoo.com.

[20] U.S. Government's Commodity Futures Trading Commission (CFTC): Was established in 1974 as an independent agency with the mandate to regulate commodity futures and option markets in the U.S., and consists of five Commssioners appointed by the President and confirmed by the Congress for five-year terms. The Commission monitors markets and market participants, trading firms and end users closely. The Office of the Chief Economist provides economic support and advice to the Commission, conducts research on the public policy issues facing the agency, and provides education and training for Commission staff; the Office of the General Counsel is the Commission's legal adviser, represents the Commission in appellate litigation and certain trial-level cases, including bankcruptcy proceedings involving futures industry professionals, and advises the Commission on the application and interpretation of the Commodity Exchange Act and other administrative status; the Office of the Executive Director formulates and interpretes the management and administrative functions of the CFTC and the agency's budget; the Office of the Chairman includes the Office of External Affairs which acts as the CFTC's liaison with news media,producer and market user groups, educational and academic groups, and the general public, provides information about the agency, etc.; the Division of Clearing and Indemnity Oversight oversees market intermediaries, including derivatives clearing organisations, financial integrity of registrants, customer fund protection, stock index margin, sales practice reviews, foreign market access by intermediaries, National Futures Association activities related to intermediaries; the Division of Enforcement investigates and prosecutes alleged

contract on the Nymex, by having to pay only 6% of the value of the contract. At today's price of $128 per barrel, that means a futures trader only has to put up about $8 for every barrel. He borrows the other $120. This extreme 'leverage' of 16 to 1 helps drive prices to wildly unrealistic levels and offsets bank losses in sub-prime and other disasters at the expense of the overall population."

- *Goldman Sachs again in the middle:[21] "The oil price today, unlike twenty years ago, is determined behind closed doors in the trading rooms of giant financial institutions like Goldman Sachs, Morgan Stanley, J.P. Morgan Chase, Citigroup, Deutsche Bank or UBS. The key exchange in the game is the London ICE Futures Exchange (formerly the International Petroleum Exchange – IPE). ICE Futures is a wholly-owned subsidiary of the Atlanta Georgia International Commodities Exchange (AGICE). ICE in Atlanta was founded in part by Goldman Sachs which also happens to run the world's most widely-used commodity index, the GSCI, which is over-weighted to oil prices...ICE was focus of a recent congressional investigation. It was named both in the Senate's Permanent Subcommittee on Investigation and House Committee on Energy & Commerce's Hearing in energy futures that looked into unregulated trading in energy futures. Through a convenient exception granted by the Bush Administration in January 2006, the ICE Futures trading of U.S., energy futures is NOT regulated by the CFTC, even though the ICE Futures U.S. oil contracts are traded in ICE affiliates in the U.S. And at Enron's request, the CFTC exempted the Over-the-Counter oil futures trades in 2000..."*

- *U.S., Senate Report:[22] "There's a few hedge fund managers out there who are masters at knowing how to exploit the peak oil theories and hot buttons of supply and demand, and by making bold predictions of shocking price advancements to come, they only add more fuel to the bullish fire in a sort of self-fulfilling prophecies."* [as quoted by CRG Associate F. William Engdahl.]

1.3.2 Oil-Consumer Countries, Companies

OIL TRADING – supply and demand – in oil consuming countries is the core responsibily of oil and petroleum companies. At the height of the Oil-and-Food-Crisis, the World Petroleum Council (WPC) held its 19th Congress in Madrid, Spain on 29 June 2008,[23] to celebrate its 75th Anniversary. WPC Congresses held every

---

violations of the Commodity Exchange Act and Commission regulations. Violations may involve commodity futures or option trading or U.S. futures exchanges or the improper marketing and sales of commodity futures products to the general public.

[21] Ibid; U.S., Permanent Subcommittee on Investigations: Staff Report, 27 June 2006; and House Committee on Energy & Commerce's Hearing, December 2007.

[22] U.S. Senate Report: *"The Role of Market Speculation in Rising Oil and Gas Prices."* June 2006.

[23] World Petroleum Congress (WPC) 19th World Petroleum Congress, Madrid, Spain 2008: *"A World in Transition: Delivering Energy for Sustainable Growth."* http://www.19wpc.com/final-newsletter.pdf; info@19wpc.com; sponsored by StatoilHydro, gasNatural, Petro-bras, ACS, and Saudi Aramco; WPC Sessions were: Natural gas: An Emerging Global Commodity" and "Advancing sustainability in the Oil and Gas Industry" both sponsored by StatoilHydro; Women's Forum: "Making the Business Case for more Women in the Industry"; "Advanced drilling and completion: Novel approaches to improve energy efficiency at refineries"; "Gas transportation challenges"; "NOCs and the IOCs – Competing or Co-sponsoring?"; "Reporting of reserves and resources logo acs linea negra"; "Unconventional petroleum resources"; "Unconventional crude oils and feedstocks to refineries"; "Natural gas as a transportation fuel"; Attracting and retaining future energy professionals"; "Motor fuel quality: regulators and industry views"; "Biofuels and oil products: Integration vs Competition"; "Reservoirs monitoring"; "Refineries in the digital era: what is next in automation, control, optimization"; "Balancing the rational gas supply & demand equations"; "Coping with the impacts of natural disasters – lessons learned"; "Coal Bed Methane: Addressing environmental constraints"; "Gas Market Integration: A Perspective from IGU" – all sponsored by ACS. Host Sponsors: Repsol YPF, CEPSA; Gold Sponsors: ANH Colombia, Qatar Petroleum, ExxonMobil, Saudi Aramco, gasNatural, Human Energy, BR PETROBRAS, TOTAL, Energy for generators; Silver Partners: ABENGOA BIOENERGY, Enil, IBM, PRICEWATERCOOPERS, ERNST YOUNG, CLH, FLOUR, nexen, Grupo CUNADO, galp energia, StatoilHydro, MOL, SHELL; Official Partners: Expansion, BNP PARIBAS,

three years, are the most influential meeting places where global oil and gas industry are discussed, and governments and oil companies, IOCs and NOCs, industry and shareholders from around the globe come together to set out the way forward for the petroleum sector. The 19th Congress launched a wide-ranging Technical Programme that was aimed at recognising and honoring the scientific, technological and professional achievements of the petroleum industry.[24] The official theme *"A World in Transition: Delivering Energy for Sustainable Growth,"* explained thus:

-
   Increasing population growth, energy intensity and globalisation has led to phenomenal rise in the use of energy. Demand for oil and gas continues to grow and will have to be met through *traditional* as well as *unconventional* sources. Past experience demonstrates that price volatility and economic cycles alternating boom and bust periods were harmful to both the industry and the consumers. The challenge for the industry in a world in transition is to ensure *continuous, affordable* and *reliable supply,* meeting society's expectations in a *sustainable, transparent, ethical* and *environmentally sound* manner."

"The main challenge all companies in the [oil & gas] sector must face," said Carlos Perez de Bricio, Vice-President of the Organising Society of the 19th WPC Congress who is also the President and CEO of CEPSA, "is balancing energy needs with a solution to the problem of climate change." Bricio named other challenges as:

-
   ... increasing the use of *clean technologies* in guaranteeing the energy supply at a reasonable price...it is possible to carry out an efficient industrial activity which *respects the environment* and offers maximum safety with regards to process and supplies. To achieve this, we must have the right technology, the necessary resources

---

PETROLEUM ECONMIST, CNBC, THE WALL STREET JOURNAL, Delotte; Bronze Sponsors: accenture, GRACE, KPMG, ACS, Schlumberger, AENOR, TECNICAS REUNION, Bureau Veritas, TEEBAY, enagas; Official Support: EU, asicma, UNIVERSIDAD AUTONOMA (UA), MINISTERIO DE ASUNTOS EXTERIORES Y DE COOPERACION, MINISTERIO DE INDUSTRIA TURISMO Y COMERCIO, La Surra de Todos, MADRID CONVENTION BUREAU, AOP, Sedigas, Universidad de Alcala, UNIVERSIDAD CARLOS III DE MADRID, UNIVERSITAS COMPLUTENSIS MATRITENSIS, POLITEC-NICA, Universidad Rey Juan Carlos; Plenary Sessions: Plenary 1 – Delivering Energy for Sustainable Growth: The European Perspective. Speakers: Andris Plebalgs, Chair The European Commission; Sergey Bogdanchikov, Pres., Rosneft; Antonio Brufau, Chairman/CEO Repsol Ypf; Tony Hayward, CE0 BP; Carlos Perez de Bricio, Pres/CEO, Cepsa; Jeroen van der Veer, Chief Executive, Royal Dutch Shell plc; Plenary 2 – Inaugural Session-Wpc Excellence Awards Presentation: H.M. Juan Carlos I De Borbon, King of Spain; Jose Luis Rodriguez Zapatero, Prime Mnister of Spain; Andris Plebgas, Energy Commissioner, The European Union; Ms Esperanza Aguirre, Pres of the Regional Gov't of Madrid; Miguel Sebastian, Spain's Minister of Industry, Trade & Labor; Alberto Ruiz Gallerdon, Mayor of Madrid; Antonio Brufau, Chairman/CEO-Repsol YPF; Randall Gossen, Pres World Petroleum Council; Plenary 3 – Deliverability Challenges: Security of Supply and Demand Perspectives. Speakers: Jesus Reyes Nevoles, General Director, Premax; Taite Costa, Pres CNE, Spanish Energy Commission; Chengdu Fu, Pres CNOOC, China Offshore Oil Corporation; Sergey Shmatko, Minister of Energy, Russian Federation; Christophe de Margerie, CEO, Total; Rex W. Tillerson, Chairman/CEO, ExxonMobil; Plenary 4 – Natural Gas: An emerging Global Commodity. Speakers: H.E. Abdullah bin Hamad al-Attiyah, Minister of Energy and Industry, Qatar; Ms Linda Cook, Execitive Director, Gas+Power, Royal dutch Shell plc; Mohamed Meziane, Pres/Director Gen., Sonatrach; Plenary 5 – Advancing Sustainability in the Oil and Gas Industry. Speakers: William Hogan, Prof Global Energy Policy, Kennedy School of Gov't, Harvard University, U.S.; Helge Lund, Pres/CEO, StatoilHydro; Francis Saville, Chair of Board, Nexen Inc.; Plenary 6 – Accessing Reserves and Investing in Infrastructure. Speakers: H.E. Shri M.S. Srinivasam, Secretary, Ministry of Petroleum & Natural Gas, India; Sergio Gabrielli, Pres, Petrobras; John Watson, Executive Vice Pres, Strategy and Development, Chevron Corporation; Plenary 7 – Societal Expectations of the oil and gas industry. Speakers: Andrew Gould, Chairman/CEO, Schlumberger; Ms Huguette Labelle, Chair, Transparency International; James J. Mulva, Chairman/Pre/CEO, ConocoPhillips.

[24] WPC 19th Congress Technical Programme: (1) "The Media Village": was a designated area planned as a single-topic exhibition on Social Responsibility located in Pavilion 7 of the IFEMA Convention Centre, where major collaboration projects, especially those related to sustainability and human rights between industry and NGOs were displayed; "Social Responsibility" featured Global Partnership projects like: BP's 'The Solar Power Technology Support Project to Agrarian Reform communities (SPOTS Projects); 'CDA Collaboration Learning Projects, Inc'; 'Corporate Engagement'(CEP); 'Engineer Against Poverty '(EAP) and AMEC; 'ExxonMobil (& Esso Angola)'; Gala-Shell: 'PROMOVER (PROMOTE) – Programme of socio-environmental capacity-building and mobilisation'; 'International Petroleum Industry Environmental Conservation Association (IPIECA) – A Human Rights Training for the Oil & Gas Industry'; 'Marathon Oil – Bioko Island Malaria Control Project – Equatorial Guinea'; 'Ptrobras: Carnauba Viva' initiative; Respol YPE's 'Trinidad & Tobago Red Cross HIV/AIDs Programme'; Total's 'Support Programme to Prevent HIV/AIDs in Truck Drivers in Morocco'.

and international relationships which favor the positive development of all countries...the only way to ensure permanence is by working in the long-term, in a responsible manner and creating a value for our shareholders and for society as a whole. In short, by living up to their expectations of us which focus mainly on supplying energy safely at a reasonable price while respecting the environment.

•

Petroleum is a finite resource, although I believe that , if we also take account of bituminous sands and extra-heavy crude oils, it will continue to be the most commonly used energy for many years. CEPSA seeks to guarantee the energy supply. In Exploration and Production, we marketed 7.2 million barrels in 2007 and performed seismic drilling operations in various blocks in a number of countries. We recently commenced oil exploration activity in Peru. Through our increasing efficient refineries and commercial network we place petroleum products at the service of society. To even further enhance our refineries we also plan to invest 1.65 billion euros in the Refinery area over the next five years. These action plans will enable us to increase production of middle distillates by 39% with only a 17% increase in the total refinery capacity, thereby contributing to minimise the market's external dependence on these products."

The evolution of oil prices will always be in the spotlight, given the influence of oil on global economy and our welfare society, according to Antonio Brufau, President of the Organising Society of the 19th WPC and President/CEO of RepsolYPF...The increase in profitability due to higher oil prices makes it possible for companies to increase their investment in new technologies. This makes deeper oil exploration possible, with more efficient energy consumption and less pollution...The important increase in demand is not only affecting refinement, but also the entire value chain of the industry, from oil extraction to oil supply, transportation and the marketing of end products. This is why companies, such as Repsol, are making large investments in refineries to try to meet the demand in the most efficient way possible.

Upstream, Downstream and LNG is Repsol's integrated strategic business area, a vision to optimise operations and organic growth through 10 key projects, represnting 60% of investments in a strategic business in the period 2008-20012, which will account for 75% of the operating results growth. Downstream is a strategy based on two concepts: improving the profitability of current assets, and investing in expansion and conversion in Spain and Portugal, in three key project areas: expansion of the refinery of Cartagena; new coker in the refinery of Bilbao; and expansion of the petrochemical site of Sines in Portugal. Upstream is a strategy based on an essentially organic growth focusing on three core areas: North of Africa, Latin America and Aguas Profundas (Brazil and the Gulf of Mexico), with five key growth projects: Shenzi/Genghis Khan, in the Gulf of Mexico; Reganne, in Algeria; I/R fields in Libyia; Carioca, in Brazil; and Bloque 39 in Peru. The LNG stragegy is to maximise the marketing competitive advantages by optimising investment levels, with two key growth projects: LNG integrate project in Peru, and regasification plant in Canaport, in Canada.

An article[25] that appeared as part of the WPC Congress's agenda, denied that oil companies were responsible for reckless speculation that resulted to hiked oil prices. It also blamed the Organisation of Petroleum Exporting Countries (OPEC) for marking time on oil supplies, as well as other market factors, explaining: "Financial flows, either away from weaker returns on equity and bonds markets, or as a flight away from a weaker U.S. dollar, are often cited as [being] responsible for an oil price inflated above some loosely defined notion of 'fair value.'" The article cited studies based on *disaggregated data* by market participants (supposedly not available to the public) issued by the U.S., Office of the Chief Economist of the Commodities Futures Trading Commission (CFTC) as concluding that:

- *... commercial players (hedgers) like oil companies, airlines or utilities seem to react first to news. At the same time, large speculators do not tend to change their positions nearly as frequently as hedgers. Changes in speculative positions rather seem to follow changes in the commercial positions, providing the liquidity necessary for risk management. Speculative activity of this sort cannot therefore be seen as a basis for setting an escalating trend in oil prices in the past five years. At the same time, price discovery and risk transfer afforded by speculators is essential to a well-functioning futures market...*

- *As far as the oil market goes, fundamentals have supported a continuous strengthening in prices since the second quarter of 2007. The oil market like other commoditiy markets boils down to a question of growth: can we supply the increment of demand as we move forward? On the basis of annual growth in global oil demand less OPEC supply (supply which is unconstrained by a directional cartel decision), the answer would appear to be 'no' until late in the second half 2008 (even when we include supply growth from recent OPEC members Angola and Ecuador). This leaves the market dependent on the marginal barrel that OPEC chooses (or not) to put on the market. OPEC[26] has removed in excess of 1 mbld year-on-year in the first three quarters of 2007, only recently raising its supply. But given a contraction year on year of non-OPEC supply in Q1 08, the recent increase from OPEC only maintains a status quo on a tightening situation. The end result, given a strong oil demand growth in energy intensive emerging markets, is that adjustments in the market will need to come from either lower oil inventories, higher prices or both. And if market participants attach a high convenience yield to production capacity, the adjustment must come mostly through price.*

- *"There are downside risks to oil demand growth in the second half of 2008. A U.S. economic downturn will affect advanced and emerging markets, while financial markets to date remain*

---

[25] Harry Tchilingguirian, Senior Oil Market Analyst, BNP Pariba – Commodity Derivatives: *"Oil Prices above $100: Funds or Fundamentals?"*

[26] The Organisation of Petroleum Exporting Countries (OPEC): Was formed in 1960 in Bagdad, Iraq by 13 oil-producing countries (mostly Arabic nations): Iran, Iraq, Kuwait, Qatar, Saudi Arabia, United Arab Emirates (UAEs), Libya, Algeria, Nigeria, Angola, Venezuela, and Ecuador – with headquarters in Vienna, Austria since 1965. Indonesia and Gabon joined but later resigned their membership, while Ecuador also withdrew its membership. Venezuela first broached the idea to form such an oil-producers' group in 1949, but the idea took form in 1960 triggered by a U.S., national security law issued by Pres Dwight Eisenhower, that forced quotas on Venezuelan and Persian Gulf oil imports in favor of the Canadian and Mexican oil industries. In 1968, Arab-OPEC members Kuwait, Libya, Saudi Arabia plus non-Arab OPEC countries Egypt and Syria, formed the Organisation of Arab Petroleum Exporting Countries (OAPEC) – after the break of the Six Day Arab-Israel War of 1967 – and imposed an Oil Embargo during the 1973 Oil Crisis. One of OPEC's goals is to determine the best means for safeguarding the organisation's interests individually and collectively. It also pursues ways and means of ensuring the stabilisation of oil prices in international oil markets with a view to eliminating harmful and unnecessary fluctuations, an efficient economic and regular supply of petroleum to consumer nations, and a fair return on their capital. OPEC nations account for two-thirds of the world's oil reserves and 35.6% of the world's oil production as of March 2008. The next largest oil-producing countries after OPEC are: members of the Organisation for Economic Cooperation & Development (OECD) with 28.8%, and former Soviet Union republics with 14.8%. OPEC-Member State use quotas in thousands of barrels/day production. The quotas are meant to mitigate *Global Warming* since fossil-fuel production produces large amounts of CO^2 and other *Greenhouse Gases* (GHGs). Current OPEC quotas (in barrels/day) are: Algeria: 894 (7/1/05), 1,360 (production 1/07), 1,430 (capacity); Angola: 1,900 (7/1/05), 1,900 (production 1/07); 1,700 (capacity); Ecuador: 520 (7/1/05), 500 (production 1/07), 500 (capacity); Iran: 4,110 (7/1/05), 3,700 (production 1/07), 3,750 (capacity); Iraq: 1,481 (production 1/05), 2,600 (capacity); Kuwait: 2,247 (7/1/05), 2,500 (production 1/07), 2,600 (capacity); Libya: 1,500 (7/05), 1,600 (production 1/07), 1,700 (capacity); Nigeria: 2,306 (7/1/05), 2,250 (production 1/07), 2,250 (capacity); Qatar: 726 (7/1/05), 810 (production 1/05), 850 (capacity); Saudi Arabia: 10,099 (7/1/05), 8,800 (production 1/05), 10,500 (capacity); UAEs: 2,444 (7/1/05), 2,500 (production 1/05), 2,600 (capacity); Venezuela: 3,225 (7/1/05), 2,340 (production 1/05), 2,450 (capacity); Total: 31,422 (7/1/05), 30,451 (production 1/05), 32,230 (capacity).

rattled by the U.S. Sub-prime crisis.[27] Inflationary pressure in emerging markets [e.g., China and India] – is a risk and could lead to tighter monetary policy and slower growth. However, supply constraints today and tomorrow sustain upward pressure on price. The oil industry is cyclical and concerns over adequate investment in future supply capacity (be it in upstream or downstream) are entrenched. These include cost evaluation, constraints in services (manpower and equipment) and competing claims on capital, all which can delay projects. Most of non-OPEC supply growth is backloaded in 2008 and can easily slip into 2009.

- *"Price is a necessary but not sufficient condition to spur investment and to bring new oil to the market. Setting aside geological cost challenges that come with complex or harder to reach reservoirs, uncertainty is growing either in the ability to access reserves or the reliability of the operating regime with rising reserve nationalism. Contractual agreements and sharing agreements signed by international oil companies are aiming to capture a growing role in the development of new oil supply. This is only likely to give pause to international oil companies before engaing vast sums of capital over long periods of time.*

- *"One can take the easy way out and choose to blame speculators for high oil prices. But that would be turning a blind eye to some real constraints in oil sector's ability to deliver future supply in light of the established emergence of new large consumers on the world scene."*

Oil commodity speculators shift the blame for the cause of high energy prices to OPEC's low production. The OPEC bounces the snowballing cause of hiked oil prices right back to the commodity speculators during the WPC Congress. The Secretary-General of OPEC, His Excellency Abdalla Salem El-Badri[28] said in his address:

- ...Some of us here have today attended the Jeddah Energy Meeting ten days ago. It was interesting to note how often the subject came up of the need for OPEC and the IEA[29] to work together closely and take the lead in *restoring calm to the oil market,* with the IEA forming the third corner of the triangle...there has long been a strengthening of relations between OPEC and IEA, on both a formal and informal basis, especially in technical areas. This includes the holding of joint high-level workshops and OPEC's participation in the IEA's *Greenhouse Gas*

---

[27] U.S. Sub-Prime Crisis: An economic problem that affected the U.S., economy in 2007- 2008 torched off by sub-prime mortgage house-lending practices or sub-prime lending. Sub-prime lending or B-paper , near-prime, non-prime or second chance lending – refers to lending practices at a higher expectation of risk than that of A-paper, and generally accompanied by higher interest rates. Sub-prime loans or leases do not meet normal lending guides, and may have less room for financial difficulties for the borrower and can often lead to late or deferred payments, and are accompanied by exhorbitant fees and hidden terms and conditions. Many sub-prime lenders are believed engaged in predatory practices and deliberate targeting of borrowers who could not understand what they were signing for, or on borrowers who could never meet the terms of their loans. The advantages of sub-prime loans to borrowers is that, it makes it possible for people who would otherwise not have access to the credit market. The U.S., sub-prime mortgage industry collapsed in March 2007 due to higher than expected home foreclosure rates. More than 25 sub-prime lenders went bankrupt because of significant losses they incurred, and 84% of New Century Financial – the largest U.S., subprime lender – plunged. Other U.S., financial intitutions, and Wall Street banks like Merryll Lynch, Lehman Brothers, City Bank, and the Government's own lending banks Freddie Mac and Fanny Mae, applied for bankruptcy protection and infusion of new cash from the Government. Banks, monetary institutions and commodity futures around the world were badly hit as well. The surge in oil prices which was at the basis of these finacial and commodity tumult meanwhile, continued its cause skywards.
[28] HE Abdalla Salem El-Badri, OPEC Secretary-General: *"Meeting the Challenges in the international oil market",* an OPEC/IEA Luncheon Address to the 19th World Petroleum Congress, Madrid, Spain, 29 June-3 July 2008. OPEC, Obere Donaustrasse 93, A-1020 Vienna, Austria.
[29] International Energy Agency / Information Energy Agency (IEA): Established in 1974, the IEA is the energy forum for 27 industrialised countries, among them: Australia, Austria, Belgium, Canada, Czech Rep, Denmark, Finland, France, Germany, Greece, Hungary, Ireland, Italy, Japan, Korea (South), Luxembourg, Netherlands, New Zealand, Norway, Portugal, Slovak Rep, Spain, Sweden, Switzerland, Turkey, UK. Founding Objectives: maintain and improve systems for coping with oil supply disruptions; promote energy policies in a global context through cooperative relations with non-member countries, industry and international organisations; operate a permanent information system in the international oil market; improve the world's energy supply and demand structure by developing alternative energy sources and increasing the efficiency of energy use; promote international collaboration on energy technology; assist in the mitigation of environmental and energy policies. IEA-member governments are committed to taking joint measures to meet oil supply emergencies, agree to share energy information, coordinate their energy policies and cooperate in the development of rational energy programmes that ensure energy security, encourage economic growth and protect the environment. IAE Executive Director: Nobuo Tanaka; www.iea.org .

R&D Programme...and both heavily involved in setting-up the IEF and the Joint Oil Data Initiator...there is much common ground between OPEC and the IEA on oil matters. And this is supportive of the continued healthy development and evolution of the industry in the years ahead.

- The most pressing of these [four issues] is *oil price volatility,* which has increased over the past ten months, with effects that are being felt, increasingly, in other parts of the world economy. Of special concern here is the impact on the least-developed countries. The volatile price behavior was discussed at length at the Jeddah meeting, which was called specifically for that purpose. There was widespread agreement that a combination of factors has been behind this behavior. OPEC's Member Countries stressed *that speculation has been playing a significant role in the volatility and rising prices. This is evident in the fact that more than 70% of oil futures contracts on the NYMEX are currently held by speculators.* This represents a dramatic rise.

- It is becoming increasing clear that some form of regulation is needed to moderate this. Industrialised nations have made some heavy way in this regard...Better regulation should eventually affect the *price behavior* of just not oil, but also other commodities, which have suffered a similar way from excessive speculation.

- "Let me state here clearly that market has no shortage of physical crude, with comfortable stock levels. Today, supply is higher than demand. OPEC output currently stands at 32.2 million barrels a day, well above the forecast average level of demand for its oil for the whole of 2008, of 31.8 million barrels per day. OPEC has played a great part in stabilising the oil market, by increasing output when needed and only by aiming to ensure that there is always a comfortable cushion of spare capacity. Indeed, OPEC supply has risen by 7.4 million barrels a day since 2002, compared with just 3.2 million barrels a day from non-OPEC...If more crude is needed, we will supply it. But what is the point of supplying more volume, if it is not needed? Does anyone behave like this in any other economic sector? The problem is not volume. It is price."

The OPEC believes that the world has plenty of oil and that recoverable reserves of conventional oil worldwide have doubled since the early 1980s and continue to rise, as technology has increased improved efficiency. There is a vast potential for expansion of non-conventional sources of oil, such as tar sands, oil shale, heavy oil, gas-to-liquids, coal-to-liquids and biofuels. OPEC also agrees that oil is a finite resource, enough of it being available to meet global needs; although rising costs of more than 120 upstream development projects is underway in OPEC countries with over USD120 billion in cumulative investment to 2012 – this will add to net capacity of around five million barrels a day. OPEC member countries are also investing heavily in downstream refinery capacity and delivery infrastructure. "This will help address the refinery bottlenecks in some consuming countries, which have been adding to the general oil price volatility," declared the OPEC Secretay-General. One of OPEC's top worries is stability and maximisation of transparency and predictability in consuming countries and economies, especially in policy-making and technology supply. Technology applies right across the supply chain, and can bring enormous gains in efficiency and output, as well as impacts the environment. The technology of Carbon Capture and Storage (CCS), and the IEA's JODI Programme[30] for example, will have the potential for combating climate change, and is receiving greater international recognition.

---

[30] Jean-Yves Garnier, Head of Energy Stattistics Division, IEA: *"The Joint Oil Data Initiative: The 5th JODI Conference."* Paris, 18 Nov., 2004. In the IEA's Joint Oil Data Initiative (JODI) Programme, Carbon Capture and Storage (CCS), has a potential being fully developed. CCS is a technique, which fully developed commercially, will capture or absorb and bury at least 80-90% of Carbon Dioxide being produced at modern conventional power plants, thus preventing such carbon emissions to be released

The need for better regulation that could affect the "price behavior of not just oil, but also other commodities which have suffered a similar way from excessive speculation" referred to by the OPEC Secretary General, is exemplified by The U.S. Democratic Presidential Campaign 2008 for the White House's pledge to: "Crack down on excessive energy speculation by Closing the 'Enron Loophole.'"[31] If elected President on 4 November 2008, Democratic Presidential Candidate Nominee, promised, thus:

- Senator Barack Obama today announced his plan to crackdown on excessive energy speculation and fully close the *'Enron Loophole'* to ease the impact of skyrocketing gas prices. The Enron Loophole was created by Republican Presidential Candidate Nominee, Senator John McCain's Campaign co-chair Phil Gramm at the behest of Enron – just one example of the special interest politics that put the interests of Big Oil and speculators ahead of interests of working people. And the American people have seen the results: *record corporate profits* while Americans pay *record prices* at the pump.

- For the past years, our energy policy in this country has been simply to let the special interests have their way – opening up loopholes for the oil companies and speculators so that they could reap record profits while the rest of us pay $4.00 a gallon," Senator Obama said. "My plan fully closes the Enron Loophole and restores commonsense regulation as part of my broader plan," spelled out thus:

- *Fully Close Down the 'Enron Loophole:* One of the reasons our energy market is particularly vulnerable to excessive speculation is the so-called Enron Loophole. This provision was slipped into law by Senator Phil Graham in late 2000 at the behest of Enron lobbyists to exempt some energy traders from the regulations and public protection applicable to exchange traded commodities. As a result, the Commodity Futures Trading Commission (CFTC) is unable to fully oversea the oil futures market and investigate cases where excessive speculation may be driving up oil prices. This regulatory gap is dangerous because: 1) *The absence of government oversight has the potential to facilitate abusive or price manipulation; and* 2) *The failure of a large derivatives dealer could trigger disruptions of supplies and prices in energy markets.* As President, Barack Obama will go beyond the changes included in the recently passed FARM BILL and fully close the Enron Loophole by *requiring that U.S., energy futures trade on regulated exchanges.* He will call for new, disaggregated data on index fund and other passive investments to increase transparency and oversight of the growing number of institutional investors participating in commodities futures markets. And he will support legislation directing the CFTC to investigate whether additional regulation is necessary to eliminate excessive speculation in U.S., commodities markets, including higher margin requirements and position limits for institutional investors.

into the atmosphere. CCS captured carbon emissions are envisaged to be properly buried either in deep geological formations, in deep ocean floors or in the form of mineral carbonates. The IEA's Joint Oil Data Initiative (JODI), pursued in cooperation with OPEC and 90 oil- producing and – consuming nations, began in 2000, to provide answers to political and technical concerns in oil production, oil storage, reporting mechanism, oil data bases and marketing oil. http://www.iea.or/Textbase/work/2004/jodi/garnier_jodi.pdf .
[31] U.S. Democratic Presidential Campaign for the White House 2008: *"Obama Release on 'Enron Loophole': Obama Announces Plan to Fully Close the Enron Loophole, Crack Down on Excessive Energy Speculation."* The Page-Politics up to the Minute by Mark Halperin, Sunday 22 June, 2008. TIME/CNN.

- *Ensure that U.S. Energy Futures* Cannot Be Traded on Unregulated Offshore Exchanges: CFTC oversight of oil market speculation is also limited by rules that allow energy traders to engage in unregulated transactions through foreign subsidiaries of U.S., exchanges. Currently, about 30% of U.S. oil futures trade fly below the regulatory radar because they are transacted on a U.S., exchange that works through a subsidiary in London. Similar arrangements are being pursued by U.S., exchanges in partnership with Dubai as well. Barack Obama would limit the price impacts of excessive speculation by *preventing* U.S., crude oil from routing their transactions through offshore markets in order to evade speculation limits and also impose reporting requirements.

- *Work with Other Countries to Coordinate Regulation of Oil Futures Markets:* As the global energy market expands and trading for oil futures increases, Barack Obama believes we must work with other countries to establish uniform approaches to avoiding excessive speculation in commodities futures markets. As President, he would work through the International Organisation of Securities Commissioners (IOSC) and other international organisations to harmonise regulations across countries. This effort will help to ensure that as the U.S., strengthens oversight and transparency in U.S., exchanges, these efforts are not undermined by overseas trading subject to lax regulations.

- *Call on the Federal Trade Commission and Department of Justice* to Vigorously Investigate Market Manipulation in Oil Futures: In 2007, Senator Obama supported legislation that gave the Federal Trade Commission (FTC) new authority to investigate and pursue price manipulation in oil markets. However, even in the face of record oil prices and growing concerns about excessive speculation, the Bush Administration has failed to utilise these new powers. Barack Obama does not believe we can afford to wait weeks and months more to vigorously investigate whether energy traders and oil companies are manipulating the market at the expense of consumers. He is calling on the FTC to immediately expedite its investigation into market manipulation, including in the oil futures markets. He is also calling on the Department of Justice to *open an investigation of energy traders who may have engaged in illegal activities that have helped drive up the price of oil and food.*

While commodity futures speculators, oil marketers and oil producer OPEC shifted the blame on one another, the price surges in oil and food and inflation, sparked off by the excessive oil speculators – maintained their course. The world's teeming billions of people and domesticated animals were restlessly saddled with increased high food prices, high fuel prices, incomes that were low and falling lower still, high costs of everything Nature provides without charge like air, water and sunlight. Who will protect them even from the ravages of climate change, global warming, and faultering agricultural input?

At the height of the oil-and-food-crisis debacle, the UN Food and Agricultural Organisation (FAO)[32] met in June 2008 to follow-up on its January and April 2008 Expert Meetings and Shareholder Consultations. They discussed the food price crisis, climate change and agriculture, and food vs fuel. They concluded that:

---

[32] UN Food and Agricultural Organisation (FAO): *"High-Level Conference on World Food Security 200: The Challenge of Climate Change and Bioenergy."* The FAO was established by the UN General Assembly in December 1974 on the recommendation of the UN World Food Conference (WFC) to serve as a coordinating body for national ministries of agriculture, to help reduce malnutrition and hunger. WFC, first held in Rome in 1974 in the wake of a devastating famine in Bangladesh, was itself officially suspended in 1993 and its functions absorbed by the FAO.

Conventional causes, falling world food stockpiles, structural changes in trade and agricultural production, agricultural price supports, subsidies in developed countries, diversions of food commodities to high imput foods and fuel, commodity market speculation, and climate change - contributed to the crisis. Global Warming brought about by Climate Change is significantly affecting agriculture, for example, temperature, precipitation and glacial runoff – which determine the carrying capacity of the biosphere to produce enough food. Rising CO_2 levels affect the weather with detrimental as well as beneficial effects on crop yield. Agricultural production itself produces greenhouse Gases (GHGs) such as CO_2, methane, nitrous oxide; these affect both weather and crops and mobility of the landcover to absorb or reflect heat and light, and thus contributing to radiative forcing, deforestation, and desertification. Storms, hurricanes, and floods also greatly impact on food production. The uses of fossil fuels are sources of major *anthropogenic* or man-made sources of CO_2.

In its bi-annual *Food Outlook*, the FAO estimated the world's import bill rose in 2007 to $745 billion, up 21% from 2006. In developing countries, costs went up by a quarter to nearly $233 billion – all caused by record oil prices, farmers switching out of cereals to grow biofuel crops, extreme weather and growing demand from countries like India and China. Hardest hit were sub-Saharan Africa, where many of the world's poorest nations depend on both high-cost energy as well as food imports. The situation was deteriorating further and cash-poor governments are forced to choose between the former which has almost always won out in the past.

The oil-and-food prices crisis clearly demonstrate the damage done to life, standards of living, lifestyles, economies, welfare, politics, weather, climates, societies and civilisation. The use and production of fossil fuels is the root cause of all these. It would have dawned on governments and leaders of the world, decision-makers, institutions and organisations of great influence and organised individual and heightened group activity, to set forth and smooth the way to transit fossil fuels to Renewable Energies. But as Colin J. Campbell[33] wrote on 29 April 2000 under *Rational National Energy Policies*, next to nothing is being seriously undertaken:

- So far as I can find out, virtually no government is properly aware of the growing energy crisis, still less making any serious plans. For some strange reason, they have all become concerned about the climate issue and have at least addressed it, howerver limited their actual response. It is not clear either that many of the senior management in oil companies truly perceived their own situation: *most of them are forced to sing to the short-term stock market. They do, however, make oblique references.* For example, BP now describes itself as a gas company also concerned with oil and *renewables* (the CEO on BBC); the CEO of ExxonMobil speaks of being excited by technology, but confesses that it does not change the raw resources; the CEO of Shell says he agrees with the USGS, which itself puts out the message in heavy camouflage. I think the Japanese are the most aware of the position. The Indians, too, are worried. Norway in earlier years did have a policy of restraint and good management, but has become corrupted by easy money like everyone else. *They pretend to be cutting production to support OPEC*, but in reality they are just watching the depletion of their oil fields. The last government fell over an issue of the CO_2 emissions from gas to generate

[33] Colin J. Campbell: *"OIL CRISIS."* http://healthandenergy.com/oil-crisis.htm

electricity. I think that the Germans are becoming aware, and may prove to lead Europe. They are however, tangled up with the Green movement wishing to close nuclear stations, and find it difficult to admit that they will need nuclear in the face of an oil crisis. I think that the U.S., UK and European Commission are all quite useless on this issue, understanding little and doing less. Not a promising picture...

- Americans use 25 barrels per capita yearly (Bcy). Japanese and Italians consume 10-Bcy. The global average is less than 4.7 Bcy. China, and to a lesser extent, India, are following the U.S., in rapidly expanding their economies using the technology of cheap oil. The high prices for oil today are creating a widespread awareness that there isn't enough oil being pumped to supply everybody. Poor Third World countries are already priced out of the market. Humans, with a few notable exceptions, *have not yet demonstrated willingness and desire to make any sacrifice to assure a successful transition to renewable energy sources before fossil fuels become extremely scarce and expensive.* The best depiction of the past 100 years of human endeavor is exponential growth of the world's population based on an exponentially growing world economy based on the *utilisation of incredibly cheap fuel.* What happens when there isn't enough oil to keep growing or maintain the current human population?"

We can no longer afford to pretend that fossil fuels are infinite. Or afford to pretend that oil shortages and oil-and-food price surges will no longer occur. Or afford to pretend that dependence on foreign oil will cease on its own; Or afford to pretend that global warming and climate change will cease if we continue to use fossil fuels; Or afford to pretend that Greenhouse gases (GHGs) - which include toxic mercury, lead, nitrous oxides, carbon dioxide, and other pollutants - will cease terrorising the atmosphere, if we continue to produce and use fossil fuels, while we continue aggressively to promote their production and expansion.

We can begin to unravel our pretenses and prepare to begin the *Renewable Energies Era* by updating and enforcing building codes that require much less energy while allowing very good indoor air quality; increase the use of solar and wind energy, fuel cells and co-generation systems in buildings to generate electricity, hot water, absorption cooling and space heating; stop subsidising the fossil oil industry so that fossil fuels can compete with renewable energies in the free market; invest the funds used to subsidise fossil fuels in the timely development of enough renewable energy systems; wean ourselves from non-renewables by implementing vigorous action now to reduce fuel consumption.

One huge reason why we must and should divest ourselves of our pretenses and deal in honesty and good faith to smooth the way from *fossil fuels* to *renewable energies'* uses and production, is climate change. Projections of future changes in climate provided by the Intergovernmental Panel on Climate Change (IPPC)[34] are more than frightening and devastating. They are our dead warrants without an alternative to drop fossil fuels like hot potatoes, and immediate embrace renewable energies to run our economies:

- For the next two decades a warming of about 0.2°C per decade is projected for a range of **SRES** emissions scenarios. Even if the concentrations of all GHGs and aerosols had been kept constant at year 2002 levels, a further warming of about

---

[34] Intergovernmental Panel on Climate Change (IPCC): *"Climate Change 2007. Synthesis Report."* http://www.ipcc.ch/pdf/assessment-report/ar4/syr/ar4_syr.pdf; IPCC-Sec@wmo.int

0.1°C per decade would be expected. Afterwards, temperature projections increasingly depend on specific emissions scenarios [Figure 2].

- Continued GHG emissions at or above current rates would cause further warming and induce many changes in the global climate system during the 21st century that would *very likely* be larger than those observed during the 20th century.

- Climate Change is expected to exacerbate current stresses on water resources from population growth and economic and land-use change, including urbanisation. On a regional scale, mountain snow packs, glaciers and small ice caps play a crucial role in freshwater availability. Widespread mass losses from glaciers and reductions in snow cover over recent decades are projected to accelerate throughout the 21st century, reducing water availability, hydropower potential, and changing seasonal flows in regions supplied by meltwater from major mountain ranges (e.g., Hindu-Kush, Himalaya, Andes), where more than one-sixth of the world population currently lives.

- "The negative impacts of climate change on freshwater systems outweigh its benefits (high confidence). Areas in which runoff is projected to decline face a reduction in the value of the services provided by water resources (very high confidence). The beneficial impacts of increased annual runoff in some areas are *likely* to be tempered by negative effects of increased precipitation variability and seasonal runoff shifts on water supply, water quality and flood risk...Generally, the irreversible phenomenon of climate change impacts on *water supply, ecosystems, food production and supply, coasts, and health,* are:

- (a) *"Water:* Increased water availability in most tropics and high latitudes; Decreasing water availability and increasing drought in mid-latitudes and semi-arid low latitudes; Hundreds of millions of people exposed to increased water stress.

- (b) *"Ecosystems:* Up to 30% of species face increasing risk of extinction or significant extinctions around the globe; Increased coral bleaching, most corals bleached; Widespread coral mortality; Terrestrial biosphere tends toward a net carbon source as – 15%-40% of ecosystems affected; Increasing species range shifts and wildfires' risk; Ecosystem changes due to weakening of the meridial over turning circulation.

- (c) *"Food:* Complex, localised negative impacts on small holders, subsistence farmers and fishers; Tendencies for cereal productivity to decrease in low latitudes; Productivity of all cereals decreases in low latitudes; Tendencies for some cereal productivity to increase at mid-to-high latitudes; Cereal productivity to decrease in some regions.

- (d) *"Coasts:* Increased damage from floods and storms; About 30% of global coast wetlands lost; Millions more people experience coastal flooding each year.

- (e) *"Health:* Increasing burden from malnutrition, diarrhoeal, cardio-respiratory and infectious diseases; Increased mortality from heat waves, floods and drought; Changed distribution of some disease vectors; Substantial burden on health services. [Figure 3.6: Examples of Impacts associated with global average temperature change; Table 3.2: Examples of possible impacts of climate change due to changes in extreme weather and climate events, based on projections to the mid-to late 21st century. These do not take into account any changes or developments in adaptive capacity. The *likely* hard estimates in columns 2 relate to the phenomena listed in column 1- Phenomenon and direction.]

- *"Virtually Certain To Occur:* Human Health: Reduced human mortality from decreased cold exposure; Industry, settlement and society: Reduced energy demand for heating; Increased demand for cooling; Declining air quality in cities; Reduced disruption to transport due to snow, ice; Effects on winter tourism.

- *"Very Certain To Occur:* Phenomenon and direction of trend will be witnessed; Over most land areas, warmer and fewer cold days and nights, warmer and more frequent hot days and nights; Agriculture, forestry and ecosystems: Increased yields in colder environments, decreased yields in warmer environments, and increased insect outbreak; Water resources: Effects on water resources relying on snowmelt, effects on some water supplies; Increased water demand; Water quality problems, (e.g., algal blooms); Water scarcity may be relieved; More waterspread water stress; Power shortages causing disruption of public water supply; Decreased freshwater availability due to saltwater intrusion.

- *"Very Likely To Occur:* Land degradation, lower yields and crops damage and failure, increased livestock deaths, increased risk of wildfire, damage to crops, windthrow (uprooting) of trees, damage to coral reefs; Salinisation of irrigation water, estuaries and freshwater systems; Reduced yields in warmer regions due to heat stress, and increased danger of wildfires, damage to crops, soil erosion, inability to cultivate land due to waterlogging of soil; Increased water demand; Adverse effects on quality of surface and groundwater; Contamination of water supply; Increased risk of heat-related mortality, especially for the elderly, critically sick, very young and socially isolated; Increased risk of deaths, injuries and infections, respiratory and skin diseases; Increased risk of food and water shortages; Increased risk of malnutrition; Increased risk of water- and food-borne diseases; Increased risk of deaths, injuries, water- and food-borne diseases; post-traumatic stress disorders; Increased risk of deaths and injuries by drowing in floods; Migration-related health effects; Reduction in quality of life for people in warm areas without appropriate housing; Impacts on the elderly, very young and poor; Disruption of settlements, commerce, transport and societies due to flooding: Pressures on urban and rural structures, loss of property; Water shortage for settlements, industry and societies; Reduced hydropower generation potentials; Disruption by floods and high winds; Withdrawal of risk coverage in vulnerable areas by private insurers; Potential for population migrations; Costs of coastal protection versus costs of land-use relocation; Potential for movement of populations and infrastructure; Also tropical cyclones and earthquakes.

- *"Likely To Occur:* Warm spells and heat waves; Frequency of increases over most land areas; Heavy precipitation events; Area affected by drought increases; Intense tropical cyclone activity increases; Increased incidents of extreme high sea level (excludes tsunami)."

Oil Consumer- and Producer-countries and companies[35] are aware of the predicament their oil wares present to its clients, economies, markets,

---

[35] Oil & Petroleum Companies: The Oil & Petroleum Industry is a world-wide phemenon. There are thousands of companies and enterprises, big, small and medium operating in both Oil-Consuming as well as Oil-Producing Countries. The largest ones country-by-country, include: **U.S.A:** Alon, Dallas; Amerada Hess Corporation, New York; Anadarko Petroleum Corporation, The Woods, Texas; Apache Corporation, Houston, Texas; Chevron Corporation, San Ramon, California; Citgo Petroleum Corporation, Houston, Texas; ConocoPhillips, Houston, Texas; Crown Central Petroleum/Crown Central LLC, Baltimore, MD; Devon Energy Corporation, Oklahoma City, Oklahoma; ExxonMobil Corporation, Irving, Texas; Koch Industries, Wichita, Kansas; Marathon Oil Corporation, Houston, Texas; Occidental Petroleum, Los Angeles; California; ENSCO International, Dallas, Texas; Shell Oil Company, Houston, Texas; Sinclair Oil, Salt Lake City, Utah; Sunoco, Philadelphia, Pennsylvania; United Refining Company, Warren, Pennsylvania; **Canada:** Canadian Natural Resources Ltd, Calgary, Alberta; Enbridge Inc, Calgary, Alberta; Encata, Calgary, Alberta; Husky Energy, Calgary, Alberta; Imperial Oil, Calgary, Alberta; Irving Oil, Saint John, new Brunswick; Petro-Canada, Calgary, Alberta; Shell Canada, Calgary, Alberta; Gulf Oil International Ltd, Luxembourg, Calgary, Alberta; Suncor

environments, and the future of the health of the planet Earth. Everyone agrees that fossil fuels could be in place for the foreseeable couple of estimated 22-52 years (up to 2030-2060 for oil and gas respectively); that oil wells will sooner or later dry up - new investments and improvements notwithstanding – and in spite of its advantages. "Cheap" oil enables economies to expand, as well as provides jobs and wealth for employees and owners of the enterprises. The global economy is literally fossil-fuel-based. *But compared to oil's disadvantages, it is a pyrrhic victory. It is more than a pyrrhic victory because: increasing volumes of GHGs blown into the atmosphere continue to defy and render this pyrrhic victory, at least, very slippery, and at worst, no victory at all!*

Attempted cosmetic repairs to oil's slippery, pyrrhic victory are numerous, but are mostly ineffectual and ineffective, as for example:

- *UN Framework Convention on Climate Change* (UNFCCC)[36] or Earth Summit: Established in

---

Energy, Calgary, Alberta; Syncrude Canada Ltd, Fort Murray, Alberta; Talisman Energy Inc, Calgary, Alberta; **Latin America/Caribbean:** *Bolivia:* Ecopetrol, Bogota; Yacimientos Petroliferos Fiscales Bolivianos (YPFB), Bogota; *Brazil:* Petrobras, Rio de Janeiro; *Cuba:* Cupet, Havana; *Trinidad & Tobago:* PETROTRIN, Trinidad & Tobago; *Mexico:* Mexicanos (PEMEX), Mexico City, Mexico; *Suriname:* State Oil Company of Suriname; *Venezuela:* Petroleos de Venezuela, S.A., Caracas; **Africa/Atlantic:** *Algeria:* Societe Nationale pour le Recherche, la Production, la Transportation, et la Commercialisation des Hydrocarbones s.p.a , Algiers; *Angola:* Sonangol Group, Luanda; *Congo-Brazz:* Snpc, Brazzaville; *Libya:* National Oil Corporation (NOC), Tripoli; *Nigeria:*Nigerian National Petroleum Corporation (NNPC), Abuja; *Tunisia:* Enterprise Tunisienne d'Activites Petrolieres (ETAP), Tunis; **Asia/Pacif/Atlantic:** Atlantic Petroleum, Torshan, Faroe Is; *China:* PetroChina Company Ltd, Beijing; Sinopec/China Petroleum and Chemical Corporation, Chaoyang District, Beijing; CNOOC Ltd, Beijing; *India:* Assam Company of India (ACL); Bharat Petroleum Corporation Ltd India, (BPCL) Mumbai; Cairn Energy; Essar Oil Ltd, Jamnagar; Petronet LNG, New Delhi; Gujarat Gas Co. Ltd; Gujarat State Petroleum Corporation Ltd; Hindustan Petroleum Corporation; Indian Oil Corporation, New Delhi; Reliance Industries Ltd, Mumbai; Oil India Ltd, New Delhi; Oil & Natural Gas Corporation (ONGC), Dehradum; *Japan:* Cosmo Oil Company, Tokyo; IB Daiwa, Tokyo; Impex Corporation, Tokyo; Japan Energy Corporation (Jomo), Tokyo; Nippon Oil Corporation, Tokyo; San-Ai-Oil, Shinagawa; *Indonesia:* Perusahaan Tambang Minyak Negara (Pertamina), Jakarta; MedcoEnergi, Jakarta; *Pakistan:* Oil & Gas Development Company Ltd; Pakistan State Oil (PSO), Karachi; *Philippines:* Petron Corporation, Lima Bataan; *Malaysia:* PETRONAS, Kuala Lumpur; *Singapore:* Singapore Petroleum Company Ltd (SPC); **Australia/New Zealand:** *Australia:* BHP Billiton, Melbourne; Woodside Petroleum Ltd, Perth, Western Australia; *New Zealand:* Todd Corporation, Wellington; **Europe:** *Austria:* Österreische Mineralölverwaltung (OMV), Vienna; *Croatia:* Industrija Nafte (INA), Zagreb; *Finland:* Neste Oil, Espoo; *Hungary:* Mol Group, Budapest; *France:* Total S.A., La Defense, Paris; *Germany:* E.ON AG, Düsseldorf; Kuwait German Petroleum Company (KGPC), Berlin; Preußische Elektrizitäts AG, (PreussenElektra) Berlin; Wintershall AG, Kassel; Rheinisch-Westfälisches Elektrizitätswerk (RWE Dea) AG, Essen; VEBA AG; *Greece:* Hellenic Petroleum, S.A., Athens; *Ireland:* Maxol Group, Dublin; *Italy:* Eni S.p.A, Rome; *Netherlands:* Royal dutch Shell plc (Shell), The Hague; *Norway:* StatoilHydro, Stavanger; *Portugal:* Galp Energia, Lisbon; *Poland:* Grupa LOTOS, Gdank; Polski Konern Nafftowy Orlen (PKN Orlen), Plock; Polskie Gornictwo Haftowe i Gazownictwo (PGNiG) – Polish Oil & Gas Company, Warsaw; *Romania:* Petrom, Bucharest; Rompetrol Group N.V., Bucharest; *Serbia:* Naftna Industrija Srbije, Novi Sad; *Spain:* Repsol YPF, Madrid; *United Kingdom:* BG Group, Reading; British Petrol, London; **Middle East:** *Iraq:* Iraq National Oil Company (INOC) Bagdad; Iraq Petroleum Company (IPC), Bagdad; North Oil Company, Kirkuk; South Oil Company, Basrah; *Iran:* National Iranian Oil Company (NIOC), Teh'ran; *Qatar:* Qatar Petroleum (QP), Dohar; *Saudi Arabia:* Saudi Arabian Oil Company (ARAMCO), Dharan; *Sultanate of Oman:* Oman Oil Company (SAOC); *United Arab Emirates* (UAE); Abu Dhabi National Oil Company (ADNOC), Abu Dhabi; **Russian Federation/Former Republics:** Lukoil, Moscow; Gazprom, Moscow; OJSC Surgutneftegas, Tyumenskaya Oblast, City of Surgut; OAO Rosneft Oil Company (Rosneft), Moscow; *Azerbaijan:* State Oil of Azerbaijan Republic (SOCCAR), Baku City; *Ukraine:* Naftohaz Ukrainy - Naftogas of Ukraine), Kiev.

[36] UN Framework Convention on Climate Change (UNFCCC): *"Objectives & Principles"*, UN 1992, FCCC/84, GE.05-62220 (E) 200707: Objectives: to achieve, in accordance with the relevant provisions…stabilisation of GHG concentrations in the atmosphere at a level that would present dangerous anthropogenic interference with the climate system. Such a level should be achieved within a time frame sufficient to allow ecosystems to adapt naturally to climate change, to ensure that food production is not threatened and to enable economic development to proceed in a sustained manner; Principles: 1. Protect the climate change system for the benefit of present and future generations of mankind, on the basis of equity and in accordance with their common but differentiated responsibilities. Acoordingly, the developed countries should take the lead in combating climate change and the adverse effects thereof; 2. the specific needs and special circumstances of developing countries, especially those that are particularly vulnerable to the adverse effects of climate change, and those that would have to bear a disproportionate or abnormal burden under the Convention, should be given full consideration; 3. take precautionary measures to anticipate, prevent or minimise the causes of climate change and mitigate its adverse effects. Where these are threats of *serious* or *irreversible* damage, lack of full scientific certainty should not be used as a reason for postponing such measures, taking into account that policies and measures to deal with climate change should be cost-effective so as to ensure global benefits at the lowest possible cost. To achieve this, such policies and measures should take into account different socio-economic contexts, be comprehensive, cover all relevant sources, sinks and reservoirs of GHG and adaptation,and comprise all economic sectors. Efforts address climate change may be carried out cooperatively by interested parties. [GHGdata@unfccc.int; secretariat@unfccc.int]

1994 with 192 countries ratifying it. It sets an overall framework for intergovernmental efforts to tackle the challenge posed by climate change – recongising that the climate change system is a shared resource whose stability can be affected by industrial and other greenhouse gases (GHGs). Governments gather and share information on GHGs, national policies and best practices; launch national strategies for addressing GHG emissions and adapting to expected impacts, including the provision of financial and technological support to developing countries; cooperate in preparing for adaptation to the impacts of climate change. The Kyoto Protocol establsihed on 11 December 2007, linked to the the FCCC – binds targets for 37 industrialised countries and the European Union (EU) for *reducing GHG emissions to an average 5%* against 1990 levels over the five-year period 2008-2012. FCCC GHG data include: *Carbon dioxide* (CO^2); *Methane* ($CH4$); *Nitrous oxide* (N^2O); *Perfluorocarbons* (PFCs); *Hydrofluorocarbons* (HFCs); *Sulphur hexafluoride* (SF); and indirect GHGs such as SO^2, *NOx, CO* and *NMVOC.* The FCCC encourages and commits the industrialised nations to stabilise GHG emissions, while the Kyoto Protocol "recognises that developed countries are principally responsible for the current high levels of GHG emissions in the atmosphere as a result of more than 150 years of industrial activity, and for this, the protocol places a heavier burden on developed nations under the priciple of 'common but differentiated responsibilities'". The detailed rules for the protocol's implementation were adopted at the COP 7 in Marrakesh, Morocco in 2001, and are called the "Marrakesh Accords".

- *Adaptation Fund* (AF) or Clean Development Mechanism (CDM): A Kyoto Protocol Agreement which allows industrialised or Annex I countries with GHG reductions commitment to invest in projects that reduce GHG emissions in developing countries as an alternative to more expensive emission reductions in their own countries. A CDM carbon-approved project establishes that its planned emission reduction would not occur without the additional incentive provided by emission reduction credits.

- *Clean Technology Fund* (CTF): This is a fund temporarily assigned to the World Bank management responsibilites by the UN to focus on making Renewable Energy cost-competitive with coal-fired power as quickly as possible. But after the UN's Copenhagen climate change conference in December 2009, the World Bank will instead re-direct CTF to the Bank's continued investment in huge coal-fired power plants (i.e., fossil fuels coal, natural gas, oil), as well as applying CTF as additional source of money for business as usual. CTF money is contributed by Multi-development Banks (MDBs) and countries purposed to bridge the financing-learning gap between now and a post-2012 global climate change agreement.[37]

- *Global Environmental Facility* (GEF)[38]: This 178-member countries' partnership organisation with international institutions, Non-Governmental Organisations (NGOs) and the private sector, is the largest that addresses global international issues as well as support natural sustainable development initiatives. It provides grants for projects related to biodiversity, climate change, international waters, land degradation, ozone layer and persistent organic pollutants (POPs). GEF finances the *FCCC* or Earth Summit, and three other major international environmental organisations: *Convention on Biological Diversity* (CBD) or Biodiversity Diversity (BD) - a treaty adopted in Rio de Janeiro in 1992 to conserve biological diversity; sustainable use of its components; fair/equitable sharing of benefits arising from genetic sources; *UN Convention to Combat Desertification* (UNCD), a direct recommendation of the Conference's *Agenda 21* – to combat desertification and mitigate the effects of drought through nationalisation programmes that incorporate long-term strategies supported by international cooperation and partnerships; *Stockholm Convention on Persistent Organic Pollutants* (POPs)[39] - organic compounds that are resistant to environmental degradation through chemical,

---

[37] Clean Technology Funds (CTF) funds are divided into CTF and the Strategic Climate Fund (SCF).
[38] Global Environmental Facility (GEF): Was established in 1991 by the UN Environmental Programme (UNEP), and has since provided $7.6 billion in grants and leveraged $30.6 billion in co-financing for over 2000 projects in some 165 countries. Ontact: Ms Maureen Shields Lorenzetti, Spokesperson, mlrenzetti@thegef.org; www.gefweb.org.
[39] Persistent Organic Pollutants (POPs): These include: *aldrin; chlorane; DDT; dieldrin; heptachlor; hexachlorobenzene; mirex; polychlorinated dibenzofurans; toxaphene;carcinogenic polycyclic aromatic hydrocarbons* (PAHs); *brominated flame-retardants* 1. minerals: *asbestos;* compounds: *aluminum hydroxide, magnesium hydroxide, antimony trioxide,* various hydrates, *red phosphorous, boron compounds;* 2. Hydroxymethyl or Tretrakis phosphorium salts used as fire retardants for textiles; 3. Synthetic materials, usually *halo carbons: organochlorines* such as *polychlorinated biphenyls* (PCBs), *chlorendic acid* derivatives and *chlorinated paraffins, organobromines* such as *polybrominated diphenyl ethers* (PBDEs) which is further broken down into *pentabromodiphynyl ether* (pentaBDE), *octabromo-diphenyl ether* (octaBDE), *decabromodiphenyl ether* (decaBDE) and

biological and photolytic process. POPs are capable of long range transport, bioaccumulate in human and animal tissue, biomagnify in food chains, and have significant fatal impacts on human health and the environment.

But leaders of the world's major economies or G8 Nations[40] in their annual 2008 Summit Meeting in Tokyo, Japan, on 08 July 2008, failed to set specific targets in GHG emissions. They merely "endorsed the idea of halving greenhouse gas emissions worldwide by 2050"[41], as follows:

- Conscious of our leadership role in meeting such challenges, we, the leaders of the world's major economies – commit to combat climate change in accordance with our common but differentiated responsibilities and respective capabilities...We support a shared vision for long-term cooperative action, including a long-term global goal for emission reductions, that assures growth, prosperity and other aspects of sustainable development...Taking in account assessments of science, technology, and economics, we recognise the essential importance of enhanced greenhouse gas mitigation that is ambitious, realistic and achievable..."

- We've made progress – significant progress – toward a comprehensive approach...In order to address climate change, all major economies must be at the table, and that's what we had here today," U.S., President Bush said, who was referring specifically to China and India – his main cause to oppose the Kyoto Protocol, as it did not include *strict emissions limits for China and India.*

Significant progress towards climate change and stricter emissions limits should concern both oil- consuming countries and companies as well as oil producing- countries and companies. The scoping meeting and Special Report on Renewable Energies held by the Intergovernmental Panel on Climate Change (IPCC) in Lübeck, Germany,[42] provided a better understanding of the Resources by region and impacts of climate change on these resources; Mitigation of potential of renewable energy sources; Linkages between renewable energy growth and co-benefits in achieving sustainable development by region; Impacts on global,

---

hexabromocyclododecaue (HBCD); organophates or halogenated phosphorous compound: *trio-cresyl phosphate, tris(2-3 dibromo-propyl) phosphate* (TRS), *bis (2, 3-dibromopropyl) phosphate, tris(1- aziridinyl)- phosphine oxide* (TEPA), etc. trio-cresyl phosphate; PAHs occur in oil, cod and tar deposits, and are produced as byproducts of fuel burning and identified as *carcinogenic,metagenic* and *teratogenic,* and they are also found in the interstellar medium, in commets and in meteorites.

[40] Group of G8 Nations: This an international forum that comprises Canada, U.S., France, Japan, Germany, UK, Russia, Italy, whose heads of state hold annual summits, and their foreign ministers also meet four times a year – to deliberate on economic, political and financial matters. The first G8 Summit was founded in November 1975 in the aftermath of the 1973 Oil Crisis. Brazil, India, South Africa and Mexico were included in the 2008 G8 Summit, but generally, these additional nations have held only sideline meetings with G8 summiteers. The Group of Seven is the Summit Meeting of the Finance Ministers from the group of 7 industrialised nations formed in 1976 who meet regularly each year to discuss economic issues. [There is also a (1) G20, 21 and 22 Developing Nations Group – which is a bloc of developing or underdeveloped countries formed during the 20 August 2003 5th Ministerial World Trade Organisation (WTO) Conference in Cancun, Mexico. These more than 150 countries classified as 'underdeveloped, developing or emerging' economies, and include: Russia, India, China and South Africa. The G20 Group formed under the Brasilia Declaration which complain about: "the protectionist policies in developed nations who at the same time pressure underdeveloped nations to pursue liberalisation of their markets."; (2) D-8 or Developing Eight: a group of developing nations who formed an economic development alliance in 1997 in Istanbul, Turkey with the objective to improve developing countries' position in the world economy. They include: Bangladesh, Egypt, Indonesia, Iran, Malaysia, Nigeria, Pakistan, Turkey; (3) Group of Eleven: a forum formed by mostly developing countries in 2005 and called on the rich developed G8 nations to ease their debt burden owed to the industrialised nations. The G11 Group includes: Jordan, Croatia, Ecuador, Georgia, El Salvador, Honduras, Indonesia, Morocco, Pakistan, Paraguay, Sri Lanka; (4) Goldman Sachs Investment Bank lists a Group of N-11 or Next Eleven Nations it has identified as having a high potential of becoming the world's largest economies: Brazil, Russia, India, China (BRICs), Bangladesh, Egypt, Indonesia, Iran, Mexico, Nigeria, Pakistan, Philippines, South Korea, Turkey, Vetnam.]

[41] *"ENVIRONMEMENT: World Leaders avoid setting greenhouse target."* TIME in Partnership with CNN, Business/07/09/ga.summit/index.html.

[42] PROPOSAL FOR AN IPCC SPECIAL REPORT ON RENEWABLE ENERGY: Was presented by Germany and supported by Austria, Belgium, Denmark, Gambia, Greece, Spain, Sweden and The Netherlands, during the Global Renewable Energy Forum (GREF) discussions held in Foz do Iguacu, Brazil, 18-21 May 2008, organised by the UN Industrial Development Organisation (UNIDO), Brazilian Ministry of Mines & Energy Electrobas, and Itaipu Binacional.

regional and national energy security; Technology and market status, future developments and projected rates of development; Options and constraints for integration into the energy supply system and other markets, including energy storage options; Economic and environmental costs, benefits, risks and impacts of development; Capacity building, technology transfer and financing in different regions; Policy options, outcomes and conditions for effectiveness; Accelerated development could be achieved in a sustainable manner.

1.3.3 Oil-Producer-Consumer Countries, Companies

MANY OIL-consuming countries also produce oil. There are OPEC and non-OPEC oil-producers and oil-consumers. They have common programmes and expectations: that consumers be sustainably, securely and regularly be supplied with oil; demand for oil to be constant to spur up more investments for further expansion of the oil and petroleum industry; construction of more upstream and downstream facilities around the globe; and, to make the industry attractive to college and university graduates. Oil merchants, (exporters and importers) coordinate their efforts with international oil companies, national oil companies, service companies, governments and academia, to make sure that the world remains *more addicted* to fossil oil dependency. The OPEC details the strategy:[43]

- With the world expected to rely on fossil fuels for many decades to come, it is critical to ensure that future energy growth that supports both economic growth and social progress, is compatible with tackling the issue of climate change as well as move towards a more carbon-constrained world. It points to the need to promote the early development and deployment of cleaner fossil fuel technologies. Carbon capture and storage (CCS) is a technology that could make a significant contribution to abate the growth of CO_2 emissions, and can be applied to large stationary sources of CO_2 emissions such as power, cement and steel plants. The Intergovernmental Panel on Climate Change (IPCC) estimates the range of economic mitigation potential for CCS to be 200-2000 gigatons of CO_2 by 2100. Its development also points to a 'win-win' option. Combining it with enhanced oil recovery where possible, could help offset part of its development costs. The oil and gas industry can offer valuable expertise and opportunities for cost reductions...

- ...On top of these challenges, the oil industry faces great uncertainties over how much to invest. For example, the U.S. Energy Security and Independence Act (ESIA) of 2007 and recent European Union (EC) proposals to address climate change and *renewables targets* could have substantial impacts upon the amount of oil that could be supplied by OPEC. The ESIA introduces changes to automobile efficiency standards, as well as a requirement to rapidly increase the contribution of alternative fuels in the transportation sector. The EU Commission package of implementation measures for climate change and renewable energy sets out a greenhouse gas emission reduction of 20% by 2020 compared to 1990 levels, and a target of 20% renewable energy by 2020, including a 10% biofuels target in road transportation.

- Scenarios show that these policy measures *could reduce* the call on OPEC oil by close to 4 million barrels per day (mb/d) by 2020. A key question that arises from this is, the extent to which *new fuel economy standards and targets* for biofuels and renewables should already be

---

[43] *"WORLD OIL OUTLOOK 2008."* OPEC's Executive Summary; ISBN-978-3-200-01253-0; OPEC Secretariat, Obere Donaustrasse 93, A-1020 Vienna, Austria. www.opec.org.

factored into future reference case projections. It is thus increasingly becoming necessary to review even reference case oil demand projections.

- Greater uncertainty exists for the required amount of OPEC oil, if we move beyond these specific policy measures. Broader scenarios for OPEC crude oil suggest that the range of uncertainty for OPEC oil is considerable. By 2020, the amount of crude oil needed is in the range 29-38 mb/d, a gap of 9 mb/d. This translates into an uncertainty gap for upstream investment needs in OPEC Member Countries of over $300 billion in 2007 dollars."

Oil-petroleum-producing countries and companies, like oil-petroleum-consuming countries and nations, are working round-the-clock producing, purchasing and investing heavily on fossil fuel's everlasting destiny and integrity – even as the production, uses, investment and establishment of renewable energies are gathering momentum globally. Central to Big Oil's stratagem to keep pace with the rapid rate of incoming alternative fuels other than fossil fuels is: Refining of crude for quality and continued move toward more stringent quality specifications; Supply capabaility and economics; Supply levels of non-crudes that essentially bypass refineries; Demand growth and mix; and, possible imposition of carbon emissions targets on refineries. It has been estimated that between 2008 and 2015, around 7.6 mb/d of new crude distillation capacity will add to the global refining system to increase it by 8.8 mb/d from 2007 levels.

Distillation additions would exceed requirements in each year from 2010-2013 as a range of new projects come onstream, to ease refining tightness and margins. Crude distillation unit addition by 2015 appears close to sufficient, but those for secondary processing units are not, and so need substantially more especially for *hydrocracking*hy and *desulphurisation*. The gap between supply and demand likely to drive price differentials towards a premium for diesel and could have an impact on the absolute level of product and crude prices for middle distillates will grow, if more diesel-oriented projects are not implemented. The proportion of non-crudes in the total supply will *rise*, while that for crudes to be processed per barrel of additional product demand will *decline*.

Total non-crudes are projected to cover more than 16 mb/d of supply in 2000 and 20 mb/d in 2030 compared to an estimated 10.5 mb/d in 2007, and this increase impacts the downstream as these downstreams are predominantly light and clean and most of them bypass refining processes. Also, as ethanol supply increases, especially in the Atlantic Basin, it exacerbates the weakness in gasoline margins globally in the medium-term, while biodiesel growth in Europe widens diesel deficits sharply. Total substantial investment of more than $320 billion in refinery processing to 2015 is required in all regions, and the 2030 figure is close to $800 billion. Inter-regional oil trade increases by more than 25 mb/d will reach the level of 77 mb/d by 2030; both crude oil, refined product exports and tanker capacity will increase around 170 million deadweight tonnes (dwt) by 2030, from 2007 levels of around 550 million dwt.

The phenomena of Climate Change, CO^2 emission reduction targets and subsequently, Global Warming, seem *coincidental* in the scheme of fossil fuel production and consumption jigsaw puzzle. Second-generation fossil fuels, no matter how 'clean' they may be polished, cannot even superficially, delay, postpone or hinder the smooth establishment of renewable energies. The 2008 price rises in food and other commodities caused by price hikes in oil by commodity futures speculators, will not be the last. The ensuing global economic disarray besides giving birth to global bankcruptcies, and intense economic pains and distress to developed and developing economies, made investment in gold futures, not in renewable energies futures, more attractive.

1.3.4 Upsurge In Gold Investment, Gold Prices

THE FAILURE of the fossil oil economy to deliver would have been replaced with a renewable energies economy and futures. Instead, the precious metal[44] gold assumed its traditional and historical role of last resort banker. It became a hard, solid investment medium during the oil-and-food-crisis. Inflation also roared parallel with the roaring commodity and energy prices. As a consequence, gold prices also dramatically went up in the principal London and Sydney Gold Stock Exchanges.

Gold (Au), along with copper (Cu), silver (Ag), is also used as money, a store of value and in jewelry. Gold is the ultimate insurance against anticipated economic and financial difficulties comparable only to a health insurance against anticipated health problems. For over 20 years, the bear market, major investors and even Wall Street disdained gold as an asset which was massively oversold. Since 2000, investment in gold has exhibited signs of a comeback with an accumulation of up to 80% in four years, and breaking out to near 17-year highs. The many commercial uses of gold include:

•
  Gold is often used in jewelry, coinage and as a standard for monetary exchange in many countries. It is an asset that central banks will increasingly use to shore up confidence in sagging fiat currencies as the excessively loose money and fiscal policies cause inflation; and the likelihood of paper money or fiat currency competitive devaluations going forward. Green gold is used in specialised jewelry. White gold serves as the substitute for platinum. Gold is made into thread and used in embroidery. It is also used in awards, and formed the basis for the Gold Standard[45] after the breakdown of the Bretton Woods System.[46].

---

[44] Precious Metal: Is a rare metallic chemical element of high economic value. Precious metals include gold (Au), copper (Cu), silver (Ag), roentgenium (Rg) - known as coinage metals; platinum (Pt), ruthenium (Ru), rhodium (Rh), palladium (Pd), osmium (Os), iridium (In), rhenium (Re) - known as the platinum group; Platinum is the most widely traded. Precious metals are less reactive chemically than most elements, have a high lustre, are softer or more ductile, and have high melting points than other metals. Precious metals have been used as *currency* and as investment and industrial commodities. For this purpose, precious metals gold, silver, platinum and palladium are each given an ISO (International Standards Organnisation) currency code or the currency code to define the names of different currencies, and the code for more currencies established by the ISO, which describes three-letter. The ISO 4217 code list is the established norm in banking and business all over the world for defining various currencies. In many countries, the codes for the more common currencies are also well known publicly along the exchange rates of different world currencies; ISO 4217 codes are also used on airline and international train tickets; The ISO, composed of representatives from various national standards organisations, was established on 23 February 1944, and promulgates worldwide propriety industrial and commercial standards. Headquaters: Geneva, Switzerland.

[45] Gold Standard: Before the collapse of the Bretton Woods system, gold formed the basis for the *Gold Standard* system – which is the International Standards Organisation (ISO) currency code; the ISO currency code of gold bullion is "XAU".

[46] Bretton Woods [Monetary] System: The Bretton Woods System was set up in 1945 by a meeting of world governments after World War II. It set up a system of rules, institutions and procedures to regulate the international monetary system. The five institutions (World Bank Group) set up by the Bretton Woods System are: International Bank for Reconstruction & Development (IBRD), known as "The World Bank"; International Monetary Fund (IMF); International Development Association (IDA); International Finance Corporation (IFC); Multilateral Investment Guarantee Agency (MICA); and International Centre for

- Gold flake is used in sweets and drinks.

- Gold is used as an industrial metal due to its high electrical conductivity and resistance to corrosion: It performs critical functions in computers, communications equipments, spacecraft, jet aircraft engines and a host of other products. It is used as thin layer electroplated on the surface of electrical connectors to make sure of good connections. Used in restorative dentistry.

- Colloidal gold (a gold *nanoparticle)* is an intensely colored solution used as gold paint on ceramics prior to firing. Chloranic acid is used in photograpphy to tone silver image. Gold (III) chloride is utilised as a catalyst in organic chemistry. Gold is used for protection coatings on many artificial satellites as it is a good reflector of infrared and visible light. The isotope of gold, Au-198, is used in some cancer treatments and for some other diseases.

Gold coins are typically produced as either 90% gold or 22% carat (92% gold) alloyed with copper and silver making up the remaining weight, usually as commemorative coins and South African Krugerrands. *Bullion gold* coins contain up to 99.999% gold. Silver coins typically contain 90% silver, or *sterling silver* with 92.5% alloyed with copper making up the remaining weight. Copper coins are often of quite high purity, around 97%, and usually alloyed with small amounts of zinc and tin. *Numismatics,* the hobby of collecting gold, copper and silver coins, includes coins, tokens and paper money. *Alloyed* metals make the resulting coins harder, less likely to become deformed, and more resilient to wear. Investments in gold are traditionally done in bullion gold bars or gold bullion coins.

A *bullion gold bar* is a gold ingot which may be produced in many different types, weights and categories. Investors buy and sell bullion bars "over the counter", or directly through ownership, or indirectly through certificates, accounts, spread betting, derivatives or shares between two parties of the major banks. *Gold certificates* of ownership held by gold investors, instead of storing the actual gold bullion, allows investors to buy and sell security without the inconvenience associated with the transfer of actual physical gold.

Gold occurs in Nature as nuggets or grains in rocks, in underground "veins" and in alluvial deposits. Gold, copper and silver are known as *coinage metals.* Major gold producing countries, in order of output, are South Africa, U.S., Australia, China, Canada, Russia, Indonesia, Peru, Uzbebekistan, Papua New Guinea, Ghana, Brazil, Chile, Philippines, Mali, Mexico, Argentina, Kyrgyzstan, Zimbawe, and Colombia. London is the great gold clearing house; New York is the home of futures trading in gold; Zurich is the physical turntable for gold; Istanbul, Dubai, Singapore and Hong Kong are the doorways to important gold consuming

---

Settlement of the Investment Disputes ICSID). These five organisations and their enormous outgrowths are responsible for providing finance and advice to countries for the purposes of economic development, and eliminating poverty. Under the Bretton Woods System, each country was obliged to adopt a *monetary policy* that maintained the *exchange rate* of its currency *within a fixed value* – plus or minus *1% - in terms of gold,* and the ability of the IMF to bridge *temporary imbalances of payments.* All went well, until 1971, when the U.S., suspended *convertibility* from dollars to gold because of increasing strain. From 1971 onwards, (instead of gold), the U.S. dollar (USD) became the *rescue currency* for all the nation-states which had signed the Bretton Woods Agreements.

regions; Tokyo is where the TOCOM sets the mood of Japan; and Mumbai is India's most liberalised gold regime. *Gold rushes* have occurred in several countries down the centuries. A *gold rush* is a period of feverish migration of workers into the area of a dramatic discovery of commercial quantities of gold. Several gold rushes took place in the 19th century in Argentina, Australia, Brazil, Canada, Chile, New Zealand, South Africa and the U.S.

Gold farming or extraction techniques used by major gold mines[47] and small-scale enterprises include:

* *Dredging:* This is done by small-scale miners using suction dredges – small machines floating on the water usually operated by one or two people. A suction dredge has a sluice box supported by pontoons attached to a suction hose controlled by the miner working beneath the water. Modern suction dredges impact the environment and fisheries significantly because their action disturbs spawning gravels and fish migration.

* *Metal Detecting:* Metal detectors use electromagnetic induction and oscillators to locate gold. A positive reading from the metre indicates the presence of gold below the surface of the scanned area.

* *Pacer Sediment Mining:* Gold panning is manually sorting gold from wide shallow pans filled with sand and gravel that may contain gold, added water, and shaking the pans to sort the gold from the gravel and other material. The gold quickly settles to the bottom as it is much more denser than rock. The silt is usually removed from stream beds, often at a bend in the stream, or resting on a bedrock bed of the stream where the weight of gold causes it to separate out of the water flow. Gold found in streams or dry streams is called *placer deposits.*

* *Open Pit:* Opencast, open-cut mining, strip mining, quarries or borrows – are similar methods used to extract gold and other minerals from rock or from the earth.

* *Sub-surface Mining:* Underground mining of hard rock or Underground mining of soft rock extracts metals like gold, copper, zinc, nickel, lead or gems such as diamonds; soft rock mining excavates the softer minerals such as coal or oil sands and oil shale.

* *Surface Mining:* By this method, overlying rock is left in tact, and the mineral is removed through shafts or tunnels. Surface mines are typically enlarged until either the mineral deposit is exhausted, or the cost of removing larger volumes of overburden makes further mining uneconomic. Heavy equipment machinery such as earthmovers first remove the overburden or the soil and rock above the deposit, followed by huge machines such as dragline excavators that extract the mineral.

* *Highwall Mining:* Another form of surface mining with much better recovery than augering that evolved from auger mining. The coal seam is penetrated by a continuous miner propelled by a hydraulic Pushbeam Transfer Mechanism (PTM).

* *Contour Stripping:* This involves removing the overburden above the mineral seam near the

---

[47] Report Buyer: *"Analysing the Major Gold Mines Worldwide."* August 2007. Product Code ARU00419, $750.00. Aruvian Research, 54 Maltings Place, London, UK. service@reportbuyer.com. The 52 major gold mines include: Argonaut; Battle Branch; Boroo; Cadia; Calhoun; Campbell; Carlin Trend; Clogau; Cortez; Crisson; Dolaucothi; Franklin-Creighton Mine; Galore Creek; Gold Ridge Mine; Goldstrike; Grasberg; Greenwood; Kalgoorlie; Kemess; Kennedy; Kisladag; Kolar; Kumtor; La Coopa; Lagunas Norte; LaRonde; Leadhills; Lihir; Lost Dutchman's; Loud; Mnogovershinnoye; Musselwhite; Naranjal; Old Hundred; Pogo; Porcupine JV; Porgera; Red Lake; Reed Gold Mine; Rosebele; Sleeping Giant; South Deep; Super Pit; Tanjianshan; Target; Timbarra; Treadwell; Tulwaka; Turquoise Ridge JV; Veladero; Yanacocha.

outcrop in hilly terrain where the mineral outcrop usually follows the contours of the land. Contour stripping is often followed by auger mining into the hill side to remove more of the mineral. Contour Stripping leaves behind terraces in mountainsides.

- *Auger Mining:* Uses a device for moving material or liquid by means of a rotating helical or smooth curve flighting, where the material is moved along the axis rotation, in the manner of a grain crop auger mounted on a tractor in harvesting in agriculture.

- *Hard Rock Mining:* Gold encased in rock or minerals rather than particles in loose sediment, is mined. Sometimes, open-pit mining is used (i.e., opencast and open-cut mining and strip mining).

- *Shrinkage Stop Mining:* Used in the upwards mining from a lower to a higher horizon, leaving broken rock in the excavation created. The brocken rock acts as a working platform and helps to stabilise the excavation by supporting the walls; method can be used for ore mining in shrinkage stopes, raising, and underground construction projects where excavations of considerable vertical height may be required, such as ore and waste bins, crusher rooms, penstocks and tailrace tunnels.

- *Room and Pillar:* Involves the removal of ore from rooms, usually for relatively flat-lying deposits, such as follow particular stratum.

- *Mountain Removal* (MTR): Method employed in coal mining that involves mass restructuring of earth in order to reach the coal seams as deep as 1000 ft below the surface. Streams near valley fills from mountaintop removal contain high levels of minerals in the water and decreased aquatic biodiversity (US Environmental Protection Agency). Like other methods of coal mining and processing, mountaintop removals generate waste slurry or coal sludge, which is usually stored behind a dam on-site. Many coal slurry impoundments like the one in West Virginia (U.S.), exceed 500 million gallons in volume; in Raleigh County (U.S.) they exceed 7 billion gallons. Impoundments can be hundreds of feet high and be in close proximity to schools or private residences. The 400 yards sludge dam impoundment above Marsk Fork Elementary School (U.S.) is permitted to hold 2.8 gallons of toxic sludge, 21 times larger than the Buffalo Creek Flood pond in Logan County, West Virginia (U.S.) which killed 125 people on 26 February 1972.

Mining generally, involves extraction of any non-renewable resources, (e.g., petroleum, natural gas, bauxite, coal, gold, silver, copper, diamonds, iron, lead, limestone, nickel, phosphates, oil shale, rock salts, tin, uranium, molybdenum, and even water) – and the resulting environmental degradation. Light aluminum is the most abundant metallic element on Earth comprising more than 8% of the Earth's crust, and also abundant on the surface of the Moon. Aluminum does not exist in pure form in Nature, but always in combination with oxygen, sand, iron, titanium, clay or other substances. The principal aluminum ore is bauxite rock or aluminum hydroxide. Nearly all rocks, particularly igneous rocks or rocks of volcanic origin, contain some aluminum in the form of aluminosilicate or aluminum plus silicon dioxide minerals.

Global gold mining on a large scale is carried on by some 74 multinationals, provided by AME Mineral Economics – Gold's "List of companies with ownership in gold operations."[48]

---

[48] AME Mineral Economics – Gold: *"List of Companies with Ownerships in Gold Operations."* AME House, 342 Kent Street,

Noble minerals are gold, silver, mercury to which the platinum group – palladium,

Sydney NSW 2000, GPO Box 362, Sydney NSW 2001, Australia. http://www.ame.com.au/Companies/Au/Companies.htm; ame@ame.com.au.: **Agnico-Eagle Mines Ltd:** Product: Gold; Gold Mine: Beaconsfield; Country: Canada; **Allstate Explorations NL :** Product: Gold; Gold Mine: Beaconsfield, Australia; **Anglo Ashanti:** Product: Copper, Gold; Gold Mines: Bibiani, Boddington, Cerro Vanguardia, Cresson//Cripple Creek, Crixas/Serra Grande, ERGO, Geita, Great Noligwa, Iduapriem, Kopanang, Moab Khotsong, Morila, Morro Velho, Mponeng, Obuasi, Sadiola, Savuka, Saguiri, Sunrise Dam, Tau Lekora, Tan Tona, Vaal River-Surface, Yatela; Countries: Argentina, Australia, Brazil, Ghana, Guinea, Mali, South Africa, Tanzania, U.S.; **Apollo Gold Corporation:** Product: Gold; Gold Mine: Florida Canyon, U.S.; **Arizona Star Resource Corp.:** Product: Gold; Gold Mine: Cerro Cascale (Aldebaran), Chile; **Ballarat Goldfields NL:** Product: Gold; Gold Mine: Ballarat, Australia; **Barrick Gold Corp:** Products: Copper, Gold, Nickel; Gold Mines: Bald Mountain, Betze-Post, Bouquet Complex, Bulyanhulu, Cortez/Pipeline, Coval, Darlot-Centenary, Eskay Creek, Golden Sunlight, Granny Smith, Hemlo, Henty, Homestake, Kalgoorlie Super Pit, Kalgoorlie West, Kanowna Belle, Kundana, Lagunas Norte, Lawlers, Marigold, McLaughlin, Meikle, Mount Charlotte, North Mara, Pascua-Lama, Pierina, Plutonic, Porgera, Round Mountain, Ruby Hill, Tulawaka, Turquoise Ridge, Veladero; Countries: Agentina, Australia, Canada, Chile, Papua New Guinea, Peru, Tanzania, U.S.; **Beaconsfield Gold NL:** Product: Gold; Gold Mine: Beaconsfield; Country: Australia; **Bema Gold Corp:** Product: Gold; Gold Mines: Cerro Cascale (Aldebaran), Golden Reefs (Grootvlei), Refugio; Countries: Chile, South Africa; **Bendigo Mining Limited:** Product: Gold; Gold Mine: Bendigo; Country: Australia; **BHP Billiton:** Products: Aluminum, Copper, Gold, Iron Ore, Lead, Metallurgical Coal, Nickel, Thermal Coal, Zinc; Gold Mine: Escondida; Country: Chile; **Buenaventura:** Product: Gold, Lead, Zinc; Gold Mine: Yanacocha; Country: Peru; **Cambior Inc:** Product: Gold; Gold Mines: Doyon, Omai, Rosebel; Country: Canada, Guyana, Suriname; **Centerra Gold:** Product: Gold; Gold Mine: Boroo; Country: Mongolia; **Croesus Mining NL:** Product: Gold; Gold Mine: Norseman; Country: Australia; **Cumberland Resources Ltd:** Product: Gold; Gold Mine: Meadowbank; Country: Canada; **Dayton Mining Corporation:** Product: Gold; Gold Mine: Andacollo Oro; Country: Chile; **DRDGOLD:** Product: Gold; Gold Mines: Blyvooruitzicht, Crown Section, ERPM (East Rand Pty Mines Ltd), Hartebeestfontein (NOW), Porgera, Tolukuma, Vatuloula; Countries: Figi, Papua New Guinea, South Africa; **Durban Roodepoort Deep:** Product: Gold; Gold Mine: Durban Deep/West Wits; Country: South Africa; **Echo Bay Mines Limited:** Product: Gold; Gold Mine: McCoy/Cove; Country: U.S.; **Eldorado Gold Corporation:** Product: Gold; Gold Mines: Kisladag, Sao Bento; Country: Brazil; **Emperor Mines Limited:** Product: Gold; Gold Mines: Porgera, Tolukuma, Vatukoula; Countries: Figi, Papua New Guinea; **Freeport-McMoRan C&G:** Products: Copper, Gold; Gold Mine: Grasberg; Country: Indonesia; **GBS Gold International Inc:** Product: Gold; Gold Mine: Union Reefs; Country: Australia; **Glamis Gold Ltd:** Product: Gold; Gold Mines: El Sauzal, Marlin, San Martin (Honduras); Countries: Guatemala, Honduras, Mexico; **Gold Fields Ltd:** Products: Copper, Gold; Gold Mines: Agnew, Beatrix, Cerro Corona, Damang/Abosso, Driefontein Consolidated, Kloof Division, Oryxe, South Deep, St. Ives, Tarkwa; Countries: Australia, Ghana, Peru, South Africa; **Goldcorp Inc:** Product: Gold; Gold Mines: Ampari, Campbell, Dome, La Copia,Los Filos, Luismin, Marigold, Musselwhite, Peak Gold, Red Lake, Wharf; Countries: Australia, Brazil, Canada, Chile, Mexico, U.S.; **Golden Resources Ltd:** Product: Gold; Gold Mines: Bogoso, Wassa; Country: Ghana; **Governments:** Products: Aluminum, Copper, Gold, Zinc; Gold Mines: Bogoso, Damang/Abosso, Obotan, Omai, Rosebel, Sapon, Tarkwa, Wassa; Countries: Ghana, Gyana, Laos, Suriname; **Harmony Gold Mining Co. Ltd:** Product: Gold; Gold Mines: Bambani, Big Bell, Elandsrand, Evander, Freegold, Harmony (Free State), Hill 50, Joel, Jubilee, Kalgold, Matjhabeng, New Celebration, Randfontein, South Kalgoorlie, St. Helena, Target, Tshepong; Countries: Australia, South Africa; **LAMGOLD Corporation:** Product: Gold; Gold Mines: Damang/Abosso, Sadiola, Tarkwa, Yatela; Countries: Ghana,Mali; **Industrias Pendes SA de CV:** Products: Copper, Gold, Lead, Zinc; Gold Mine: La Herradura; Country: Mexico; **Immet Mining Corp:** Products: Gold, Zinc; Gold Mines: Ok Tedi, Troilus; Countries: Canada, Papua New Guinea; **International Finance Corp** (IFC): Products: Copper, Nickel; Gold Mines: Escondida, Syama, Yanacocha; Countries: Chile, Mali, Peru; **Ivanhoe Mines Ltd:** Products: Copper, Gold; Gold Mines: Turquoise Hill (Oyu Tolgor); Country: Mongolia; **Khumalo Bathong:** Product: Gold; Gold Mines: Crown Section, ERPM (East Rand Pty Mines Ltd); Country: South Africa; **Kingsgate Consolidated NL:** Product: Gold; Gold Mine: Chatree; Country: Thailand; **Kinross Gold Corporation:** Product: Gold; Gold Mines: Crixas/Serra Grande, Fort Knox, Kettle River,La Coipa, Lupin, Morro do Ouro/Brasilia/Paracatu, Refugio, Round Mountain; Countries: Brazil, Canada, Chile, U.S.; **Leviathan Resources Limited:** Product: Gold; Gold Mine: Stawell; Country: Australia; **Leyshon Resources Limited:** Product: Gold; Gold Mine: Mount Leyshon; Country: Australia; **LionOre Mining International Ltd:** Products: Gold, Nickel; Gold Mine: Thunderbox; Country: Australia; **Mali Government:** Product: Gold; Gold Mines: Loulo, Syama, Yatela; Country: Mali; **Meridian Gold Company:** Product: Gold; Gold Mine: El Penon; Country: Chile; **Mitsubishi Corporation:** Products: Aluminum, Copper, Gold, Iron Ore: Gold Mine: Escondida; Country: Chile; **Mitsubishi Materials Corporation:** Products: Aluminum, Copper, Gold, Iron Ore, Metallurgical Coal, Thermal Coal, Zinc; Gold Mine: Escondida; Country: Chile; **Mwana Africa Holdings:** Product: Gold; Gold Mine: Freda/Rebecca; Country: Zimbabwe; **Nawcrest Mining Limited:** Products: Copper, Gold; Gold Mines: Cadia, Cadia-Ridgeway, Cracaw, Gosowong, Telfer; Countries: Australia, Indonesia; **Newmont Mining Corporation:** Products: Copper, Gold; Gold Mines: Ahafo (Yamfo-Sefwi), Batu Hijau, Boddington, Golden Giant, Granites, The (Tanami Operations), Holloway, Holt-McDermott, Jundee, Kalgoorlie Super Pit, La Herradura, Leeville, Martha Hill, Mishasa, Mount Charlotte, Mount Leyshon, Nevada Complex, Pajingo (Vera-Nancy), Phoenix, Turquoise Ridge, Yanacocha, Yandal; Countries: Australia, Canada, Ghana, Indonesia, Mexico, New Zealand, Peru, U.S.; **NipponMining & Metals:** Products: Copper, Gold, Lead, Zinc; Gold Mine: Escondida; Country: Chile; **Northern Mining Exploration Ltd:** Product: Gold; Gold Mine: Tulawaka; Country: Tanzania; **Northgate Minerals Corporation:** Products: Copper, Gold; Gold Mine: Kemess; Country: Canada; **OceanaGold:** Product: Gold; Gold Mine: Macraes; Country: New Zealand; **Orogen Minerals Ltd:** Product: Gold, Nickel; Gold Mine: Misima; Country: Papua New Guinea; **Other-Public Companies, Governments, Minor Parties:** Products: Copper, Gold; Gold Mines: Batu Hijau, Cerro Vanguardia, Iduaprien, Kemess, Kidston, Lihir, Morila, Ok Tedi, Sadiola, Siguiri; Countries: Argentina, Australia, Canada, Ghana, Indonesia, Mali, Papua New Guinea, Guinea; **Oxiana Limited:** Products: Copper, Gold, Zinc; Gold Mines: Prominent Hill, Sepon; Countries: Australia, Laos; **Pacific Rim Mining Corp:** Product: Gold; Gold Mine: Rawhide; Country: U.S.; **Perilya Limited:** Products: Gold, Lead, Zinc; Gold Mine: Fortnum; Country: Australia; **Peter Hambro Mining Plc:** Product: Gold; Gold Mine: Pokrovskiy Rudnik; Country: Russia; **Placer Donne Inc:** Product: Copper, Gold; Gold Mines: Kidston, Misima, Musselwhite; Countries: Australia, Canada, Papua New Guinea; **PT Aneka Tambang:** Products: Gold, Lead, Nickel, Zinc; Gold Mine: Gosowong; Country: Indonesia; **PT Pakuafu Inda:** Products: Copper, Gold; Gold Mine: Batu Hijau; Country: Indonesia; **Queenstake Resources Ltd:** Product: Gold; Gold Mine: Jernitt Canyon; Country: U.S.; **Randgold Resources Ltd:** Product: Gold; Gold Mines: Loulo, Morila; Country: Mali; **Resolute Limited:** Product: Gold; Gold Mines: Golden Pride, Obotan, Syama; Country: Ghana, Mali, Tanzania; **Rio Tinto:** Products: Aluminum, Copper, Gold, Iron Ore, Lead, Metallurgical Coal, Nickel, Thermal Coal, Zinc; Countries: Chile, Indonesia, Mongolia, Papua New Guinea, U.S.; **Sedimentary Holdings Ltd:** Product: Gold; Gold Mine: Cracow: Country: Australia; **St. Barbara Mines Limited:** Product: Gold; Gold Mines: Carosue Dam, Marvel Loch/Yilgarn-Southern Cross Region, Sons of Gwalia-Leonora, Tarmoola; Country: Australia; **Straits Resources Ltd:** Products: Copper, Gold, Thermal Coal; Gold Mine: Mount Muro; Country: Indonesia; **Sumitomo Corporation:** Products: Aluminum, Copper, Gold, Lead, Metallurgical Coal, Thermal Coal, Zinc; Gold Mine: Batu Hijau; Country: Indonesia; **Teck Cominco Ltd:**

iridium, rhodium, ruthenium, and osmium, belongs. Noble metals are highly resistant to chemical reaction or corrosion (i.e., they will not easily dissolve in a solvent or rust.) All other metals are called *base metals* because they will more easily undergo chemical changes.

1.4. Rising Transportation, Agri-Produce Prices

TRANSPORTATION AND agricultural prices rose dramatically in harmony with oil and food price rises. This applied also to the mining and extraction industry. This is because transportation and agricultural systems rely heavily on fossil fuels and the automobile industry which also heavily relies on fossil fuels. There is increasing demand for fossil fuels by an ever-increasing number of vehicles especially in emerging economies like India and China. Transportation is the art of moving people and goods from one near or far location to another. The transport industry does this as it also provides the means, methodology, expertise and technology to accomplish this task through infrastructure, vehicles and operation policy and management. *Infrastructure* consists of roads, railways, airways, waterways, canals, pipelines, terminals, rail and bus stations, sea and river ports, and pedestrian alleys. *Vehicles*[49] include horses and wheelcarts,

---

Product: Copper, Gold, Lead, Metallurgical Coal, Zinc; Gold Mine: Hemlo; Country: Canada; **Viceroy Resources Corporation:** Product: Gold; Gold Mine: Bounty; Country: Australia; **View Resources:** Product: Gold; Gold Mine: Bronzewing; Country: Australia; **Western Areas Ltd:** Product: Gold; Gold Mine: South Deep; Country: South Africa; **Xstrata:** Products: Copper, Gold, Lead, Metallurgical Coal, Nickel, Thermal Coal, Zinc; Gold Mines: Alumbrera, Ernest Henry; Countries: Argentina, Australia.

[49] Dan Neil, Pulitzer Prize-winning automative critic and syndicated columnist for the Los Angeles Times, and TIME look at the greatest lessons of the automative industry on the 50th anniversary of the Ford Edsel: *"The 50 Worst Cars of All Time – 1899-2007."* Monday 05 November 2007, TIME/CNN: 1899 Horsey Horseless: Somewhere between an early car and the head-in-the-bed scene in *The Godfather*, the Horsey Horseless, the brainfart of inventor Uriah Smith of Battle Creek, Mich., was intended to soothe the skittish nerves of our equine servants. A wooden horse head was attached to the front of the chuffing bug in order to make it resemble a horse and a carriage (Smith recommended the horse head be hollow to contain volatile fuel – another great idea). "The live horse would be thinking of another horse," said Smith, "and before he could discover his error and see he had been fooled, the the strange carriage would be passed." Stupid horse! It's not clear if the Horsey Horseless was ever actually built or if it is a chimera of auto history, but it reminds us just what a radical, hard-to-conceptualise thing a horseless carriage was; 1909 Ford Model T: Uh-oh. Here comes trouble. Let's stipulate that the Model T did everything that the history books say: It put America on wheels, supercharged the nation's economy and transformed the landscape in unimagined when the first Tin Lizzy rolled out of the factory. Well, that's just the problem, isn't it? The Model T - whose mass production technique was the work of engineer Wialliam C. Klann, who had visited a slaughterhouse's disassembly line" – conferred to Americans the notion of automobility as something akin to natural law, a right endowed by our Creator A century later, the consequences of putting every soul on gas-powered wheels are piling up, from the air over our cities to the sand under our soldiers' boots. And by the way, with its blacksmithed body panels and crude instruments, the Model T was a piece of junk, the Yugo of its day; 1911 Overland OctoAuto: Milton Reeves had a very hard head and, apparently, very poor eyesight. While the general conformation of the automobile was largely sorted out in the first decade of the 20th Century – particularly that business about four wheels – Reeves thought perhaps eight or a minimum of six wheels might provide a smoother ride. Welding in some bits to a 1910 Overland and adding two more axles and four more guncart-style wheels, Reeves created the OctoAuto, proudly displaying it at the inaugural Indianapolis 500. Like its Marvel Comics-worth name, the car was a bit of a monster, measuring over 20ft. long. Talk about scaring the horses. Zero orders for the patently ugly and silly OctoAuto apparently didn't discourage Reeves, who tried again the next year with the Sextauto (six wheels, single front axle design). Reeves is remembered today as the inventor of the muffles, which is far from ignomity; 1913 Scrpps-Booth Bi-Autogo: A 3,200-lb. Motorcycle with training wheels, V8 engine and enough copper tubing to provide every hillbilly in the Ozarks with a still, the Scripps-Booth Bi-Autogo was the daft experiment of James Scripps-Booth, an heir of the Scripps publishing fortune and a self-taught – or untaught – auto engineer. The Bi-Autogo was essentially a two-wheeled vehicle, carrying its considerable heft on 37-in. wooden wheels. At low speeds, the driver could lower small wheels on outriggers to stabilise the vehicle so it wouldn't plop over. This is not a case of the advantage of hindsight; this was obviously a crazy idea, even in 1913. The Bi-Autogo does enjoy the historical distinction of being the first V8-powered vehicle ever built in Detroit, so you could argue it is the beginning of an even greater folly; 1920 Briggs and Stratton Flyer: By 1920, the automotive was no longer a primitive experiment. Companies such as Rolls-Royce, Cadillac, hispano-Suiza and Voisin were making potent and luxurious automobiles, the technical achievement of the age. And then there was this, the Flyer, which is no more than a motorised park bench on bicycle wheels. No suspension, no body work, no windshield.It was actually a five wheeler, with the dinky 2-hp Briggs and Stratton engine driving a traction wheel on the back, like a boat's outboard motor. The Flyer represents something we'll see several times on this list: The drive to make the absolute cheapest, most minimal automobile possible; 1933 Fuller Dymaxion: Designer-genius R. Buckminster Fuller was one of the century's great auto nutjobs, a walking unorthodoxy who originally conceived of the Dymaxion as a flying automobile, or drivable plane, with jet engines and inflatable wings. It would be one link in his vagely totalitarian plan for the people to live in mass-produced homes deposited on the landscape by dirigibles. *Okayyyy...* Deprived of wings, the Dymaxion was a three-wheel, ground-bound zeppelin, with a huge levered A-arm carrying the

rear wheel, which swivelled like the tail wheel of an airplane. The first prototype had a wicked death wobble in the rear wheel. The next two Dymaxions were bigger, heavier, and only marginally more drivable. The third car had a stabilizer fin on top, which did nothing to cure the Dymaxion's acute instability in crosswinds. A fatal accident involving the car – cause unknown - doomed its acceptance. Though unworkable, this three-wheeled suppository was the boldest of a series of futuristic, rear-engined cars of the 1930s, including the Tatra, the Highway Corporation's "Fascination" car and, everybody's favorite, the Nazi's KdF-wagen; 1934 Chrysler/Desoto Airflow: The Airflow's "worst"-ness derives from its spectacularly bad timing. Twenty years later, the car's many design and engine.ring innovations – the aerodynamic singlet-style fuselage, steel-spaceframe construction, near 50-50 front-rear weight distribution and light weight – would have been celebrated. As it was, in 1934, the car's dynamic streamliner styling antagonized Americans on some deep level, almost as if it were designed by Bolsheviks.It didn't help that a few early Airflows had major, engine-falling-out-type problems that stemmed from the radical construction techniques required. Chrysler, and even more hapless Desoto, tried to devolve the Airflow stylistically, giving it more conventional grill and raising the trunk into a kind of busle (some later models were named Atrstream), but the damage was done. Sales were abysmal. It wouldn't be the last time American car buyers looked at the future and said, "no thanks."; 1949 Crosley Hotshot: The first sports car produced in postwar America was a major hunk of a junk. Actually, at 1,100 lbs 145 in. long, the Crosley Hotshot was a minor hunk of a junk, but at least it was slow and dangerous. A wonderously mangled and compacted Hotshot can be glimpsed in the 1961 driver's ed scare film *Mechanized Death*. The Hotshot was the work of consumer products pioneer Powel Crosley Jr., of Cincinnati, he of Crosley radio fame. But what he really wanted to do was build cars, while he did it middling failure until the doors closed in 1952. A Hotshot actually won the "index of performance" – an honor for the best speed for its displacement – at the 1950 Six Hours of Sebring, puttering around at an average of 52 mph. What killed the Hotshot was its engine, a dual-overhead cm.75-litre four cylinder, not cast in iron brazed together from pieces of stamped tin. When these brazed welds let go, as they often did, things quickly got noisy, and hot; 1956 Renault Dauphine: The most ineffective bit of French engineering since the Maginot Line, the Renault Dauphine was originally named the Corvette, *tres ironie*. It was, in fact, a rickety, paper-thin scandal of a car that, if you stood beside it, you could actually hear rusing. Its most salient feature was its slowness, a rate of acceleration you could measure with a calendar. It took the drivers at *Road and Track 32 seconds* to reach 60 mph, which would put Dauphine at a severe disadvantage in any drag race involving farm equipment. The fact that ultra-cheap, super-sketchy Dauphine sold over 2 million copies around the world is an index of how desperately people wanted cars. Any cars; 1957 King Midget Model III: The King Midget story reminds us what a middle-class nation the U.S., was in the '50s. Claud Dry and Dale Orcutt, of Athens, Ohio, buddies from the Civil Air Patrol, wanted to sell bare-boned utility car that anybody could afforf, unlike that bloody elitist peacenik Henry Ford with his fancy Model T. King Midget's cars made the Model T look like a Bugatti Royale. In the late 1940s, they began offering the single-seat Model I as a home-built, $500 kit, containing the front frame, axles and sheet metal patterns, so that the body panels could be fabricated by local tradesmen. Any single-cylinder engine would power it. The result was a truly crap-tastic little vehicle, the four-wheel equivalent to those Briggs-and-Stratton powered minibikes. Amazingly, Midget Motors continued to develop and sell mini-cars until the late 1960s. The crown jewel was the Model III, introduced in 1957, a little folded-steel crackerbox powered by a 9-hp motor. Government safety standards, at long last, put the King Midget out of our misery; 1957 Waterman Aerobile: Waldo Waterman wanted avaiation piuoneer Glenn Curtis to like him in the worst way. Inspired by what was apparently Curtiss' casual remark about driving an airplane away from the field, Waterman spent years developing a roadable airplane. In 1934, he flew his first successful prototype, the "Arrowplane," a high-wing monoplane with tricycle wheels. On the ground, the wings folding against the fueselage like those of a fly (now would be a good time to note that Waterman must have been crazy to get airborne in such a contraption). Nonetheless, the Arrowplane goes down as the first real flying car. Two decades later, Waterman finally perfected, if that's the word, what he then called the Aerobile, configured as a swept-wing "pusher" (prop in the back). There were few customers with so consumate a death wish as to order their own Aerobile, and Waterman's one working car-plane eventually wound up in the Smithsonian, where it can't kill anyone; 1958 Ford Edsel: That's why we're all here, right? To celebrate E Day, the date 50 years ago when Ford took one of the autodom's most hilarious pratfalls. But why? It really wasn't that bad a car. True, the car was a kind of homely, fuel thirsty and too expensive, particularly at the outset of the late '50s' recession. But what else? It was the first victim of Madison Avenue hyper-hype. Ford's marketing mavens had led the public to expect some plutonium-powered, pancake-making wondercar; What they got was a Mercury. Cultural critics speculated that the car was a flop because the verical grill looked like a vagina.maybe. America in the '50s was certainly phobic about the female business. How did the Edsel come to be synonymous with failure? All of the above, consolidated into an irrational groundthink and pressurized by a joyous cathy media. Interestingly, it was Ford President Robert McNamara who convinced the board to bail out of the Edsel project; a decade later, it was McNamara, then Secretary of Defense, who couldn't bring himself to quit the disaster of Vietnam, even though he knew a lemon when he saw one; 1958 Lotus Elite: Fibreglass was the '50s carbon fibre – tough, versatile, lighter than steel and more affordable than aluminum. The Kaiser Darrin and Corvette sports cars were wrapped in fibreglass bodies, for instance. Colin Chapman, the founding engineer of Lotus, was bonkers for weight savings. It was inevitable that he would be drawn to the material. And so, the Elite. Weighing just 1,100 lbs and powered by a punchy, 75-hp Coventry Climax engine, the Elite (Type 14) was a successful race car, winning its class at the 24 Hours of Le Mans six times. It was also a lovely little Coupe, which made the moment when the suspension mounts punched through the stressed-skin monocoque all the more pathetic. The unreinforced fibreglass couldn't take the structural strain. In Chapman's cars, failure was always an option; 1958 MGA Twin Cam: A point of personal privilege. I own a 1960 MGA that I restored with my own two hands, and it is a fantastic British sports car, with lovely lines panned by Syd Enever, a stiff chassis, and a floggable Character. The car was introduced in 1955 as a replacement for the venerable TD and was itself replaced by the MGB in 1962. Along the way, somebody decided my little car was anemic – hey! I resent that! – so MG offered an optional high-performance engine with dual overhead cams, thus the "twin cam." It was a leaking, piston-burning, plug-fouling nightmare of a motor that required absolute devotion to things like ignition timing, fuel octane and rpm limits, less the whole shebang vomit connecting rods and oil all over the road. Many years after the engine was taken out of service, it was discovered that the problem lay in the carburetors. At certain rpm, resonant frequencies would cause the fuel mixture to froth, leaning out the fuel and burning the pistons. I've never had any such trouble with my iron-block, pushrod, lawn tractor engine. I'm just saying; 1958 Zunndap Janus: Built in Nuremburg, Germany by the well-established motorcycle firm during a downturn in the two-wheeler market, this push-me-pull-you was based on a Dornier prototype and powered by a 250-cc, 14-horsepower engine, giving it a top speed of only 50 mph, assuming you had that kind of time. Its unique feature was the rear-facing bench seat, which meant passengers could watch in horror as traffic threatened this rolloing roadblock of acar. Soon it became clear – "Auch Du lieber!" – that the Janus was a disaster, coming and going; 1961 Amphicar: A vehicle that promised to revolutionise drowning, the Amphicar was the peacetime descendant of the Nazi Schwimmwagen (say it out loud – it's fun!). The standard line is that the Amphicar was both a lousy car and a lousy boat, but it certainly had its merits. It was reasonably agile on land, considering, and fairly maneuverable on water, if painfully slow, with a top speed of 7 mph. Its single greatest demerit – and this is a great one – was that it wasn't particularly watertight. Its flotation was entirely dependent on whether the bilge pump could keep up with the leakage. If not, the Amphicar became the world's most aerodynamic anchor. Even so, a large number of the early 4,000 cars built between 1961 and 1968 are still on the road/water. In animals. *Operations* and services provided by the transport industry deal with the

fact, during the recent floods in Britain, an Amphicar enthusiast served as a water taxi, bringing water and grocesies to agroup of stranded school kids. Bully!; 1961 Corvair: Rear-engine cars are fun to drive and even more fun to crash. While rear engine packaging offers enormous advantages, putting the vehicle's heaviest components behind the rear axle gives cars a disticnt tendency to spin out, sort of like an arrow weighted at the end. During World War II, Nazi officers in occupied Czechoslovakia were banned from driving the speedy rear-engined Tatras because so many had been killed behind the wheel. Chevrolet execs knew the Corvair – a little and lovely car with an air-cooled, flat-six in the back, a la the VW Beetle – was a hardful, but they declined to spend the few dollars per car to make the swing-axle rear suspension more manageable. Ohhh, they came to regret that. Ralph Nader put the smackdown on GN in his book *Unsafe at Any Speed,* also noting that the Corvair's single-piece steering column could impale the driver in a front collision. Ouch! Meanwhile, the Corvair had other problems. It leaked oil like a derelict tanker. Its heating system tended to pump noxious fumes into the cabin. It was offered for a while with a gasoline-burner heater located in the front "trunk," a commonbut dangerously dumb accossory at that time. Even so, my family had a Corvair, white with red interior, and we loved it; 1966 Peel Trident: Less a car than a 5th-grade science project on seed germination, the Peel Trident was designed and built on the Isle of Man in the 1960s for reasons yet undetermined, kind of like Stonehenge. The Trident was the evolution of the P-50, which at 4-ft, 2-in. in length could justify its claim as the world's smallest car,or fastest barstool.The Trident is a good example of why all those futuristic bubbletop cars of GM's Motorama period would never work. The sun would cook you alive under the Plexiglass. We in the car business call the phenomeno "solar gain." – You have to love the heroic name: *Trident/* More like Doofus on the half-shell; 1970 AMC Gremlin: American Motors designer Richard Teague – remember that name – was responsible for some of the coolest cars of the era. The Gremlin wasn't one of them. AMC was professionally in the weeds at the time, and the Gremlin was the company's attempt to beat Ford and GM to the subcompact punch. To save time and money, Teague's design team basically whacked off the rear of the AMCHornet with a cleaver. The result was one of the worst curiously proportioned cars ever, with a long low snout, long front overhang and a truncated tail, like the tail snapped off a salamander. Cheap and incredibly deprived – with vacuum-operated windshield wipers, no less – the Gremlin was also awful to drive, with a heavy six-cylinder motor and choppy, unhappy hardling due to the loss of suspension travel in the back. The Gremlin was quicker than other subcompacts, but alas, that only meant you heard the jeers and laughter that much sooner; 1970 Triumph Stag: You could put all the names of all the British Leyland cars of the late '60s in a hat and you'd be guaranteed to pull out a despicable, rotten-to-the core mockery of a car. So consider the Triumph Stag merely representive. Like its classmates, it had great style (penned by Giovanni Michelotti) ruined by some half-hearted, half-witted, utterly temporized engineering: To give the body structure greater stiffness, a T-bar connected to the roll hoop to the windscreen, and the windows were framed in eye-catching chrome. The effect was to put the driver in a shiny aquarium. The Stag was lively fun to drive, as long as it ran. The 3.0-litre Triumph V8 was a monumental failure, an engine that utterly refused to confine its combustion to the internal side. The timing chains broke, the aluminum heads warped like mad, the main bearings would seize and the water pump wood poop the bed – Ka-POW! Oh, that piston through the bonnet, that is a spot of bother. We'll not hear the last of Triumph on this list; 1971 Chrysler Imperial LeBaron Two-Door Hardtop: The glamorous Imperial marque was, by the late '60s, reduced to a trashy pseudo-luxury harlot walking the streets for its pimp, the Chrysler Corporation. By 1971, only the Imperial LaBaron was left and it shared the monstrous slab-sided "fueslage" styling of corporate sibblings like the Chrysler New Yorker and the Dodge Monaco. Appearing to have been hewn from solid blocks of mediocrity, the Imperial LaBaron two-door is memorable for having some of the longest fenders in history. It was powered by Chrysler's silly-big 440-cu.-in. V8 and measured over 19fr. Long. The interior looked like a third-world casino. Here we are approaching the nadir of American car building – obese, under-engineered, horribly ugly. Or, it would be the nadir, except for the abysmal 1980 Chrysler Imperial, which had an engine cursed by God. The Imperial name was finally overthrown in 1983; 1971 Ford Pinto: They shoot horses, don't they? Well, this is fish in a barrel.Of course, the Pinto goes on the Worst list, but not because it was a particularly bad car – not particularly – but because it had a rather volatile nature. The cartendedto erupt in flame in rear-end collisions. The Pinto is at the end of one of autodom's most notorious paper trails, the Ford Pinto memo, which ruthlessly calculates the cost of reinforcing the rear end ($121 million) versus the potential payout to victims ($50 million). Conclusion? Let'em burn; 1974 Jaguar XK-EV12 Series III: The 1961 Jaguar E-Type was heavenly, dead-sexy, 150-mph supercar, a stilleto heel to the heart of any car-loving man. By 1974, it had morphosed into this, this *thing.* In order to compensate for power-sapping emissions controls required in the U.S., the car's primary export market, Jaguar discontinued the reliable 4.2-litre six for an anchor-heavy 5.3-litre V12, which was a total bitch to try to keep in tune and made the car nose-heavy besides. Jaguar also discontinued the elegant fixed-head coupe and offered the car only as a long wheel base 2+2 on convertible. Imagine taking one of the world's most beauutiful cars and sticking it in a taffy puller.Not finished running the lines, Jag plumed up the fenders, spoiling the smooth, aero-sleek contours of the original. The piece de resistance, Jag offered hideous rubber bumpers – Dagmars, really – in a lame attempt to meet 5-mph bumper standards. To which car enthusiasts can only say, "You *bastards;* 1975 Bricklin SV1: The Bricklin I ever sat in caught on fire and burned to the axles. This is notably ironic, since the car's creator – the smooth-talking Malcolm Bricklin – didn't include an ashtray or lighter in the car, to discourage smoking. Despite its hand-removing, 100 lb.gullwing doors, the SV1 was supposed to exemplify the safer car of the future;the name stands for "Safety Vehicle 1." The bodies were made of brightly coloured, dent-resistant plastic, like PlaySkool furniture. Another safety feature: incredible, crust-of-the-Earth-cooling slowness. All those resin panels and compressible bumpers added hundreds of pounds that the emissions-limited V8s couldn't handle. This thing couldn't outrun the Rose Bowl Parade.Less than 3,000 of the wedgy coupes were built, but Malcolm Bricklin was far from through, as we'll see; 1975 Morgan Plus 8 Propane: The venerable, and I do mean venerable, Morgan Company of Malvern, Warwickshire, has been making cars the oldfashioned way since it was radical and high-tech. With wing fenders, wooden-frame bodies, and sliding-pillar front suspensions, Morgan are mailed to us direct from 1935. But in the early 1970s, new U.S., emissions and safety requirements caused Morgan to pull out of the market. To the rescue came Bill Fink, a San Francisco Moggie-phile and dealer who managed to get the car certified by running its Buick/Rover V8 on propane. For years, small numbers of these bouncy little roadsters had tanks of liquid propane hung perilously behind the rear bumper. And people gave the Pinto grief?; 1975 Triumph TR 7: "The shape of things to come" quickly became the shape that came and went, in a great cloud of "good riddance." The doorstop-shaped TR7, and its rare V8-powered sibling TR8, were the last Triumphs sold in America and among the last the company made before it folded its tents in 1984. The trouble was not necessarily the engineering, or even the peculiar design, which looked fit to split firewood. It was that the cars were so horribly made. The thing had more short-circuits than a mixing board with a bong spilled on it. The carburetors had to be constantly romamnced to stay in balance. Timing chains snapped. Oil and water pumps refused to pump, only suck. The sunroof leaked and the concealeable headlights refused to open their peepers. One owner reports that the rear axle fell out. How does that happen? It was as if British Leyland's workers were trying to sabotage the country's balance of trade. Oh yeah; 1975 Trabant: This is the car that gave Communism a bad name. Powered by a two-stroke pollution generators maxed out at an ear-splitting 18 hp, the Trabant was a hollow lie of a car constructed of recycled worthlessness (actually, the body was made of fibreglass-like Duroplast, reinforced with recycled fibres like cotton and wool). A virtual antique when it was designed in the 1950s, the Trabant was East Germany's answer to the VW Beetle – a "people's car," as if the people didn't have enough to worry about. Trabants smoked like an Iraqi oil fire, when they ran at all, and often lacked even the most basic amenities, like brake lights or turn signals. But history has been kind to the Trabi. Thousands of East Germans drove their Trabants over the border when the Wall fell, which made it a kind of automative liberator. Once across the border, the none-too-sensational Ostdeutschlanders immediately abandoned their cars. Ich bin Junk!; 1976 Aston Martin Lagonda: In the disco days of the 1970s, even supercars were cocaine-thin. Meet the Aston Martin Lagonda, a four-door exotic that lived on dinner mints and hot water. Designed by A M penman William Towns – undoubtedly wearing a very large cravat at the time – the Lagonda was as beautiful a car as ever resembled a penscil box. Mechanically, it was a constant catastrophe, Aston Martin's Dunkirk. The company decided to build the Lagonda with a brace of cutting-edge, computer-driven electronics and cathode-ray displays, which would have been very impressive if any of them ever worked. NASA couldn't have built this car, much less the heirs to Joseph Lucas, the British electronics famous "Prince of Darkness." Still, I'd kill to have one of these cars, and the O-scope and multi-metre to fix it; 1976 Chevy Chevette: I include the Chevy Chevette only to note that even the most unloved and unlovely cars have had their partisans. There are Pacer fan clubs and Yugo fan clubs, and if there is a Chevette fan club, let it begin with me. My girlfriend in college had a diaper-brown Chevette three-door hatchback, as bare bones as an exhibit at the natural history museum. It had a 51-hp engine and a four-speed manual transmission and not much else. It was loud and it was tinny, but we drove that car across the country three times and it never failed us. Once I got a 85-mph speeding ticket in it. That was on the down slope of the Appalachians, but still. The last time I saw that Chevette it was still plugging along. Vaya con Dios, old paint; 1978 AMC Pacer: A recent poll by Hagerty Insurance asked enthusiasts to name the worst car design of all times: This glassline bolus of dorkiness is the pathetic winner. Remember Richard Teague, designer of the amputated Gremlin? Him again. But, cone on, the Pacer, its Wayne and garth's Mirth-mobile, for Heaven's sake! You can't hate on that.Indeed, my family owned a dark green Pacer with that Navajo-blanket upholstery, and it worked just fine until I drove it through a ditch, after which the heavy doord hung off their hinges like beagle ears. What I remember of this car is that, in the summer, it was like being an ant under a mean kid's magnifying glass. The air conditioning was non-existent. You could actually see fumes of volatile petrochemicals out-gassing from the plastic dash. Wayne, I feel woozy; 1980 Corvette 305 "California": Federal emissions requirements of the 1970s took a big neutering knife to American musclecars, and no car bled more than the Corvette. The worst of it came in California – dang hippy *liberals!* – where stricter state regs required that the barely adequate 350 cu.-in. small block in the 1980 Corvette be replaced with a wholly inadequate 305 V8, putting out 180 hp of pure shame. On top of that, the "California" Corvette sucked its pitiful rivulet of horsepower through the straw of tonque-sapping three–speed automative transmission. That gave Corvette – the very totem of hairy-chest, disco machismo – acceleration comparable to a very hot Vespa. These were dark days indeed; 1980 Ferrari Mondial 8: Even the legendary Italian sports car company whiffs once in a while, and the first Ferrari Mondial was a big red disaster. Based on the 308 chassis, this large and relatively heavy 242 couple had a mere 214 hp on tap from its transversely mounted mid-engine V8, and its transistor–based electronics had more bugs than a Barstow model rollway. Eventually, every single system would fail, not infrequently accompanied by the smell of burning wires. The factory-authorised service, meanwhile, was more like factory-authorised extortion.It hasn't helped the Mondial reputation that it was one of the "cheap" Ferraris, within reach of a reasonably successful orthodontist. Mondials eventually got much better. They could hardly get worse; 1981 Cadillac Fleetwood V-8-6-4: These days, cylinder deactivation, or variable displacement, is relatively common – the 2008 Honda Accord V6 has it, for instance. And it's a beautiful idea. When the engine is running at light loads, it's logical to shut down unneeded cylinders to save fuel, like turning off lights in unused rooms. But in 1981, when semiconductors and on-board computers still in their infancy, varible displacement was a huge technical challenge. GM deserves credit for trying, but the V-8-6-4 was the Titanic of engine programmes. The cars jerked, bucked, stalled, made rude noises and generally misbehaved until wild-eyed owners took the cars to have the system discontinued. For some it was the last time they ever saw the inside of a cadillac dealership; 1981 De Lorean DMC-12: Automatic icon, snappy dresser and FBI target John Z. Delorean left the building in 2005, leaving behind 8,582 stainless steel Deloreans and one time-travelling hotrod. Few car projects were more maledicted than the DMC-12. By the time John Z. got the factory in Northern Ireland up and running – and what could possibly go wrong there? – the losses were piling up fast. The car was heavy, underpowered (the 2.8-litre Peugeot V6 never had a chance) and overpriced. And De Lorean was having a few dramas of his own, resulting in one of law enforcement's more memorable hidden-camera tableaux: the former GM executive sitting in a hotel room with suitcases on money, discussing the supply-and-demand of nose candy. The Giugiaro-designed DMC-12 sure was cool looking, though. In August of this year, the Texas company that controls the rights to the name annouced it will build a small number of new DMC-12s. How's that for time travel?; 1982 Cadillac Cimarron: The horror. The horror. Everything that was wrong, lazy and mendacious about GM in the 1980s was crystallised in this flagrant insult to the good name and fine customers of Cadillac. Spooked by the success of premium small cars from Mercedes-Benz, GM elected to rebadge its awful mass-market J-platform sedans, load them up with chintzy fabrics and accessories and call them "Cimarron, by Cadillac." Wha...? Who? Seeking an even hotter circle of hell, GM priced these pseudo.caddies (with four-speed manual transmissions, no less) thousands more than their Chevy Cavalier sibblings. This bit of temporising nearly killed Cadillac and remains its biggest shame; 1982 Camaro Iron Duke: There was time when 90 horsepower was a lot, and that time was 1932. Fifty years later, it was bupkis, especially under the hood of Chevy's bloved Mustang-fighter, the Camaro. As the base engine for the redesigned 1982 Camaro (and Pontiac Firebird), the 2.5-litre, four-cylinder "Iron Duke" was the smallest, least powerful, most un-Camaro-like engine that could be and, like the California Corvette, it was connected to a low-tech three-speed slushbox. So equipped, the Iron Camaro had o-60 mph acceleration of around 20 seconds, which left Camaro owners to drum their fingers while school buses rocketed past in a blur of yellow; 1984 Maserati Biturbo: "Biturbo" is, of course, Italian for "expensive junk." At least, it is now, after Maserati tried to pass off this bitter heartbreak-on-wheels as a proper grand touring sedan. The Biturbo was the product of a desperate, under-funded company circling the drain of bankruptcy, and it shows. Everything that could leak, burn, snap or rupture did so with the regularity of the Anvil Chorus. The collected service advisories would look like the Gutenberg Bible. The only greater ignominy was the early 1990s Maserati TC, a version of the Chrysler Le Baron (a flaccid, front-drive, four cylinder loser-mobile) with the pound Mazzer Trident on the nose. Finally, sir, have you no shame?; 1985 Mosler Consulier GTP: Warren Mosler, a brilliant economist and investor, built his sports racer out of bits and parts that fell off the Big Three's table – a steering wheel from a minivan here, a Chrysler engine there, some mismatched gauges – but mostly wheat he did was to add lightness. The resulting fibreglass-bodied car had a marvelous power-to-weight ration and did so well in racing that it was eventually banned. Or it might have been that the course workers were suffering from post-traumatic stress from the sight of the thing. Mosler had thought of every thing but a stylist, and the pride and joy of this arch-capitalist looked like something from an East german kit-car company. Truly one of the ugliest cars ever, the Consulier GTP proved once and for all that building a car is harder than it looks; 1985 Yugo GV: Malcolm Bricklin, he of the Bricklin SV1, wouldn't be satisfied until he had forced every American towalk to work. To that end, in 1985,he began importing the Yugo GV, which turned out to be the Mona Lisa of bad cars. Built in Soviet-bloc Yugoslavia, the Yugo had the distinct feeling of something assembled at gunpoint. Interestingly, in a car where "carpet" was listed as a standard feature, the Yugo had a rear-window defroster – reportedly to keep your hands warm while you pushed it. The engines went ka-booey, the electrical system – such as it was – would sizzle, and things would just fall off Yugo. Or not; 1986 Lamborghini LM002: This V12-powered super dune buggy gets on the list – well, my list anyway – purely because of its appalling clientele. The "Rambo Lambo" was the civilian version of a military vehicle that Lamborghini sold to those beacons of democracy, Saudi Arabia and

infrastructure. Transportation and the transport industry consume the lion share

Libya, among others. The luxurious LM002 appealed to spoiled young Saudi sheks wanting to cross the sand to survey their oil field holdings. Uday Hussein, son of Saddam [Hussein], had one, which the U.S., military cheerfully blew up in 2004 during a "test" to stimulate the effects of a car bomb. The LM002 is the forerunner of another large and unnecessary SUV that signals pure contempt for one's fellow man, the Hummer H2. Read on; <u>1995 Ford Explorer:</u> How could the best-selling passenger vehicle in America 14 years running, the mother of all mom-mobils, the beloved suburban schlepper of misllions, wind up on this list? Forget about the whole Firestone tyre controversy. In its very success, the Ford's Explorer is responsible for setting this country on the spiral of vehicular obesity that we are still contending with today. People, particularly women drivers, discovered that they liked sitting up high. Even though more fuel-efficient minivans do the kid- and cargo-hauling duties better, people came to prefer the outdoor, go-anywhere image of SUVs. In other words, people became addicted to the pose. And, as vehicles got bigger and heavier, buyers sought out even bigger vehicles to make themselves feel safe. *Helloooo* Hummer. All of that we can lay at the overachieving feet of the Explorer; <u>1997 GM EV1:</u> The EV1 was a marvel of engineering, absolutely the best electric vehicle anyone had ever seen. Built by GM tocomply with California's zero-emissions-vehicle mandate, the EV1, was quick, fun and reliable. It held out the promise that soon electric cars – charged from the grid will all sorts of groovy sources, like wind and solar – could replace the smelly old internal combustion vehicle. And therein lies the problem: the promise. In fact, battery technology at the time was nowhere near ready to replace the piston-powered engine. The early car's lead-acid bats, and even the later nickel-metal hybride batteries, couldn't supply the range or durability required by the mass market. The car itself was a tiny, super-light two-seater, not exactly what American consumers were looking for. And the EV1 was horrifically expensive to build, which was why GM's execs terminated the programme – handing detractors yet another stick to beat them with. GM, the company that had done more to advance EV technology than any pther, became the company that "killed the electric car."; <u>1997 Plymouth Prowler:</u> by the mid-1990s, car designers had powerful new computer tools at their disposal, allowing them to pursue low-volume, high-zoot projects that before would never have recovered the development costs. The Prowler was one such project. Inspired, if not plagiarised, by a retro-roadster design by Chip Foose, the Prowler looked like a dry-lake speedster from the 22nd century, with an open-wheel front end and low-slung hotrod fuselage. Except they forgot to make it a hotrod. Intent on containing costs, Chyrsler stuck its standard-issue 3.5-litre V6 under hood, good for a rather less than spectacular 250 hp. The Prowler didn't even have a manual transmission, which made it almost impossible to lay down the requisite stripes of hot rubber. The result was a flaccid little jerk of a car that threatened much but delivered little; <u>1998 Fiat Multipla:</u> "Multipla" is a time-honored name for Fiat. The company made an adorable microvan by that name in the '50s and '60s, based on the Fiat 600, The Multipla that appeared in 1998 was anything but adorable. With its strange high.beam lenses situated at the bottom of the A-pillars (base of the windshield), the Multipla looked like it had several sets of eyes, like an irradiated tadpole. It had this weired probosis out front and a bulky, glass cabin in back, and the whole thing was situated on dwarfish wheels. I rented one of these in Europe and it worked beautifully, but it was just so tragic to look at. The Multipla (and the Aztek and the Consulier GTP) reminds us that cars cannot just work beautifully. They have to be beautiful. At least they can't look like this; <u>2000 Ford Excursion:</u> GM had its H2. Ford had the Excursion, a Mount Rushmore-sized SUV based on the company's Super Duty truck platform. Dubbed the Ford "Valdez" by the Sierra Club, the Excursion was a passenger vehicle of gob-smacking proportions. It weighed 7,000 lbs, measured almost 19ft. long and stood 6.5ft tall. At the time, Ford argued that many customers – ranches, farmers, um, tugbaot enthusiasts –needed a vehicle this big with over 10,000-lb. Towing capacity. Maybe that was true, but that didn't keep Suzy Homemakers from driving them to the mall. To its dubious credit, the Excursion pioneered the use of the blocker bar, a kind of under-vehicle roll bar designed to keep the Excursion from rolling over anything unfortunate enough to be hit by it. *The Simpsons* wrote the Excursion's cultural obituary in the episode where Marge buys the "Canyonero." *"Can you name the truck with four wheel drive, smells like a steak and seats thirty-five...Canyoner- oooo!"*; <u>2001 Jaguar X-Type:</u> A business case is not the same as wisdom. Certainly, Jaguar needed an entry-luxury model to compete against the BMW 3—series and Mercedes-Banz C-class. Yes, the company, owned by Ford, had access to a very successful world car platform, the Mondeo, which Americans knew as the Ford Contour. There was money to be saved. But in the attempt to turn the front-drive compact car into the limits of platforms engineering. The result was an English version of the Cadillac Cimarron, a tarted-up insult to a once-proud marque and a financial disaster for the company. It hardly matters that the X-Type was not that bad a car. Young affluenced buyers had the feeling they were somehow being grifted. They were; <u>2001 Pontiac Aztek:</u> I was in the audience at the Detroit auto show the day GM unvieled the Pontiac Aztek and I will never forget the gasp that audience made. Holy Hell! This car could not have been more instantly hated if it had a Swastika-tatoo on its forehead. In later interviews with GM designers – who, for decency's sake, will remain unnamed – it emerged that the Aztek design had been fiddled with, fussed over, cost-shaved and otherwise compromised until the tough, cool-looking concept had been reduced to a bully, plastic-clad mess. A classy case of losing the plot. The Atzek violates one of the principal rules of car design: We like cars that look like us. With its multiple eyes and supernumerary nostrils, the Atzek looks like deformed and scary, something that dogs bark at and cathedrals employ to ring bells (cf. Fiat Multipla). The shame is, under all that ugliness, there was a useful, competent crossone; <u>2002 BMW 7 series:</u> The Munich company's flagship sedan was nothing less than everything the company knew about car building, and that was quite a lot. Perfectly constructed, astonishingly fast and utterly besotted with technology, the big, gracious 7-series had but two flaws: The first was something called iDrive, a rotary dial/joystick controller situated on the centre console,through which drivers adjusted dozens of vehicle settings, from climate, navigation and audio functions to things like the sound of the door chime. The reason for iDrive and simulated systems is that designers were running out of room for switches and instruments. The trouble was that the iDrive was hard to work. Damn near impossible, in fact. Drivers spent many hairsplitting minutes driving to figure out how to add radio presets, for example, or turn up the air conditioning. When confronted with complaints, BMW engineers said, with barely disguised contempt: *Ze system werks perfectly. Dis is no problem.* Since 2002, BMW has gradually improved iDrive to make it more intuitive, but it's still a pain. The other flaw? The silly bubble-butt, called the Bangle Bustle, after lead designer Chris Bangle; <u>2003 Hummer H2:</u> One struggles to think of a worse vehicle at a worse time. Introduced shortly after 9/11 – an event whose causes were tangled in America's unquenchable thirst for oil – the Hummer H2 sent all the wrong signals. It was/is arrogantly huge, overtly militaristic, openly scornful of the common good. As a vehicle choice, the H2 was a spiteful reactionary riposte to notions that, you know, maybe we all shouldn't be driving tanks that get 10 miles per gallon. Not surprisingly, the green-niks struck back. A Hummer dealership was torched in Southern California. The H2 was also a PR catastrophe for GM, who happened to be repossessing and crushing the few EV1 electric cars at the time. It all contributed to GM's emerging image as the Dick Cheney of car companies; <u>2004 Chevy SSR:</u> It's surprising, considering that Chrysler and GM are in the same town, that GM didn't learn from the Plymouth Prowler episode. When GM decided to kick upsome custom retro mojo, it commissioned the Chevy SSR, an awsome-looking hotrod pickup truck with complete body panels and a slick convertible top. Alas, the chassis and mechanicals for the SSR were borrowed fromGM's corporate midsize SUV programme, making the putative performance machine heavy, underpowered and unforgivably lazy. It was no more hotrod than Britney to the next Hellen Mirren.In the next couple of years, Chevy amped up the SSR but by then the credibility was gone. The SSR also violated a principle of hotrodding.Hotrods are homemade subversions of the existing order, mechanical folk art. There is no such thing as a factory hotrod. Seems obvious, in retrospect.

of energy generated from petroleum, a fossil fuel. Apart from walking, cycling, wheelcarting, horse-riding, or sporting on foot, *all transportation activities involve the burning of fossil fuels and nuclear energies.*

Transportation economic costs are classified into *Vehicle, Travel time, Road and parking facility, Congestion, Traffic crashes, Environmental, Fuel externalities, Impacts on Non-motorised travel, Land use impacts* and *Equity impacts.* The TDM Encyclopedia,[50] says why these costs are studied:

- *Economic Impacts* refer to costs and benefits. Costs (benefits) reduce (increase) scarce resources such as money, time, land, health, environmental quality or any other item of value. Costs and benefits have a mirror image relationship: a cost can be as a reduction in benefits and and a benefit can be defined as a reduction in costs. Transportation benefits are often measured in terms of reduced transportation costs. For example, congestion regulation reduction benefits consist of reductions in travel time and vehicle operating costs. Calculating costs is therefore, the basis for calculating benefits.

- *Costs and Benefits:* Some people are comfortable with the idea that motorised transportation imposes 'costs' on society, because it seems arbitrary and judgmental. Virtually any human activity can be considered to impose costs when viewed from some perspectives. What, they ask, is the reference case for transportation that imposes no costs? Doesn't this focus on costs ignore the benefits of motorised transportation?...certain types of transportation activity impose *higher* costs on society than others. Although motor vehicle travel provides benefits, these benefits are largely *internal,* enjoyed directly by users. It is particularly important to identify *external* impacts (benefits and costs your neighbors have on you.) Rather than focusing on costs, some people may be more comfortable focusing on benefits...In some situations, you would probably prefer that your neighbors reduce their automobile use and rely on *alternative* forms of transportation, and all else being equal, you would prefer public policies that encourage more efficient and balanced transportation, and help reduce the transportation costs you bear.

- *Vehicle Costs:* Vehicle costs are *direct* user financial expenses for vehicles. These are often divided into *vehicle ownership* (fixed) and *vehicle operating* (variable) costs. Several organisations publish typical vehicle purchase, ownership and operating cost estimates.[51] *Vehicle Operating Cost:* These are usually defined to include short-term (or "out-of-pocket) expenses, such as fuel and oil, tyre wear, tolls and short-term parking fees, and sometimes also includes a portion of vehicle maintenance. Fuel is usually the largest portion of vehicle operating costs. Fuel price and consumption data are available from:[52]

[50] TDM Encyclopedia:*"Transportation Costs & Benefits: Resources for Measuring Transportatin Costs and Benefits."* 22 July 2008. Published by Victoria Transport Policy Institute (VTPI), 1250 Rudlin Street, Victoria, BC, V8V 3R7, Canada. info@vtpi.org; www.vtpi.org./tca. Economists have developed estimates of many transportation costs for use in economic analysis, including vehicle expenses, travel time, road, parking facility, crash and environmental costs.

[51] Among such organisations are: The American Automobile Association's *"Your Driving Costs Booklet"* (Public Affairs Office: www.aaapublicaffairs.com) provides estimates of typical annualised ownership costs for several types of vehicles during their first 5 years of operation; The Canadian Automobile Association's *"Driving Costs"* (www.caa.ca) PDF/3708-EN-2005.pdf), provides estimates of typical annualised vehicle ownership/operating costs for several types of vehicles during their first 4 years; Rundzheimer International (www.runzheimer.com) sells estimates of typical annualised ownership/operating costs for several types of vehicles, which is the basis for the values published by automobile associations; The Black Book, National Auto Research Division of Hearst Business Media Corporation (www.blackbookusa.com and www.canadianblackbook.com), provides wholesale and retail price estimates for new and used vehicles, taking into account model, age, condition,mileage, accessories and geographic location; also available at (www.cars.com); Intellichoice (www.intellichoice.com), provides price estimates for new and used vehicles; UK Automobile Association (www.theaa.co.uk), provides estimates of typical annualised ownership/operating costs for several types of vehicles; The Way To Go Seattle Ca Cost Worksheet (www.cityofseattle.net/carsmart/carcostworksheet.htm) calculates your car cost and compares it to other transportation options.

[52] International Energy Agency (IEA): www.iea.org; The American Petroleum Institute (API): www.api.org; Canadian Petroleum Communication Foundation (CPCF): www.pcf.ab.ca; Canadian Petroleum Products Institute (CPPI): www.cppi.ca; Transportation Energy data Book (TEDB)(: www.ott.doe.gov.

- Travel Time Costs: Travel time is one of the largest transportation costs, and travel time savings are often the greatest *potential* benefits of transport improvements. Various studies have calculated travel time values relative to wage rates based on traveller behavior, and several time value schedules have been developed based on such studies (Wardman, 1998; Small, et al., 1999).[53] Many specific attributes of travel, such as comfort, safety and prestige, can be reflected in travel costs. Some of the main factors affecting travel costs are:

- Commercial Vehicle Costs: Driver's wages and oversight costs; costs for the value of freight (particularly perishables), and sometimes costs for delays beyond a critical delivery time.

- The Cost of Personal travel is usually estimated at one-quarter to one-half of prevailing wage rates; *Travel Time Costs* tend to be particularly high for unexpected delays; *Travel Time Costs per minute* tend to increase for long commutes (more than 20 minutes).

- Under Pleasant Conditions, walking and cycling can have positive value, but under unpleasant or unsafe condition (for example, walking along a busy highway or waiting for a bus in an area that seems dirty and unsafe), time spent walking, cycling and waiting for transit has cost two to three times higher than time spent travelling .

- Travel Time Costs tend to increase with income, and tend to be lower for children and people who are retired or unemployed. People with full-time jobs tend to have more demands on their time, and tend to be willing to pay more for travel time savings.

- Personal Needs and preferences vary. Some people place a relatively high cost on time spent driving in congestion, and place low value on time spent as a transit passenger, while others have the opposite preferences.

- A Certain Amount of time has a low cost or positive value because consumers enjoy the experience. Under certain conditions, walking, cycling, driving, train travel and air travel are considered enjoyable and desirable, although under certain conditions, the same type of travel is considered undesirable and costly.

- Road and Facility Costs: Roads and parking are usually provided *free* or bundled with facility costs (for example, parking is usually included with housing purchases or rents) so most consumers have little idea of what a road or parking space costs to produce. Typical cost data: Roadway Construction/Maintenance Costs; Roadway Land Value; Roadway Cost Allocation; Traffic Services; Parking Costs; Incremental (Marginal) Costs of Accommodating Urban Peak Automobile Trips.

- Congestion Costs: These consist of the incremental delay, stress, vehicle operating costs and pollution that results from each additional vehicle to the traffic stream. It is an *externality* in terms of economic efficiency, and to some degree in terms of equity due to differences in the cost per passenger mile imposed by different modes. Several approaches are used to calculate congestion costs (TRB, 1997)[54]...Most congestion studies consider only costs imposed on motor vehicle users. The delay and accident risk costs that vehicle traffic and highways impose on non-motorised travel is called "barrier effect" or "severance." Some studies have quantified this cost in terms of travel delay and non-motorised trips foregone. Such costs can be significant, particularly in urban areas.

---

[53] Wardman, 1998: *"The Value of Travel Time: A Review of British Evidence."* Journal of Transport Economics & Policy, vol. 32, No. 3, Sept 1998, pp. 285-316; Small/Noland/Chu/Lewis, 1999: *"Valuation of Travel-Time Savings and Predictability in Congested Conditions for Highway Users- Cost Estimation."* NCHRP 431, Transport research Board (www.nas.edu/trb).
[54] TRB, 1997: *"Quantifying Congestion."* TRB (www.trb.org), NCHRP Project 7-13.

- *Traffic Crashes:* Traffic crash costs include deaths, injuries, pain, grief, disabilities, lost productivity, material damage, and crash expenses. Many road safety experts prefer the term "crash" to "accident" because accident implies that the event was random, without cause or responsibility. Traffic crash statistics are available at:[55]

- *Environmental:* Environmental pollution costs include air, noise, water, waste disposal, and environmental impacts associated with transportation facilities, such as loss of wildlife habitats. Air pollution is one of the most obvious environmental costs of motor vehicle use. Per mile emissions for many pollutants have declined over time due to emission control strategies. It is common to hear claims that *"automobile emissions have declined by 90% or more over the last few decades"*, but this is an exaggeration. Engine and fuel improvements have significantly reduced *tailpipe* emission rates under design conditions, but a *significant portion of driving* occurs under *non-design conditions* and non-tailpipe emissions are *not controlled* by these technologies. Motor vehicles produce several potentially harmful air pollutants, including *carbon monoxide (CO)*, *carbon dioxide (CO_2)*, *particulates (PM)*, *nitrogen oxides (Nox)*, *volatile organic compounds (VOCs)*, *hydrocarbons (HCs) reactive organic compounds (ROCs)*, *sulphur oxides (Sox)*, *methane (CH_4)*, *road dust,* and *toxic gases* such as *benzene.* Most air pollution cost studies focus on just a few impacts and give an incomplete estimate of total pollution costs. Information on vehicle emissions is available at:[56]

- *Climate Change Emissions:* There is particular uncertainty about climate change emission costs. An increasing body of scientific evidence indicates that climate change is a significant risk. For example, the American Geophysical Union concluded that the "present level of scientific uncertainty does not justify inaction in the migration of human-induced climate change" (AGU, 1998, Table 12).[57] AGU summarises one estimate of greenhouse emission costs that indicates a greenhouse gas cost of $0.18 cents to $0, 56 cents per gallon of gasoline, or about $0. 9 cents to $0. 2.8 cents per mile; Carbon Dioxide: ECU/tonne carbon units, 74 low, 152 Midpoint, 230 high; Carbon Dioxide: ECU/tonne CO_2 units, 20 low, 42 Midpoint, 63 high; Methane, ECU/tonne CH_4, 370 low, 540 Midpoint, 710 high; Nitrous Oxide: ECU/tonne N_2O, 6,800 low, 21, 400 Midpoint, 36,000 high.

- *Noise Pollution:* Motor vehicle traffic imposes noise pollution. Traffic noise tends to increase with traffic speed, accelerations, the portion of heavy vehicles and motor cycles, and development density. Noise costs tend to be much higher on local urban roads, where traffic tends to be closer to residences. [Information on noise costs are available at: Noise Pollution ClearingHouse (www.noise.org); and FHWA (1997b).]

- *Water Pollution & Hydrologic Impacts:* Roads and motor vehicles use also contributes to water pollution, hydrologic impacts and waste disposal, such as used tyres, which impose a variety of costs on society (USEPA, 1999; Bein, 1997).[58] [information hydrologic impacts is available at: NEMO Foundation: www.canr.uconn.edu/ces/nemo; and center for Watershed Protection: www.pipeline.com/~mrunoff.]

- *Waste Disposal:* Motor vehicles produce a number of harmful products that impose

[55] Bureau of Transportation Statistics (www.bts.gov); National Highway Traffic safety Administration (www..nhtsa.dot.gov); National Center for Statistics/Analysis (www.nhtsa.dot.gov/people/ncsa); Transport Canada (www.tc.gc.ca/roadsafety); Eurostat (www.europa.eu.int); European Conference of Ministers of Transport (www.oecd.org/cem/stat); G-7 Transportation Highways (www.bts,gov/itt/G7HighwaysNov99/G-7book.pdf); International Road Traffic/Accident Database (www.bast.de/htdocs/fachthemen/irtad//english/we2.html).
[56] USEPA Transportation Air Quality Center (www.epa.gov/otag); Environmental valuation Reference Inventory (www.evri.ca); Transportation Energy Data Book (ORNL, 2000) (www.ott.doe.gov); European Environmental Agency (EEA) (www.eea.eu.int); Energy Conversation and Emission Reduction Strategies (ECERS); Several studies provide monetised estimates of the environmental costs of transportation (Bein, 1997); USEPA, 1999; Delucchi, 2000; Litman, 2001.
[57] AGU, 1998: *"Climate Change and Greenhouse Gases."* American Union (www.agu.org).
[58] USEPA, 1999:*"Indicators of Environmental Impacts of Transportation."*Office of Policy & Planning." USEPA (www.itre.ncsu.edu/cte); Bein, 1997: *"Monetisation of Environmental Impacts of Roads."* Highway Planning & Policy Branch, Ministry of Transportation & Highways (www.gov.bc.ca/tran; www..geocities.com/davefergus/Transportation/0ExecutiveSummary.htm).

externalities, including used tyres, batteries, junked cars, oil and other semi-hazardous materials resulting from motor vehicle production and maintenance. These wastes impose a variety of environmental, human health aesthetics, and financial costs, through improper disposal, residential impact even when proper disposal is observed, and because some disposal efforts are subsidised by general taxes. Some new laws and policies are intended to internalise these costs. Crankase oil recycling is encouraged, vendors are required to recycle used car batteries, and in some U.S., states a type of tax is dedicated to tyre disposal.

●

Fuel Externalities: Fuel production and consumption can impose various external costs, including national security risks and macroeconomic impacts on individual economies that import fuel, depletion of non-renewable resources, various financial subsidies, and environmental damages (including greenhouse gas emissions). Put another way, there may be benefits to society from increased energy efficiency and conservation. Greene and Tischchishya (2000)[59] estimate that oil market upheavals of the past 30 years have cost the U.S., economy $7 trillion (net present value) in reduced output, with a range of $1.5 to $14.6 trillion. These estimates do not include military, strategic or political costs associated with U.S., and world dependence on oil imports. They point out that each of the major price shocks during this time periods has preceded a major economic recession, and that higher petroleum import prices reduce national GDP.

●

Impacts On Non-Motorised Travel: Changes in the design of roads and parking facilities, vehicle traffic volumes and speeds, and a quality of the pedestrian environment can affect convenience, safety and comfort of walking and cycling. The "Barrier Effect" or Severance, refers to the tendency of roads and traffic to create a barrier to non-motorised travel. It represents a degradation of the pedestrian and cyclist environment that reduces the viability of these modes, often leading to increased driving.

●

Land Use Impacts: A number of studies indicate that lower density, urban periphery, automobile oriented development patterns (commonly called "sprawl") impose a number of economic, social and environmental costs (Table 13: "Sprawl Costs and Benefits) based on Burchell and others, 1998.[60] Some researchers argue that a more balanced transportation system can increase community cohesion, reduce crime and increase employment opportunities for disadvantaged populations. Untermann/Vernez Moudon 1998,[61] studied traffic impacts on neighborhoods, and and conclude: "A deeper issue than the functional problems caused by road widening and traffic buildup is the loss of sense of community in many districts. Sense of community traditionally evolves through easy foot access – people meet and talk on foot, which helps them develop contacts, friendships, trust and commitment to their community. When everyone is in cars there can be no social contact between neighbors, and social contact is essential to developing commitment to neighborhoods." Although they are difficult to measure, these indicate that there are likely to be benefits from reducing traffic impacts in neighborhoods, reducing the amount of land paved for roads and parking facilities, and preserving greenspace.

●

Equity Impacts: Transportation policies can have a variety of equity impacts. Some people argue that transportation policies that favor automobile travel are inequitable because they *favor motorists over people* who use *alternative* forms of transportation, who are often economically, socially and physically disadvantaged. To the degree that automobile use makes motorists relatively better off compared with non-motorists, it can be considered to impose equity[62] costs. The demand for mobility, and for motorised travel in particular, has

---

[59] Green/Tishchishya, 2000: *"Costs of Oil Dependence: A 2000 Update."* Oak Ridge National Laboratory, ORNL-TM-2000/152 (www.osti.gov/bridge).
[60] Robert Burchell, et. al., 1998: *"The Costs of Sprawl – Revisited."* TCRP Report 39, Transportation Research Board (www.trb.org).
[61] Richard Untermann/ Anne Vernez Moudon, 1989: *"Street Design: Reassessing the Saftety, Sociability, and Economies of Streets."* University of Washington at Seattle, U.S.
[62] Equity: Refers to the distribution of resources and opportunities. Transportation decisions can have significant equity impacts because: Transportation represents a major portion of consumer, business and government expenditures, including taxes and public land; transportation activities have external-noise and ait pollution, crash risk and barrier effects – that affect the quality of community and natural environments, and personal safety; transportation determines where people can live, shop, work, go to

increased over the last century. In previous generations, most communities were organised to allow residents to *walk* or *cycle* to neighborhood stores, schools and recreational activities. Work trips tended to be relatively short and centralised. Now, transportation systems and land use patterns are more *automobile-dependent,* increasing the need to travel and reducing travel choices, particularly for *non-drivers.* (Sanches/Brenman, 2007).[63] And Johnson, 1993[64] comments: "For those too young, too old, too poor or too infirm to drive, the paucity of mobility *alternatives* severely limits their opportunity for education and their ability to share in their essential everyday activities. Moreover, as more employers move to suburbs, more jobs require car mobility."

Transportation infrastructure across the globe shows great disparities and major differences between the 5% rich and the 5% income poor countries (Tasso Adamopoulos).[65] The five richest countries of the world have 24.5 km of total road and 1.8 km of railroad for every 1000 people, while the five poorest countries have 1.5 km of total road and 0.05 km of railroad for every 1000 people – factor differences of 14 and 30 respectively.

Furthermore, there are large differences in the efficiency and quality of these networks. In the five richest countries, 61% of the total road network is paved, while in the five poorest countries only 16% is paved. Out of the paved network in the poor countries 33% is in fair condition, while 20% is in poor condition. This leads to poor productivity, differences in income, low aggregate output per worker; transportation total factor productivity (TFP) increases the cost of transporting final goods and inter-regional inputs used in farming (chemical fertilizers, pesticides, processed seeds, fuel, energy) across regions; higher freight costs operate as natural barriers that distort the efficient spatial distribution of products; higher freight costs also have an additional adverse impact on agricultural productivity because they reduce farmers' access to the technology embodied in the modern intermediate imputs. Transportation economic costs – *travel time, road/parking facilities, congestion, traffic crashes, fuel externalities, vehicle ownership;* and *environmental, non-motorised, land use* and *equity impacts* – rose stiffly and remain high with the oil-and-food-crisis. These transportation costs affect agriculture – the basis of all economies, societies and lives on planet Earth. Agriculture, energy supply and transportation are the *make* or *break* of the survival of the world and its economies.

Agriculture is the production of goods through the growing of plants and fungi, aquaculture, horticulture, gardening, free farming; and the raising of domesticated animals and hunting of wild animals. A wide variety of specialty agriculture is done to improve quantity and quality of harvests, such as cultivation on arable land, pastoral herding of livestock on rangeland, sustainable agriculture, intensive farming, modern agronomy, plant breeding, selective breeding, irrigation, and modern animal husbandry.

---

school and recreate, and their opportunities to life; adequate mobility is essential for people to participate in socity as citizens, employees, consumers and community members.; transportation affects peoples' ability to obtain education, employment, medical service and other critical goods.

[63] Thomas W. Sanches/ Marc Brenman, 2007: *"The Right To Transportation: Moving To Equity."* Planners Press (www.planning.org).

[64] Elmer Johnson, 1993: *"Avoiding the Collision of Cities and cars."* American Academy of Arts & Sciences (Chicago, U.S.); Institute for Science & Technology Policy, Murdoch University (http://wwwistp:murdoch.edu.au).

[65] Tasso Adamopoulos:*"Transportation Costs, Agricultural Productibity and Cross-Country Income Differences."* Abstract and Introduction, Nov., 2005, York University, 4700 Keele Street, Toronto, Ontario, Canada, M3J1P3 (www.yorku.ca).

Major agricultural products include:

Foods – cereals, fruits, vegetables, meat; *Fibres* - cotton, wool, hemp, silk, flax; *fuels* – biofuels (methane from biomass, ethanol, biodiesel, cut flowers, nursery plants, tropical fruits, and birds); *Raw materials* - lumber, bamboo; *Pharmaceuticals* and *biopharmaceuticals; Drugs* - tobacco, marijuana, opium, cocaine, digitalis, curare, eugenol, reserpine, pyrethrins, taxol, resins. *Fertilizers* – organic, inorganic chemical compounds or minerals – are fed to plants to promote growth, usually applied either through the soil for plants roots' uptake or by foliar feeding for leaves' uptake. Fertilizers are *organic*, if composed of organic matter, or *inorganic* if composed of simple inorganic chemicals or minerals. *Chemical compounds* can be naturally-occurring, such as peat or mineral deposits; or manufactured through natural processes, such as the Haber-Bosch-process,[66] to produce ammonia[67] and ammonia fertilizers on a commercial scale, that in turn enable farmers to grow food crops, feed, fibre and bioenergy needs. The International Fertilizer Industry Association (IFA)[68] while acknowledging the high cost of fertilizers brought on by the oil-and-food-crisis, states the reasons for fertilizer prices' rise highs, and why they remain high:

- *Strong Community Agricultural Prices:* Because farmers generally buy fertilizers on credit that is repaid only after their harvests are sold, investing in fertilizers to increase yield and crop quality carries some risks. Therefore, the price that farmers expect to receive for their output influences their decision to invest in fertilizers.

- *Market Forces Determine Global Fertilizer Prices:* Trade in fertilizers is truly global. IFA's nearly 200 fertilizer producers whose market shares of even the biggest is 7% of total world fertilizer production. Current fertilizer prices are illustrative of the basic principles of supply and demand, and reflect a properly functioning, but very tight, global market; the underlying reason for increased fertilizer prices is: *fertilizer demand is growing in developing countries,* which trigger better diets and changed food preferences, thus increasing the demand for agricultural goods. South and East Asia countries currently account for two-thirds increase in fertilizer use; Latin America, especially Brazil, account for a significant share increase; in Africa, numerous transaction costs make fertilizers more expensive, where soil fertility is declining at alarming rates – than anywhere elses in the world. That means that fewer farmers in the region can afford to replenish the nutrients removed from their fields for each crop and lost to erosion.

- *Exchange Rates Influence Global Market Prices:* Cross-border fertilizer transactions are denominated in U.S., dollars, which has pushed quoted prices of fertilizers upwards.

---

[66] Haber-Bosch-Process or Nitrogen Fixation Reaction: A process where nitrogen is passed over an iron catalyst, to produce ammonia to produce ammonia on an industrial scale, first accomplished by the German firm BASF's Oppau plant, Germany in 1913. BASF chemists Fritz Habers and and carl Bosch developed the Haber-Bosch-Process which bears their names, for which both men earned Nobel prizes in 1918 and 1931. To begin the Haber-Bosch-Process to produce ammonia, hydrogen is isolated from methane, a natural gas, using heterogeneous catalysis. Heteregeneous catalysts provide a surface for the chemical reaction to take place on.

[67] Ammonia (NH^3) gas occurring in Nature is a colorless with a characteristic pungent smell, very soluble in water, giving alkaline solution, and is both caustic and hazardous; it contributes significantly to the nutritional needs of terestrial organisms by serving as a percursor to foodstuffs and fertilisers as its salts or solutions. The fertilizer generated by ammonia is responsible for sustaining one-third of the world's population. Ammonia is used commercially as *anhydrous ammonia*, and directly or indirectly, is a building block for the synthesis of many pharmaceuticals. Liquid ammonia dissolves the alkali metals – lithium (Li), sodium (Na), potassium (K), rubidium (Rb), caesium (Cs), francium (Fr), hydrogen (H^2=), and calcium, strontium, barium,europium and ytterbium – known as electropositive metals. In World War II, ammonia was used as fuel to power buses in Belgium, as well as in engines and solar energy applications.

[68] International Fertilizer Industry Association (IFA), a non-profit organisation, represents the global fertilizer industry, whose members serve farmers everywhere with fertilzers and growing food, feed, fibre and bioenergy needs in a sustainable manner. IFA helps ensure that farmers have all the crop nutrients needed. [28, rue Marbeuf, 75008 Paris, France; http://www.fertilizer.org/.]

- *Delivery Prices Depend on Many Factors:* The delivery price that farmers pay for the fertilizers depends on a number of factors, many of which are *not* directly related to fertilizers: *transport costs, taxes, administrative costs, mark-ups* by intermediaries and many others. Where these additional charges are determined as a percentage of the fertilizer price rather than a flat rate, they amplify the effect of higher fertilizer prices.

- *Energy and Mineral Commodities Sectors:* It is important to note that these sectors have an impact on fertilizer costs because nitrogen fertilizer production is energy-intensive and *relies on hydrocarbon feedstocks* especially natural gas. So high fuel costs drive up fertilizer prices. Energy prices also affect the cost of shipping bulky fertilizers and their raw materials around the globe, as does robust demands for limited freight capacity.

Fertilizers provide varying proportions of three major plant nutrients required by plants to grow healthy, and thus provide abundant yields of high quality crops. There are 13-20 different elements that may be needed for this, depending on the plant. Plants need large quantities of "macronutrients" and "micronutrients", which traditionally or primarily are: nitrogen (N), phosphorous (P), potassium (K), and sulphur (S) often added; secondary – calcium, sulphur, magnesium; and third - sometimes elements or micronutrients of boron, chlorine, manganese, iron, zinc copper, molybdenum, and selenium. The nitrogen in fertilizers comes from the air, which is 78% nitrogen. The chemically inert form of nitrogen that we breathe and cannot be used by plants, needs large amounts of energy to convert it into a form that is useful to plants. Phosphorous in the form of phosphates is mined from deposits that were once sediment on the bottom of the ancient seas; the potassium used to make fertilizers is in a salt form "potash", and potash deposits are derived from evaporated sea water; most of the sulphur used by the fertilizer industry is a by-product of other industrial processes, while secondary and micronutrients come from mineral deposits. Fertilizer, pesticide and herbicide uses go hand-in-hand in agriculture. While fertilizers provide nutrients to soils, pesticides and herbicides destroy plants or weed and insects that compete with crops for food welfare.

Common fertilizer products and intermediates provided by the IFA,[69] are herbicide products, and some major worldwide fertilizer and pesticide

---

[69] International Fertilizer Industry Association (IFA): *"What are some common fertilizer products and intermediatess?"*: A: <u>Common Fertilizer Products & Derivatives:</u> **Nitrogen Fertilizers:** Ammonium sulphate (AS); Ammonium nitrate (AN); Calcium ammonium nitrate (CAN); Urea; **Phosphate Fertilizers:** Single superphosphates (SSP); Triple superphosphates (TSP); Diamonium phosphate (DAP); Monoammonium phosphate (MAP); Ground phosphate rock; **Potash Fertilizers:** Muriate of potash (MOP), also called Potassium chloride; Sulphate of potash; Sulphate of potash magnesia; **Magnesium Fertilizers:** Kieserite; Epson salts; **Complex Fertilizers:** NPK fertilizers ; NP fertilizers ; NK fertilizers ; PK fertilizers; B: <u>Major Herbicides In Use Today:</u> **2, 4-D:** A broadleaf herbicide in the phenoxy group used in turf and in no-Till Crop production, the most widely used herbicide in the world, now mainly used in a blend with other herbicides that allow lower rates of herbicides; it is an example of auxin (plant) hormone; **Aminopyralid:** A broadleaf herbicide in the pyridine group, used to controlbroadleaf weeds on grassland, such as docks, thistles and nettles, notorious for its ability to persist in compost; **Atrazine:** Atrazine herbicides are used in corn and sorghum for control of broadleaf weeds and grasses; it's still used because of its low cost and because it works extraordinarily well on a broad spectrum of weeds common in the U.S., corn belt; atrazine is also commonly used with other herbicides to reduce the overall rate of atrazine and to lower the potential groundwater contamination; it is a photosystem II inhibitor; **Clopyralid:** Is a broadleaf herbicide in the pyridine group, used mainly in turf, rangeland, and for control of noxious thistles; it's notorious for its ability to persist in compost, and it's another example of a synthetic auxin; **Dicamba:** Is a post-emergent broadleaf herbicide with some soil activity, used on turf and field corn, another example of a synthetic auxin; **Glufosinate ammonium:** A broadleaf-spectrum contact herbicide, used to control weeds after the crop emerges or for total vegetation control on land not used for cultivation; **Fluroxypyr:** A sytstemic, selective herbicide used for control of broadleafed weeds in small grain cereals, maize, partners, rangeland and turf; it is a synthetic auxin; in cereal growing, fluroxypyr's key importance is control of cleaners, such as galium aparine; **Glyphosate:** Is a systemic, non-selective herbicide used in no-till burndown and for weed control in crops that are genetically modified (GMOs) to resist its effects; it's an example of an EPSPS inhibitor; **Imazapyr:** Is a non-selective herbicide used for the control of a broad range of weeds, including terrestrial annual and perennial broadleaf herbs, woody species, and riparian and emergent aquatic species; **Imazapic:** A selective herbicide for both

organisations include:[70] *Fertilisers* are either *organic, manufactured organic, inorganic* or *synthetic.*

Organic fertilizers include: naturally-occurring manure, slurry, worm castings, peat, seaweed, sewage, guano, green crops manure, minerals such as mine rock phosphate, sulphate of potash, and stone, described thus:

---

pre- and post-emergent control of some annual and perennial grasses, and some broadleaf weeds; imazapic kills plants by inhibiting the production of branched chain amino acids (valine, leucine, isoleucine), which are necessary for protein synthesis and cell growth; **Linuron:** A non-selective herbicide used in the control of grasses and broadleaf weeds, and it works by inhibiting photosynthesis; **Metolachlor:** A pre-emergent herbicide widely used for control of annual grasses in corn and sorghum, and it has displaced some of the atrazine in these uses; **Paraquat:** A non-selective contact herbicide used for no-till burndown and in aerial destruction of marijuana and coca plantings; it's more acutely toxic to people than to any other herbicide in widespread commercial use; **Penimethalin:** A pre-emergent herbicide widely used to control annual grasses and some broadleaf weeds in a very wide range of crops, including corn, sorghum, wheat, cotton, many tree and vine crops, and turfgrass species; **Picloram:** A pyridine herbicide mainly used to control unwanted trees in pastures and adges of fields, a synthetic auxin; **Triclopyr:** A systemic, foliar herbicide in the pyridine group used to control broadleaf weeds while leaving grasses and conifers unaffected; C: Fertilizer Organisations:: *Agricultural Industries Confederation* (AIC), Confederation House, Peterborough PE2 6XE, UK; enquiries@agindustries.org.uk; http://www.agindustries.org.uk/content.template/30/30/Home/Home.mspx; *China National Chemical Information Center* (CNCIC), No. 53, Xiaoguan Avenue, Andingmenwai, Beijing, 10002, China; chenl@cheninfo.go.cn; http://www.china-fertinfo.com.cn/; *British Sulphur Consultants,* A Division of CRU International Ltd, 31 Mount Pleasant, London WCIX OAD, UK; http://www.crugroup.com/Pages/default.aspx; 9517 Dunas Court, Wake Forest, NC 27587-6522, UK; *Fertecon Ltd,* Royal Victoria House, The Pantiles, Turnbridge Wells, Kent TN2 5TE, UK; http://www.fertecon.com/; *FMB Publications,* FMB House, 6 Windmill Road, Hampton Hill, Middlesex TW12 1RH, UK; *ICIS,* Quadrant Hose, The Quadrant, Sutton, Surrey, SM2 5AS, UK; *Integer Research Ltd,* London ECIM 3JB, UK; info@integer-research.com; http://www.integer-research.com/; *IFDC- An International Center for Soil Fertility & Agricultural Development,* P.O. Box 2040, Muscle Shoals, AL 35662, U.S.; http://www.ifdc.org/; IFDC's Africa Division,BP 4483, Lome, Togo; *Nexant ChemSystems,* Griffin House, 1st Floor South 161, Hammersmith Road, London W6 8BS, UK; http://www.nextant.org/.

[70] Fertilizer and Pesticide Organisations and their functions: **National Pesticide Information Center (NPIC):** Provides toll-free telephone service information to any caller in the U.S., Puerto Rico or the Virgin Islands; it is a cooperative agreement between Oregon State University and the U.S., Environmental Protection Agency (EPA); Oregon State University, 333 Weniger Hall, Corvallis, OR97331-6502; ace.ace.orst.edu; **American Crop Protection Association (ACPA):** Is a non-profit organisation representing manufacturers, formulators and distributors of crop protection and pest-control products, and the premier national association representing the concrete pavement industry, make portland cement concrete the material of choice for airports, highways, street, and local road pavements; www.acpa.org; 5420 Old Orchard Road,Suite A-100, Stokie, Illinois 60077-1059; 500 New Jersey Ave., NW 7th Floor, Washington, D.C. 20001; **Agricultural Container Research Council (ACRC):** aAnon-profit organisation that promotes/supports collection/recycling of properly raised high-density polyethylene (HDPE) crop protection production containers, facilitates the collection and recycling of one-way rigid HDPE plastic agricultural crop protection, specialty pest control, micronutrients/fertilizer, and/or adjuvant product containers throughout member-funding of cost-effective programmes that foster public health and safety, environmental protection, resource conservation and end-user convenience; Exco Dir.: Ron Perkins, 1156 15th Street NW, Suite 400, Washington, D.C. 20005; rperkins@acrecyde.org.; **Armed Forces Pest Management Board (AFPMB) Hotline:** Provides pesticide labels; pesticide selection; integrated pest management (IPM); pesticide regulation; pesticide application; environmental concerns assistance; Fact sheets; plant/animal identification services; and DOD expert locator assistance; AFPMB, WRAMC Glen Annex, BLDG 172, Forney Road, Silver Spring MD 20910-1230, U.S.; www.afpmb.org; **Asociacion Nacional De Controladores De Plaga Urbanas Mexico (ANCPU):** - National Association of Mexico for the control of insect pests; Donato Guerra No. 1 Desp... 307 Col. Juarez C.P. 06600,Mexico, D.F.; ancpuac@prodigy.net.mx; www.ancpu.org; **Plant Protection & Quarantine (PPQ):** Is a programme within the Animal/Plant Health Inspection Service (APHI.PPQ) of the USDA safeguards agriculture and national resources from the risks associated with the entry, establishment, or spread of animal and plant pests and noxious weeds to ensure an abundant high-quality, and varied food supply; LPA headquarters, Office of the Deputy Administrator, Washington, D.C.; www.aphis.usda.gov; **British Pest Control Association (BPCA):** Is a trade association representing all those professionally involved with the eradication of public health nuisance pests; ensures the provision of a legislative environment conducive to the profitability of member companies businesses; www.bpca.org.uk; tammy@bpca.org.uk; **European Fertilizer Manufacturers Association (EFMA):** Provides information and other resources about the fertilizer provided by the EFMA; responds to the needs of agriculture and society by providing, in accordance with the principles of Responsible Care, a dependable and competitive supply of high quality mineral fertilizers; www.efma.org; http://cms.efma.org/; **Entente Internationale pour le Demoustication du Litoral Mediterraneen:** Provider of mosquito control services and products; 165 avenue Paul-Rimbaud, 34184 Montpellier Cedex 4, France; eid.com@eid-med.org; www.eid-med.org; http://www.eid-med.org; **Citizen's Guide to Pest Control and Pesticide Safety:** A from the U.S., Environmental Protection Agency (EPA) that teaches consumers how to control pests in and around the home; www.epa.edu; **American Association of Pesticide Safety Educators (AAPSE):** Provides science-based pesticide safetyeducation programmes through tribal and government agencies and the land-grant university cooperative extension services, and seeks to protect human health and environment through education; organisation is composed of a grpoup of national leaders in pesticide safety education, training and certification; **FAO Integrated Pest Management (IPM):** This is part of the IPM group of Plant Protection Service of FAO, and deals with the implementation of the IPM projects from multinational to farmer level, and increases the sustainability of farming systems; improves ecological sustainability, as it relies primarily on environmentally-benign processes, including the use of pest resistant varieties, the actions of natural enemies and cultural control, and improves social stability; http://www.fao.org/WAICENT/FAOINFO/AGRICULT/AG/IPM/Welcome.htm; **Entomological Society of British Columbia (ESBC):** A a scientific society for the advancement of entomological knowledge in the province of British Columbia; www.harbour.com; **Far West Fertilizer:** Is dedicated to efficient plant production with a commitment to professionalism, environmental protection and safety; www.fwaa.org; **Pesticide Network North America (PANNA):** An international organisation providing information and resources on the use of pesticides and its impact on health, consumer, labour, environment, progressive agriculture and public interest in Canada and the U.S.; www.panna.org; **National Pest Management Association's PestWorld:** An organisation representing pest management firms worldwide; www.pestworld.org.

- *Manure:* Organic matter used as fertiliser to contribute to the soil's fertility by adding organic matter and nutrients, such as nitrogen that is trapped by bacteria in the soil on which higher organisms feed on the fungi formed and bacteria, in a chain of life that comprises the soil food website.

- *Slurry:* A thick suspension of solids in a liquid mixture of water and animal waste used as fertiliser.

- *Worm Castings, Vermincompost, Worm Compost, Worm Humus:* This is a breakdown of organic matter by some species of earthworms (microdriles or megadriles) commonly found in organic-rich soils. Earthworms together with bacteria are the major catalyst for decomposition that takes place in a healthy vermicomposting. Other soil species that play a part in vermicomposting are insects and molds.

- *Peat:* This is an accumulation of partially decayed vegetation matter that forms wetlands, peatlands, bogs, moors, muskeys, pocosins, mires, peat swamp or forests.

- *Seaweed:* Include multicellular, benthic, green, red and brown algae, used as fertiliser, food, medicine and biofuel.

- *Sewage:* This is mainly liquid waste consisting of solids produced by humans, typically comprising of washing water, feces, urine, laundry waste – that drain down from drains and toilets from households and industry – and is a potential source of pollution especially in urban areas. Water pollution by feces is estimated the largest cause of deaths worldwide. Sewage services prevent sewage pollution by managing the collection, treatment and recycling or safe disposal of sewage. About 850 billion gallons of raw sewage were dumped into waterways every year since 2004.[71] Untreated sewage carries a dangerous cargo of infectious *bacteria, viruses, parasites* and *toxic chemicals.* When it ends up in our recreational facilities and drinking water, in groundwater and in basements of our homes, it takes a severe toll on human health and the environment: each year, 1.8 million to 3.5 million illnesses are caused by swimming in water contaminated by sewage overflows, and an additional 500,000 from drinking contaminated water; U.S., medical costs associated with eating sewage-contaminated shellfish range from $2.5 million to $22 million each year.

- *Guano:* This feces excreted by seabirds, bats and seals, is used as fertiliser and gunpowder ingredients, because of guano's high levels of phosphorous and nitrogen content, and lack of odor. Superphosphate made from guano is used for aerial topdressing, (i.e., spreading of fertilisers over farmland.)

- *Green Manure:* This is a type of over-crop grown primarily to add nutrients and organic matter to the soil, to improve and protect the soil. Some green manure adds to the percentage of organic matter or biomass; others improve water retention, aeration, or deep-root nutrients for shallow-rooted organisms; others suppress weeds from growing and prevent soil erosion;

---

[71] Natural Resources Defense Council (NRDC): *"ISSUES: WATER: Sewage Pollution Threatens Public Health: Ageing sewer systems and rollbacks of environmental law are compounding the problem.,"* reports that: The U.S., Environmental Protection Agency (EPA) estimates that every year, in each county across the U.S., the amount of untreated sewage that enters the environment is enough to fill both the Empire State Building and Madison Square Garden. And *"Swimming in Sewage",* a February 2004 report by NRDC/Environmental Integrity Project, shows that sewage overflows – legal, some not – are creating an environmental and public health crisis: in Hamilton County, Ohio, a single sewer discharges 75 gallons of untreated sewage into Mill Creek each year, including during the summer months when children swim in the river; in Indianapolis, more than 1 billion gallons of untreated sewage are discharged into the environment each year because treatment plants cannot handle the flow during wet weather; in Michigan, 2000 homes sustained sewer-related damages in 1999 and 2000; in Washington, D.C., a half-inch of rain can make sewers overflow into the Anacostia River, which runs through the heart of the city; NRDC, 40 West 20th Street, New York, NY 10011; webmaster@nrdc.org.

some green manure crops, when allowed to flower, provide forage for pollinating insects. Green manure crops and plants include: legumes, oats or rye, fava, vicia, broad, faba, horse, field, itic bean; mustard; clover, trefoil, fenugreek; maithray, methi, mithi, methyada soppu, ventayam, menthulu, hilbeh, ullura bean which is cultivated worldwide and used also for curry; lupin, lupine, sweet lupins beans; sun hemp; alfalfa; tyfon; buckwheat; ferus of the azolla genus; velvet bean.

- *Manufactured Organic Fertilizers:* These include: Compost; Bloodmeal; Bonemeal; Fish Hydrolysate; Seaweed Extracts. *Compost* is brown manure which is the aerobically decomposed remnants of organic matter used in landscaping, horticulture, agriculture, soil conditioner and fertiliser, soil erosion control, stream reclamation, wetland construction, and landfill cover; *Bloodmeal* is dried powdered blood, completely soluble and can be mixed with water to be used as high-nitrogen fertiliser, liquid fertiliser, food supplement, compost or rabbit deterrant. Its raw material comes from slaughterhouse by-product; *Bonemeal* is a mixture of crushed coarsely ground bones used as an organic fertiliser for plants and formerly as animal feed, once used as human dietary calcium-supplement, and is frequently used in preparation for blowing bulbs; *Fish Hydrolysate* is groundup fish carcasses (i.e., guts, bones, cartilage, scales, meat) usually put into water and groundup, or even dried. Fish hydrolysate is quite different from fishmeal; *Fishmeal* or *fish meal* is a brown powder or cake obtained from fish, bones and offal from processed fish, by pressing the whole fish or fish trimmings to remove the fish oil. Fishmeal is predominantly used as a high-protein supplement in agricultural feed.

- *Aquatic Seaweed Extracts, Aquatic Plant Extracts* (Crops): Are made from seaweed, for example, *Kelp Extracts*[72]. Baldwin, 2001[73] describes Kelp extracts as composed of chemicals naturally found in aquatic plants, as well as breakdown products that are formed during the manufacturing process. Common sea plants as sources of aquatic extracts include: *knotted wrack* (Ascophyllum nodosum), and *Sea Bamboo* (Ecklonia maxima); aquatic plants used as soil amendments are most commonly derived from kelp (Ascophyllum ssp), and other seaweed harvested from the North Atlantic. Most commercial aquatic plant extract products do *not* provide or list specific chemical ingredient information. But aquatic plants contain *proteins, lipid sugars, amino acids, nutrients, vitamins, plant hormones,* and other biochemicals. Aquatic plant extracts are brownish-black powder or brown liquid with marine or fish odor, soluble in water and ethanol, stable and not subject to hazardous polymerization, and non-flammable. Aquatic plants are used as: foliar fertilizers on all crops or as soil conditioners; as feed; transplant solution (in combination as foliar and soil); seed treatment; source of micronutrients and growth promoters; a rooting solution for transplants and cuttings; for cold hardness in tomato, citrus fruits, and cabbage; and to reduce pest damage, such as nematodes in tomato and okra, mites in strawberries, peaches and apples. Aquatic plant extracts are sometimes applied as foliar spray by farmers seeking a natural source of micronutrients. None of the micronutrient levels in kelp extracts are high enough to correct a deficiency, but they are used as a "tonic" producing a broad array of micronutrients and other trace elements in organic farming (Hall/Sullivan, 2001).[74] Some aquatic plant products are supplemented with *synthetic* nutrients. In the European Union (EU), aquatic plant extracts are allowed (following Annex IIB-Seaweed plants), when these are directly obtained by *physical process,* including *dehydration, freezing* and *grinding; extraction with water* or *aqueous* and/or *alkaline solutions;* and *fermentation,* according to the Organic Trade Association.[75]

---

[72] Original TAP Databases Form 1995: *"Aquatic Plant Extracts",* from NOSB National Database: Kelp Extracts: Trade names for Kelp Extracts: *Alg-A-Mic; Stress-X Powder; AcadianTM Organic Powder 0.5-0.3-14 (OMRI, 2005); CAS Numbers: 8477-78-0; Extracts of* Ascophyllum nadosum . [The National Organic Standards Board (NOSB) is one of the programmes administered by the U.S. Department of Agriculture's (USDA) Agricultural Marketing Service (AMS). NOSB assists in the development of standards for substances to be used in organic production, advises the Agricultural Secretary concerning the National Organic Programme (NOP) and the consensus of the organic community. The AMS itself includes six commodity programmes: Cotton, Dairy, Fruits and Vegetables, Livestock and Seed, Poultry, and Tobacco which provide standardisation, grading and market news services for those commodities, etc.]

[73] K.R. Baldwin, 2001: *"Soil fertility for organic farming."* http://www.ncsu.edu/organic_farming_systems/news/soil_fertility.PDF .

[74] B. Hall/P. Sullivan, 2001: *"Alternative Soil Amendments: Appropriate Technology Transfer for Rural Areas."* http://attra.ncat.org/attra-pub/PDF/altsoil.pdf.

[75] Organic Trade Association, 2002:*"Comparative Analysis of the U.S., National Organic Programme (7 CFR 205) and the EU Organic Legislation* (EEC2092/91&Amendments." http://www.ota.com/pics/documents/NoPEUunifiedreport.pdf.

-

Nonsynthetic Aquatic Plant Extracts: Aquatic plant extracts can be derived naturally or nonsynthetically. For example, kelp can be dehydrated after harvest by *sun-drying,* and then ground up into a meal product, and can then be sprinkled directly on the soil, or diluted with water and either sprayed on plant foliage as foliar spray, or poured directly into the ground as a drench. Nonsynthetic products may also be produced using mechanical disruption, or freezing, pulverization, and clarification of the thawed slurry. The relative *alkali-extracted* versus *non-alkali-extracted* products has been consistently demonstrated, perhaps partly as a result of a lack of understanding of the mechanism by which aquatic plant extracts exert any purported beneficial effect (Henry, 2005).[76]

-

Inorganic, Synthetic Fertilizers: Inorganic or synthetic fertilizers (mineral fertilizers) include: *Chilean sodium nitrate* ($NaNO^3$), a salt; *Chilean saltpeter* (Potassium nitrate – KNO^3), a white solid; ordinary *saltpeter,* a white solid; mined *rock phosphate;* and *limestone,* calcium source. An example of inorganic or synthetic fertilisers, is *Urea,* (carbamide, resin, isourea, carbonyl diamide, or carbonyldiamide ($NH^2)^2CO$). Urea is actually a waste product found and extracted from *urine* from humans and *uric acid* from birds and saurcain reptiles; mammals excrete *urea;* tadpoles excrete *ammonia;* and urea is used as a *nitrogen-release* fertilizer. Synthetic fertilizers are the most used fertilizers because they are inexpensive, easy to use, and easily absorbed by plants. Most crops grown in industrialised countries receive *more* nitrogen[77] than they can use, which affects the global nitrogen cycle. The *overuse* of synthetic *nitrate* fertilizers causes environmental problems like *air* and *water* pollution, *health* problems, such as *respiratory* ailments, *heart disease,* and *cancer.* Nitrogen fertilizers also allow nitrates to *accumulate* in vegetables, causing health risks to consumers; nitrates in *drinking water* have been associated with an increased risk of *bladder cancer.* When excessive nitrates leach into the water, it greatly affects our ecosystem: a large percentage of our rivers and lakes are polluted because of agricultural runoff; high concentrations of plant nutrients causes large algal blooms to grow, which consume a lot of oxygen and block the sun light, causing other organisms to die; regions of water with low concentrations of oxygen are called *"hydpoxia regions",* often referred to as "dead zones", as for example, the Gulf of Mexico, Mississippi River, and the Black Sea. Another problem with synthetic or inorganic fertilizers, is that, they only feed the plants, not the soil: this causes a gradual decrease in organic matter in the soil; once the organic matter is used up, the soil will become compacted, and a lack of organic matter also leads to a reduction in soil organisms – making it more susceptible to insect or disease infestations; many commercial fertilizers also contain *toxic metals,* including *arsenic, mercury, lead, dioxin, chromium* and *cadmium.*

Eco-pollution by fertilizers is complemented by the use of pesticides and herbicides to control undesirable plants. Pesticides are chemical compounds or biological agents farmers apply to control weeds and other undesirable crops and plants, insects and fungi. A pesticide may be a chemical substance or biological agent, such as a *virus* or *bacterium,* antimicrobial, disinfectant or a device used against any pest. Pests include insects, plant pathogens (i.e., infectious agent, or germ with longest or most persistent potential for harboring a pathogen), weeds, mollusks, birds, mammals, fish, nematodes or roundworms, and microbes or microorganisms. Food crops are believed to compete with over 30,000 species of weeds, 3,000 species of nematodes, 10,000 species of plant-eating insects, and 20-40% of potential food production is still lost every year to pests, according to CropLife International.[78] *Biopesticides* or biological pesticides are synthetic pesticides, among them, being:[79]

[76]   Eric   Henry,   2005:   *"Report   of   Alkaline   Extraction   of   Aquatic Plants."*http://www.omri.org/AdvisoryCouncil/Aquatic_plant_extracts-2004-02-14.pdf.
[77] Nitrogen is a vital element essential to living ecosystems, and is a primary nutrient for all green plants. But when *excessive* nitrates, combined with or without a combination of toxic metals, leach into the water, it greatly impacts the ecosystem negatively.
[78] CropLife International is a global federation formed to represent the Plant Science Industry, whose membership includes BASF, Bayer CropScience, Dow Agrosciences, DuPont, Monsanto, Sumitomo and Syngenta. The source of CropLife International work includes: *chemical crop protection* (pesticides); *agricultural biotechnology* (GMO); and sustainable agriculture. Contacts:

Alternatives to the use of pesticides to avoid the negative impacts of pesticide to ecosystems and health, include:

- *Methods of Cultivation:* The use of *Polyculture* or the growing of multiple types of plantings; Crop rotation or the planting of crops in areas where the pests that damage them do not live; Timing planting according to when pests will be least problematic; The use of trap crops that attract pests away from the real crop; Spraying crops with hot water at the cost that's about the same as pesticide spraying.

- *Release of Other Organisms:* The release of other organisms, such as natural predators that fight the pests; Biological pesticides based on entomological fungi (i.e., fungi that can act as a parasite of insects) and kill or seriously disable the pests, bacteria and viruses or cause disease in the pest species.

- *Interferring Insects' Reproduction:* Interferring with insect pests' reproduction can be accomplished by sterilizing the males of the target species, and releasing the castrated males to mate with females who then cannot produce offspring. Pest insect males, such as the screw worm fly, medfly, tsetse fly, gypsy moth – have been succesfully castrated and released to mate with unsuspecting females.

- *Halfing Pesticide Use:* Halfing pesticide use, such as has been done in Sweden and Indonesia – does not lead to reduced crop yield, but to pesticides' negative impact on the environment.

- *Non-Fertilizer Initiatives:*[80] The Fertilizer Industry has undertaken some non-fertilizer initiatives aimed at addressing the Climate Change issue, thus: Several fertilizer companies have branched out into related products that can help reduce greenhouse gas (GHG) emissions, such as: *Turning Nitrogen Losses* into additional feed grass: when applied to urine patches in pastures, a compound developed by scientists at the Linclon University in New Zealand helps to prevent both the formation of GHGs and nutrient leaching. Keeping nutrients in the soil improves pasture growth and provides an inexpensive form of additional feed. The Ravensdown Fertilizer Cooperative Limited, New Zealand, has commercialised this method, which is hoped to dramatically reduce New Zealand's GHG emissions, while reducing nitrate run-off into watercourses by between 30-60%, and hold potential for other temperate zones; *Neutralising Nitrogen Oxide Emissions from Vehicles:* IFA member company Yara International ASA, Europe's largest producer of nitrogen oxides (Nox), is reducing nitrogen oxide agents based on urea and ammonia in vehicle catalysts. These products convert Nox into inert nitrogen and water vapor. Yara has also played a key role in creating awareness in the EU and amongst other shareholders, of the availability of Nox emission reduction systems for vehicles. Nitrogen oxide is generally associated with local air pollution and is not considered a GHG, but nitrogen compounds can cycle easily from one reactive form to another. Therefore, reducing Nox emissions helps indirectly to reduce the overall amounts of nitrogen oxides (N^2O) in the atmosphere.

CropLife International, aisbl, Avenue Louise 326, Box 35, B-1050 Brussels, Belgium; c/o CropLife America Offices, 1156th Street NW, Suite 400 Washington, D.C. 20005, U.S.; www.croplife.org.

[79] Types of Pesticides: These include: Algicides or Algaecides for the control of algae; Avicides for the control of birds; Bactericides for the control of bacteria; Fungicides for the control of fungi and oomycetes (a water weed); Herbicides for the control of weeds; Insecticides for the control of insects: Ovicides to kill eggs, Larvicides to kill larvae, and Adulticides to kill insects; Milticides and Acaricides for the control of mites; Molluscides for the control of slugs and snails; Nematicides for the control of namatodes; Rodenticides for the control of rodents; Virucides for the control of viruses, such as H5N1 bird flu; Biological pesticides or Biopesticides or synthetic pesticides.

[80] International Fertilizer Industry Association (IFA) Key sources: *"Global Estimates of gaceous emissions of NH^3, NO and N^2O from agricultural land, voluntary initiatives versus regulation."* FAO Rome, Italy/IFA, Paris, 2001; "Fertilizer Best Mangaement Practices: general Principles…, IFA/, 2007, Paris, France; "Sustainable Management of the Nitrogen Cycle in Agriculture and Mitigation of Reactive Nitrogen Side Effects." 1st edition, IFA, Paris, France; G. Kongshaug, 1998 : "Energy Consumption and Greenhouse Gas Emissions in Fertilizer Production." IFA Technical Conference Marrakesh, Morocco, September/October 1998; A.R. Mosier, et.al.: "Agriculture and the environment." Island Press, Washington, D.C. U.S.; 28, rue Marbeuf, 75008 Paris, France. http://www.fertilizer.org.

Fertilizer Industry Improved Efficiency: The fertilizer industry, very awake and aware of the vicious negatives caused by their products on the environment, ecosystems, health and nutrition, is gradually improving its efficiency, by reducing energy use and related emissions. Recent ammonia factories use some 30% less energy per tonne of nitrogen than those factories designed around the 1970s, thus reducing GHGs by at least 20%. In 1998, the global fertilizer industry's GHGs were calculated to be 287 million tonnes CO^2 equivalent [(CO^2-eq) per year (134 million tonnes CO^2-eq, as flue gas from energy production, 74 million tonnes CO^2-eq., as nitrous oxide from nitric acid production, and 75 million tonnes as pure CO^2).]

Increased Ammonia Production But Less Emissions: Post-1998 ammonia production has increased by about 16%, but emissions have not increased in a linear fashion because: many inefficient production sites have been closed; Older facilities have been revamped, increasing energy efficiency by some 15% at a cost of between $8-20 million per site; management innovations have improved the efficiency of many facilities; since nitric acid production is often a component of integrated fertilizer manufacturing plants, new technologies are under development to control nitrogen oxides (N^2O) emissions from this process, while a number of factories have benefitted from carbon-trading mechanisms in order to abate the N^2O emissions, and others are expected to follow.

Herbicides are pesticides that control superfluous plants, used to clear waste ground, industrial sites railways and railway embarkments. Herbicides are non-selectiove (i.e., they kill all plant material which they come in contact with); they are applied in total vegetation control (TVC) programmes for maintenance of highways and railroads, and widely used in agriculture and in landscape turf management. Smaller quatities of herbicides are used in forestry, pasture systems, and management areas set aside as wildlife habitat. Herbicides are classified as follows:

Contact Herbicides: These destroy only the plant tissue in contact with the chemical, and are the fastest acting, less effective on perennial plants which are able to grow from rhizomes, roots or tubers.

Systematic Herbicides: These are translocated through the plant either from foliar application down to the roots, or from soil application up to the leaves, and are capable of controlling perennial plants, slower acting but ultimately more effective than contact herbicides.

Soil-Applied Herbicides: These are soil-applied prior to planting and mechanically incorporated into the soil - the object of incorporation being, to prevent dissipation through photo decomposition and/or volatility.

Pre-Emergent Herbicides: These prevent the germination of seeds by inhibiting a key enzyme applied to the soil before the crop emerges and prevent germination or early growth of weeds.

Post-Emergent Herbicides: Are applied after the crop has emerged.

Mechanism of Action (MOA) *Herbicides:* These herbicides indicate the first enzyme, protein, or biochemical step affected in the plant with *ACCase, ALS, EPSPS Inhibitors,* as follows:[81]

---

[81] Enzyme Growth Inhibition: Three methods are used to inhibit the growth of growth enzymes in plants, be they weeds or in

- *Synthetic Auxim Herbicides:* The synthetic auxin is an artificial plant hormone that mimics the real plant hormone, and its application began the era of organic herbicides. These synthetic auxin herbicides mimic several points of action on the cell membrane, and in effective in control of dicot plants; 2.4-D, is an example of a synthetic auxin herbicide.

- *Photosystem II Inhibitor Herbicides:* These herbicides reduce electron flow from water to $NADPH^{2+}$ at the photochemical step in photosythesis, by binding the *Qb site* on the DI protein, and thus prevent quinone from building to this site, thus enabling electrons to accumulate on chlorophyll molecules. Because of this, oxidation reactions in excess of those normally tolerated by the cell occur, and the plant dies; the triazine herbicides, which include atrazine and urea derivatives (diuron) are photosystem II inhibitors.

- *Organic Herbicides:* These herbicides are classified (and believed) as suitable to be used in farming, are expensive to obtain, generally used along with cultural and mechanical weed control practices, and may not be affordable for commercial production. Also, they are much less effective than synthetic herbicides. Organic herbicides may include *Spices:* now effectively used in patented herbicides; *Vinegar:* used effectively for 5-20% solutions of acetic acid with higher concentrations mostly effective but mainly destroying surface growth, and so spraying to treat regrowth is needed; resistant plants generally succumb when weakened by respraying; *Steam:* Steam, commercially applied as an organic herbicide is considered uneconomic and inadequate; steam kills surface growth but not underground growth and respraying to treat regrowth of perennials is required; *Flame:* is considered a more effective organic herbicide but suffers the same difficulties as steam; *D-limonene* (Citric Acid): Is the active ingredient in Nature's Avenger Organic Herbicide, a natural degreasing agent that strips the waxy skin or cuticle from weeds, causing dehydration and ultimately, death. It's environmentally safe, registered and approved by the U.S., EPA for use in organic production, and listed by the USDA as Organic Materials Review Institute (OMRI).

Anthropogenic or man-made air and environmental pollution caused by transportation and the transportation industry, include the production of the principal greenhouse gases (GHGs) nitrous oxides (N^2O)[82], carbon dioxide (CO^2), particulates (PM) or fine dust or aerosols. The mobbying effect of these GHGs in the stratosphere is the source of many hazardous diseases and infections, besides their weather-dysfuntioning bad attitudes and manners. Agriculture and

genetically modified organisms (GMOs): Mechanism of Action (MOA) method: *ACCase Inhibitors*: Are compounds that kill grasses. Acetyl Coenzyme A carboxylace (ACCase) is part of the first step of lopid synthesis. ACCase Inhibitors affect cell membrane production in the meristems; a meristem is a tissue in all plants consisting of undifferentiated cells found in zones of the plant where growth takes place in the plant; the ACCases of grasses are sensitive to these herbicides, whereas the ACCases of *dicot* plants are not; dicot plants are a group of flowering plants whose seed typically has two embryonic leaves or cotyledons; *ALS Inhibitors*: The synthesis of the branched-chain amino acids (valine, leucine, isoleucine) begins with acetolactate synthase (ALS) or acetohydroxyacid synthase (AHAS) enzyme, and the herbicide slowly starves affected parts of these amino acids which eventually leads to the inhibition of DNA synthesis; both grasses and dicot plants are affected; the ALS Inhibitor family includes: sulfonylureas (Sus), imidazolimones (IMIs), triazolopyrimideines (TBs), pyrimidinyl oxybenzoates (POBs), and Sulfonylamino carbonyl triazolimones (SCT). ALS is a biological pathway that exsts only in plants and not in animals, and so the safest herbicide among others; *EPSPS Inhibitors:* Enolpyruvylshikimete 3-phosphate synthase (EPSPS) is used in the synthesis of the amino acids tryptophau, phenylalamine, trysosine, which affect grasses and dicot plants alike; Glyphosate (Roundup) is a systematic EPSPS inhibitor which is inactivated by soil contact.

[82] Nitrous Oxide or 'Laughing Gas': This is a colorless non-inflammable greenhouse gas (GHG) used in surgery and dentistry because of its anesthetic, analgesic and euphoric effects when inhaled, and used in motor racing as an oxidizer to increase the power output of engines. Nitrous oxide is emitted by bacteria in soils and oceans, and agricultural use of nitrogen fertilizers is the main source of anthropogenic nitrous oxides; animal waste handling can stimulate naturally-occurring bacteria to produce more nitrous oxide, and livestock such as cows, chickens and pigs produce about 65% of human-related nitrous oxide. Nitrous oxide reacts with ozone in the stratosphere to regulate stratosphere ozone to pose as one of the major GHG; Particulates (PM) or fine particles are fine particles of solid or liquid suspended in a gas, where an aerosol contains both gas and particles together. Some particulates occur naturally in/from volcanoes, dust storms, forest and grassland fires, living vegetation and sea spray; burning fossil fuels in vehicles, power plants and various industrial processes generate large amounts of aerosols which account for 10% of the total aerosols in Earth's atmosphere. Increased levels of fine particles in the air are linked to health hazards, such heart disease, altered lung function, and lung cancer. Natural and human-made aerosols affect climate functioning by changing the way electromagnetic radiations (EM), which carry energy and momentum transmitted through the atmosphere; carbon dioxide is the single largest GHGs contributor to global warming, traffic congestion, and automobile-oriented urban sprawl, which consumes natural habitats and agricultural lands.

the agriculture-food and transport industries together pollute water, air, soil and land, and are vital chains in the oil-and-food crisis.

Environmental and health impacts caused by agriculture, pesticides and herbicides include:

- *Target Overreach:* Over 98% of sprayed insecticides and 95% of herbicides reach a destination other than their intended target species, including non-target species, air, water, bottom sediments, and food. Pesticide drift occurs when some pesticides are persistent organic pollutants (POPs) and contribute to soil contamination.

- *Loss of Biodiversity:* Agricultural practices in association with fertilizers, pesticides and herbicides lead to loss of biodiversity, surplus nitrogen and phosphorous in rivers and lakes; convert ecosystems of all types into arable land, consolidates diverse biomass into a few species, and expose these to erosion, deforestation, and depletion of minerals in the soil; particulate matter (PM), including ammonia and ammonium off-gassing from animal waste, air pollution from farm equipment and transportation powered by fossil fuels, carbon dioxide release from agricultural lime, industrial output (fertilizers) and output (food, fuel, fibre) - contribute to air and water pollution.

- *Health Dangers:* Pesticides can present dangers to consumers, bystanders or workers during manufacture, transport or during and after use. The American Medical Association (AMA) recommends limiting exposure to pesticides and using safer alternatives.[83] Pesticide exposure can cause both short-term (acute) and long-term (chronic) health problems for animals and humans. Studies have shown strong associations between chemical pesticides and health problems, including *fertility* problems, *birth defects, brain tumors, breast, brain and prostrate cancers, childhood leukemia,* and *non-Hodgkin's lymphoma.* Children and young animals are especially prone to the toxic effects of pesticides: their bodies, brains, nervous, immune and reproductive systems are developing and their detoxication systems are either immature or not yet functional. The World Health Organisation (WHO), estimates that 200,000 people are killed worldwide each year as a direct result of pesticide poisoning,[84] and that 3 million people are poisoned each year, a large number of these being children; a study in England and Wales showed that 50% pesticide poisoning involved children under the age of 10.

- *Fertilizer Risk:* High solubilities of chemical fertilizers exacerbate their tendency to degrade ecosystems, particularly, through *eutrophication.*[85]

- *Pesticide Residues In Food:* When pesticides are sprayed over crops, residues can remain on the plants after the plants are harvested. These residues can cause acute long-term toxic effects in humans and animals. Fresh fruits and vegetables which are *consistently* either in highest or lowest *pesticide residues,* according to Vegan Peace "Farm Santuary",[86] are *highest*

---

[83] American Medical Association (AMA): *"Pesticide-Related Illnesses Associated with the Use of a Plant Growth Regulator – Italy, 2001."* JAMA, 2001, 286 pp; *"Surveillance for Acute Pesticide-Related Illnesses During Medfly Eradication Programme – Florida, 1998."* JAMA, 1999, 282 pp.; AMA, 515 N. State Street, Chicago, IL 60610, U.S.; http://webapps.ama.assn.org/.

[84] Pesticide Poisoning: Occurs when chemicals intended to control a pest affect non-target organisms, such as wildlife, humans, bees; pesticide poisoning through pesticide misuse, which is: when use of a pesticide violates laws regulating use of pesticides that engers humans or the environment, such as *inconsistent labelling, selling* or *using* an unregistered pesticide or one whose registration has been *revoked,* the sale or use of an *adulterated* or *misbranded* pesticide, or *alter* or *remove* pesticide labels, or sell *restricted* pesticides to an *uncertified* applicator, or *fail* to keep sales and use records of restricted pesticides.

[85] Eutrophication: This is a process which results in an increase in chemical nutrients, typically, nitrogen or phosphorus – that induces *excessive* plant growth and decay, *lack* of oxygen and *severe* reductions in water quality, fish and other animal populations; in effect, eutrophication is a *nutrient pollution,* which occurs through release of sewage effluent and run-off from lawn fertilizers into natural waters, or where nutrients accumulate systems on an ephemeral basis.

[86] Vegan Peace "Farm Sanctuary": Run by Wanda Embar, began in 1990 in wisconsin, U.S., and strives to peacefully share our Earth, end cruelty to animals, promote compassionate living through rescue, education and advocacy;

in pesticide residues: apples, bell peppers, celery, cherries, grapes (imported), nectarines, peaches, pears, potatoes, red raspberries, spinach, strawberries; *lowest in pesticide residues:* asparagus, avocadoes, bananas, broccoli, cauliflower, corn (sweet), kiwi, mangoes, onions, papaya, pineapples, pease (sweet).

- *Agribusiness[87] and Factoring Farming:* Factoring farming is a practice of raising farm animals in confinement at high stock density, where a farm operates as a factory, to produce the highest output at the *lowest* cost by relying on economies of scale, modern machinery, biotechnology[88] and global trade. *Confinement at high density* requires *antibiotics* and *pesticides* to mitigate the spread of disease and *pestilence* exacerbated by these *crowded living conditions.* Large-scale industrialised, vertically controlled farming is synonymous with corporate farming; it is a style of management that employs factoring farming. *Corporate farming* or factoring farming, is used by megacorporations involved in food production on a very large scale, that consists of the farm itself and the entire chain of agriculture-related business, including *seed supply, agrochemicals, food processing, machinery, storage, transport, distribution, marketing, advertising,* and *retail sales.* Agribusinesses or corporate farmers or megacorporations, influence education, research and public policy through their educational funding and government efforts. *Contract Farming* is a form of vertical integration where the farmer is contractually-bound to supply a given quantity and quality of product to a processing or marketing enterprise – also an integral part of factoring farming system.

- *Results of Factoring Farming On Health, Disease, Pollution, Ethics, Biodiversity Destruction:* Factoring Farming contributes greatly to food safety and low cost for consumers, standardisation and efficiency for the producers, and economically for the country. Factoring Farming also contributes the following:

- *Factoring Farming Pollution:* Large quantities and concentrations of waste are produced: lakes, rivers and groundwater are at risk when animal waste is improperly recycled, and pollutant gases are also emitted. Concentration of animals can produce unaccepted levels of smells as opposed to the tolerable odors of the countryside. In less intense conditions, natural processes can break down potential pollutants. Large farms can maintain and operate sophisticated systems to control waste products, but smaller farms are unable to maintain the same standards of pollution control. Farms can efficiently manage wastes by consolidating waste products. The production of animal products uses large resources such as fossil fuels, water and land.

- *Factoring Farming Spreads Diseases:* The use of intensive farming makes it more likely to evolve harmful diseases, as many communicable diseases spread rapidly under such conditions. Animals raised on *antibiotics* may develop antibiotic resistant strains of *pathogenic* bacteria or *superfungi.* Use of animal vaccines can create new *viruses* that kill people and cause *flu pandemic* threats. *H1N1* and *H5N1* are examples where this might have already occurred.

- *Factoring Farming Destroys Biodiversity:* Factoring Farming uses monoculture adaptation of breeds in both arable and animal farming which results in uniform product design for high yields at the risk of increased susceptibility to disease. The loss of locally adapted breeds

---

http://www.veganpeace.com/about/about.htm.

[87] Wikipedia, the free encyclopedia: "Agribusiness: Refers to various businesses involved in food production, including farming, seed supply, agrichemicals, farm machinery, wholesale and distribution, processing, marketing and retail sales. Within the agricultural industry, agribusiness is widely used simply as a convenient portmanteau of agriculture and business, referring to the range of activities and disiplines encompassed by modern food production. There are academic degrees in and departments of agribusiness, in agribusiness trade associations, agribusiness publications, worldwide.

[88] CropLife International (CI): *"What is agbiotechnology?"* CI is a federation of network of the Plant Science Industry, describes *Biotechnology* as: *Plant* Biotechnology, or *Agricultural Biotechnology,* or *Genetically Modified Organisms (GMOs).* CI explains: "Biotech crops continue to increase their share of global agriculture – over 90 million hectares were grown in 2005, with the rate of uptake being highest in developing countries... Since the first commercial biotech crops were grown in 1996, plant biotechnology has proved to be the most rapidly adopted new technology by farmers ever [?] All indications are that this growth will continue, particularly in developing countries [?] "

reduces the resilience of the agricultural system. The loss of the gene pool of domesticated animals limits the ability to adapt to future problems.

- *Factoring Farming Practices Cruelty To Animals:* Crowding, drugging, and performing surgery on animals pracised on some farms, like *debeaking* of young chicks, violates animal treatment welfare legislation. Herding pigs in barren environments leads to physical problems, such as *oseoporosis* and joint pains, boredom and frustration, evidenced by repetitive or self-destructive actions known as stereotypes.

- *Viciousness of Intensive Farming:* Intensive farming leads to a vicious cycle of exhaustion of soil fertility and decline of agricultural yields, and about 40% of the world's agricultural land is seriously degraded. The pursuance of a *Sane Farm Policy,*[89] is one of five items Michael Grunwald has recommended the new Barack Obama Administration to do on Day One: "[Present] U.S., agriculture policy is a jumble, but the basic goal is simple: *redistribute money to big commodity farmers.* The median farmer's net worth is five times the median American's, and the *top one-tenth* of farmers get *three-fourths* of the commodities. It's a *welfare programme for megafarms* that use the most fuel, water and pesticides; emit the most greenhouse gases; grow the most fattening crops; hire the most illegals; and depopulate rural America. *Antiobesity, antipoverty, free-trade, balanced budget,* and *environmental activists* have clamored for reform, but nobody works farm policy harder than the farm lobby and farm-state politicians – including Obama – have protected the status quo. Still, Obama's agri-pandering didn't win him the Farm Bureau endorsement, even though McCain opposed farm giveaways. And Obama has suggested that he's open to more sensible policies that would promote *less-intensive* agriculture. How about *repealing* the $307 billion farm bill and slashing subsidies – especially the *for-no-apparent-reason 'direct payments'* we send to commodity farmers in good times and bad. Farm lobbyists will squeal, but *60% of U.S., farms receive no subsidies.* Instead, Obama can increase conservation subsidies for farmers who adopt *green practices.* He should also *repeal* the *counter-productive mandates* that will require the production of 36 billion gal. (136 billion L) of biofuels by 2022. Biofuels like corn ethanol sound great, and Obama supports them, but they accelerate global warming, because shifting production from food to fuel leads to massive emissions from deforestation, when farmers expand to grow more food. The biofuel boom is also *jacking up* the price of grain, which is increasing food prices and triggering food riots in countries like Yemen, Haiti and Pakistan. The farm lobby and its water carriers in Congress are *long-overdue* for a smackdown. But sensible farm policies could still include goodies for farmers. For example, Obama could ditch the *preposterous* ban on subsidised farmers' growing healthy fruits and vegetables. He should expand purchase for the successful school-lunch programme while shifting the menus away from *fattening crud.* And he can expand markets for farmers and other American exporters by *ending* the *humiliatingly futile Cuban embargo,* which has been forcing the Castros out of power for 46 years now.

Factoring farming, rampant use of fertilizers and herbicides; soil, air, land and water pollutions; roaring inflation; and the elusive oil-and-food surges have stayed their courses, even as investors have taken cover in gold investments. These environmental and economic factors have heavily weighed down on the prices of other commodities and services globalwide. As the archangels of economies, any dislocation in the chain-link on energy, agriculture and transportation impacts adversely on the whole global economy. The slightest rises in energy, transportation, and agricultural production are passed on to consumers' pockets. This feeds the viral of inflation. The raging inflation monster breeds rabid bankcruptcies around the globe, with auto- and airline sectors recording punishing losses as well.

1.4.1 Rising Inflation, Bankruptcies, Recessions

---

[89] Michael Grunwald: *"Obama's Agenda: Get America Back On Track: First Things First."* TIME's Commemorative Issue: President-Elect Barack Obama; 17 November 2008; TIME/CNN; http://www.time.com/.

THE OIL-and-food-crisis, bankrolled by spiked energy, food and commodities prices, sparked a spate of economic bankcruptcies in the banking and transportation branches in economies throughout the world. It began with the inflationary plague torched off in the United States of America, the world's economic and military Superpower. The U.S., is burdened with two wars in Iraq[90] and Afghanistan,[91] the subprime lending caper, deregulation of financial and banking institutions exemplified by the Wall Street[92] banking metropolis' practices, a weak dollar,[93] and the dollar index[94] downward trend. The result from all this has been a multitude of financial and fiscal scandals, hyper-unemployment,[95] inflation-, and bankcruptcy-rates, hyper-house foreclosures and

---

[90] U.S. Invasion of Iraq, 2003: The U.S., President George W. Bush Republican Administration unilaterally ordered the invasion of Iraq on 20 March 2003 presumably as a reaction against the al-Qaeda Islamist terrorist suicide air attacks in New York (now known as "Ground Zero") on 11 September 2001. The war has cost the U.S., up to $10 trillion, $10 bn daily, and 4200 dead Americans. Al-Qaeda or al-Qa'ida used four hijacked airline jets and intentionally crashed the airliners into the Twin Towers of the World Trade Center in New York City; a third airliner crashed into the Pentagon in Virginia, and the fourth plane crashed into a field near Shanksville in rural Sommerset County, Pennsylvania; 2,993 people including 19 hijackers, and another 24 missing, perished. According to Burt Hall's *"Bush Mishandled Threat Leading up to 9/11 Disaster.* For months, Pres Bush and key members of his national security team *ignored* top expert advice and urgent warnings of an imminent attack, including the use of hijacked aircraft as weapons. The *9/11 Commission* omitted this information from its Report and did not make an overall assessment of the U.S., preparedness, as required by law. Disclosure of the urgent warnings probably would have cost Bush a second term and avoided the follow-on Katrina disaster. Missing from the 9/11 Commission Report was: *Evidence that incriminated the Pres and his national security team and showed a reckless disregard for expert advice and specific warnings of the upcoming attacks; An assessment of the U.S., preparedness for the attacks, as required by law; No attempt was made to prevent the attackes or to make them more difficult to carry out. Offensive and defensive measures to thwart the attacks were readily available to the Pres, but he did not choose to use them.* To this day, we do not know why our nation was left unprepared. Maybe the reader can figure it out." (Members of the Bush national security team and 9/11 Commission have declined comment on these comments.) Burt Hall: *"State of the Nation."* Daily Kos, 11 September 2009.

[91] Afghanistan War: The U.S., Military under UN authorisation, on 7 October 2001, lead the "Operation Enduring Freedom" invasion of the Taliban (Teleban) –led government in Kabul, Afghanistan, in retaliation against the 11 September 2001 suicide terror attacks on New York by al-Qaeda. Al-Qaeda and Taliban – both fundamentalist Sunni Islamic Terror Organisations – operated from Afghanistan before their ouster and from both Afghanistan and Pakistan after the Taliban ouster from Afghanistan. Al-Qaeda, al-Qaeda-linked organisations and al-Qaeda sympathetic individuals undergo training in one of al-Qaeda's training camps in Afganistan or in Sudan, before launching their "holy wars" or jihad against "infidels" or unbelievers around the world, with the objectives: *"to end foreign influence in Muslim countries, and creation of a new Islamic Caliphate."* The Taleban or Taliban religious movement is predominantly Pashtun tribespeople sprawled across the Afgha-Pakistan lengthy borders. The Taleban originated from Pakistan's more than 118 Koran Schools or Madrasas, invaded and overthrew the Afghan government, and were in turn ransacked by the coalition forces (NATO/US) led by the U.S.

[92] Wall Street: The Wall Street in New York's (U.S) lower Manhattan, between Broadway and South Street, runs through the centre of the Financial District where many of New York's major financial institutions, including the New York Stock Exchange (NYSE), and the American Stock Exchange (AMEX), NASDAQ, NYMEX and NYBOT are located. Wall Street or Big Business Interests is contrasted against Main Street or Small Business or Middle Class Interests. The NYSE is the largest in the world by $25 trillion dollar volume (as of 31 December 2006) and 2,764 listed securities; it lists the 4th biggest with 3,200 companies behind the Bombay Stock Exchange, London Stock Exchange, and NASDAQ in terms of company listings. NYSE is operatd by NYSE Euronext, a company formed by NYSE's merger with fully electronic Euronext located at Wall Street No. 11. Other **Financial Districts** in the U.S., and around the world include: Tokyo's Marunouchi Group (Japan); Singapore's Shenton Way at Raffles Place (Singapore); Chicago's Chicago Loop U.S.); San Francisco's Financial District (U.S.); Boston's Financial District, Boston (Massachusetts, U.S.); Hong Kong's Central (China); Shanghai's Lujiazui, Pudong (China); Paris's La Defense (France); Frankfurt's Bankenviertel (Germany); Toronto's Bay Street (Canada); Seoul's Teheranno (South Korea); Athen's Sofokleous Street (Greece); Bombay's Dalal Street (India); Dhaka's Motijheel (Bangladesh); Sydney's Martin Place (Australia); Karachi's Ibrahim Ismail Chundrigar Road (Pakistan); Sao paulo's Paulista Avenue (Brazil); Mexico City's Paseo de la Reforma (Mexico); Zurich's Paradeplatz (Switzerland).

[93] Weak Dollar: In 1971, the U.S., dollar (USD) has been the *rescue currency* for all the nation-states which signed the Bretton Woods system Agreements in 1945. In other words, the USD *replaced gold* as the exchange medium of all currencies. Gold was the *exchange medium* under the Bretton Woods System until 1971. The USD is the *basis* used by all world currencies to *exchange* their currencies to dollar worth in international business transactions and currency reserves. The power, health and prowess for any world currency depends on the power, health and prowess of the U.S., dollar.

[94] Dollar Index (USDX): Since the U.S., dollar replaced gold as rescue currency for all other currencies in 1971, the *exchange rate convertibility* uses the dollar index to relate all world currencies to the U.S., dollar. It is an index or measure of the value of the USD *relative* to all foreign currencies; the index was started in 1973, soon after dismantling of the Bretton Woods System that did this. The index started with the U.S., dollar being 100,000, and has traded since as high as the mid-60s and the low-70s; from *October 2008,* the USDX was trading in the mid-80s, and on 6 March 2008, it touched 72.89, its *lowest since* its inception in 1973, and continued its downward trend, and reached *70.698* on 16 March 2008. [USDX is updated daily seven days a week, listed on ICE Futures Exchange or New York Board of Trade (NYBOT).

[95] Associated Press (AP) Writer Jeanne Aversa on U.S. Unemployment:*"Jobless Ranks Hit 10 million, 25-year High."* TIME/CNN, 07 November 2008. Unemployment rate climed from 6.1% to 6.5% and may end to up to 8% or 8.5% by the end of 2009, before tailing off. About 10.1 million people were unemployed in October 2008, the most since the fall of 1983; more people will have jobs now, since the population has grown, but it's still a staggering jobless figure; employers are slashing jobs

hyper-greenhouse gas emissions, which phenomena have infected whole world economies. High, inflated commodity prices topped by oil, energy and food, fathered a spiral of inflationary atmosphere. The very unpopular[96] Bush Administration zigzagged from one crisis to the next to propose and implement damage-control measures against subprime lending and commercial banks, and three automakers' losses and collapses: *Bear Stern, Merril Lynch, Fannie Mae, Freddy Mac, American Insurance Group* (AIG), *American Mutual Bank, Lehman Brothers, General Motors* (GM), *Ford Motors,* and *Chrysler LLC.* The U.S., Senate and Congress passed a $700 billion[97] rescue fund package for Wall Street and Main Street. World economies and the currencies[98] issued by each country's

every month so far in 2008 and some 1.2 million positions have disappeared, over half during August-October.

[96] Mark Halperin: CNN/Opinion Research Corp. Data: *"More from the November 6-9 CNN Poll."* Relatively easy transition from Bush to Obama: 57% total, 63% whites, 48% blacks; Relatively difficult transition from Bush to Obasma: 39% total, 34% whites, 475 blacks; Approval of Bush: 24% approve, 76% disapprove; Approval of Bush by race: 27% whites, 5% blacks; Disapproval of Bush by race: 72% whites; 94% blacks. [CNN/Opinion Research Corp. Data, The Page – Politics up to the Minute, Tuesday, 11 November 2008.]

[97] Emergency Economic Stabilisation Act 2008: *"Bailout of the U.S., Financial System"* Authorises the U.S. Secretary of the Treasury to spend up to $700 billion to purchase distressed assets, especially mortgage-backed securities from the nation's banks, stabilise the economy, improve liquidity, repay tax-payers, end house foreclosures. The upcoming Obama Administration and Congress has promised to eanact a new stabilisation act to shore up mddle class businesses to create new jobs. This may include rescue pachkages for the suffering auto industry. Governments across the globe have put similar measures in place.

[98] World Governments' Currencies: *USD/Cents, Other Dollars :* United States (USD); Anguilla (East Caribbean Dollar – XCD); Antigua/Barbuda (East Caribbean Dollar – XCD); British Indian Territory (USD); British Virgin Islands (USD; Brunei (Brunei Dollar – BND; Singapore Dollar – SGD); Australian (Australian Dollar – AUD); Bahamas (Bahamian Dollar – BSD); Barbados (Barbadian Dollar – BBD); Belize (Belize Dollar – BZD); Bemuda (Bermudian Dollar – BMD); Canada (Canadian Dollar – CAD); Cayman Islands (Cayman Islands Dollar – KYD); Cook Islands (New Zealand Dollar – NZD; Cook Islands Dollar); Cocos/Keeling Islands Dollar; Dominica (East Caribbean Dollar – XCD); East Timor (USD); Ecuador (USD); El Salvador (USD); Fiji (Fijian Dollar – FJD); Grenada (East Caribbean Dollar – XCD); Guyana (Guyanese Dollar – GYD); Hong Kong (Hong Kong Dollar – HKD/Ho); Jamaican (Jamaican Dollar –JMD); Kiribati (Australian Dollar – AUD; Kiribati Dollar); Liberia (Liberian Dollar – LRD); Marshall Islands (USD); Micronesia (Micronesian Dollar), (USD); Montserrat (East Caribbean Dollar – XCD); Namibia (Namibian Dollar – NAD; South African Rand); Nauru (Australian Dollar – AUD; Nauruan Dollar); New Zealand (New Zealand Dollar – NZD); Niue (New Zealand Dollar – NZD; Niuean Dollar); Northern Mariana Islands (Northern Mariana Islands Dollar; USD); Palau (Palauan Dollar; USD); Panama (USD; Panamanian Balboa – PAB/Centesimo); Pitcairn Islands (New Zealand Dollar – NZD); Saint Kitts/Nevis (East Caribbean Dollar – XCD); Saint Lucia (East Caribbean Dollar – XCD); Saint Vincent/Grenadines (East Caribbean Dollar – XCD); Singapore Dollar – SGD; Brunei Dollar – BND/Sen); Solo,on Islands (Solomon Islands Dollar – SBD); Suriname Dollar – SRD); Taiwan (New Taiwanese Dollar – TWD); Trinidad/Tobago (Trinidad/Tobago Dollar – TTD); Turks/Caicos Islands (USD); Tuvalu (Australian Dollars – AUD; Tuvaluan Dollars); Zimbabwe (New Zimbabwean Dollar – ZWD; Old Zimbabwean Dollar – ZWD); *EU Euros* (EUR/Cents) : Akrotini/Dhekelia; Andorra; Austria; Belgium; Cyprus; Finland; France; Germany; Greece; Ireland; Italy; Kosovo (Serbian Dinar – RSD/Para); Luxembourg; Malta; Monaco; Montenegro; Netherlands; Portugal; San Marino; Slovakia (since 1 Jan., 2009); Slovenia; Spain; Vatican City; *West /Central African CFA Franc/Centimes :* Benin (XOF); Burkina Faso (XOF); Burundi (Burundian Franc – BIF); Cameroon (XAF); Central African Republic (XAF); Chad (XAF); Comoros (Comoran Franc – KMF); Congo DR (Congolese Franc – CDF); Congo-Braz (XAF); Cote d'Ivoire (XOF); Djibouti (Djiboutian Franc – DJF); Equatorial Guinea (XAF); French Polynesia (XPF); Gabon (XAP); Guinea (Guinean Franc – GNF); Guinea-Bissau (XOF); Mali (XOF); New Caledonia (CFP Franc); Niger (XOF); Rwanda (Rwandan Franc – RWF); Senegal (XOF); Togo (XOF); Wallis/Futuna (CFP Franc – XPF); *British Pound Sterling* (GBP/Penny), *Shillings, Florin :* United Kingdom (GBP); Alderney (GBP; Alderney Pound; Guernsey Pound;) Aruba (Aruban Florin – AWG/Cent); Ascension Island (SaintHelenian Pound – SHP; Egypt (Egyptian Pound – EGP/Piastre); Falkland Islands (Falkland Pound – FKP); Guernsey (GBP); Isle of Man (GBP; Many Pound); Jersey (GBP; Jersey Pound); Kenya (Kenyan Shilling – KES/Cent); Lebanon (Lebanese Pound – LBP/Piastre); Somalia (Somali Shilling – SOS/Cent); Somaliland (Somaliland Shilling/Cent); South georgia/South Sandwich Islands (GBP; South Georgia/Sandwich Islands Pound); Sudan (Sudanese Pound – SDG/Piastre); Syria (Syrian Pound – SYP/Piastre); Tanzania (Tanzanian Shilling – TZS); Uganda (Ugandan Shilling – UGX/Cents); Tristan da Cunha (Saint Helenian Pound – SHP; Tristan da Cunha Pound); *Russian Ruble* (RUB/Kopek): Russia, Transnistria (RUB); Abkhazia (RUB; Georgian Lavr); Belarus (Belarus ruble – BYR/Kapyeyka); Gibralta Pound – GIP); South Ossetia (RUB; Georgian Lari – GEL/Tetri); *Dinar/Centime, Dirham :* Algeria (Algerian Dinar – DZD); Bharain (Bharain Dinar – BHD/Fils); Jordan (Jordanian Dinar – JOD/Piastre); Kuwait (Kuwaiti Dinar – KWD/Fils); Libyan (Libyan Dinar – LYD/Dirham); Macedonia (Macedonian Denar – MKD/Deni); Morocco (Moroccan Dirham – MAD); Serbia (Serbian Dinar – RSD/Para); Tunisia (Tunisian Dinar – TND/Millime); United Arab Emirates (United Emirates Dirham – AED/Fils); *Peso/Centavo :* Argenina (Argentinian Peso – ARS); Chile (Chilean Peso – CLP); Colombia (Colombian Peso – COP); Cuba (Cuban Peso – CUP); Mexico (Mexican Peso – MXXI); Philippines (Philippine Peso – PHP); Uruguay (Uruguan Peso – UYU/Centissimo); Dominican Rep (Dominican Peso – DOP); *Lira/Real/Riyal :* Brazil (Brazilian Real – BRL/Cantavo); Iran (Iranian Rial – IRR/Dinar); Iraq (Iraqi Dinar – IQD/Fils); Saudi Arabia (Saudi Arabian Riyal – SAR/Hallalar); Qatar Riyal – QAR/Dirham); Turkey, Northern Cyprus (Turkish Lira – TRY/New kurus); Western Sahara (Moroccan Dirham – MAD); Yemen (Yemeni Rial – YER/Fils); Oman (Omani Rial – OMR/Baisa); India: *Indian Rupee :* (INR/Paisa); Mauritius: *Mauritian Rupee :* (Mauritian Rupee – MUR/Cent); Nepal: *Nepalese Rupee :* (NPR); Pakistan: *Pakistan Rupee :* (PKR); Seychelles: *Seychellois Rupee :* (SCR/Cent); Sri Lanka: *Sri Lankan Rupee :* (LKR); South Africa: *South African Rand :* (ZAR/Cent); Lesotho (ZAR; Lesotho Loti – LSL/Sente); *Swiss Frank/Rappenu :* Switzerland (Swiss Franc – CHF); Liechtenstein (Swiss Franc – CHF); Afghanistan: *Afghan Afghani :* (AFN/Pul); Albania: *Albanian Lek :* (ALL/Qintar); Angola: *Angolan Kwanza :* (AOA/Centimo); Armenia: *Armenian Dram :* (AMD/Luma); Azerbaijan: *Azerbaijani manat :* (AZN/Qapile; Bangladesh: *Bangladeshi Tala :* (BDT/Paisa); Bhutan: *Bhutanese Ngultrum :* (BTN/Chertrum); Bolivia: *Bolivian Boliviano :* (BOB/Centavo); Bosnia/Hercegovina: *Bosnia/Hercegovina Convertible Mark :* (BAM/Fening); Botswana: *Botswana Pula :* (BWP/Thebe); Bulgaria: *Bulgarian Lev :* (BGN/Stotinko); Cambodia: *Cambodian Riel :* (KHR/Sen); Cape Verde: *Cape Verdan Escudo :* (CVE/Centavo); China: *Chinese Yuan :*

government caught the inflationary flu and reeled. The very basis of capitalism and market economy[99] seemed threatened. They had served their time, and showed signs of exhaustion by exposing their vulnerable underbellies and fallibility propensities. It was time to move beyond *relying absolutely, blindly and unquestioning* on "market solutions"' near divine dictates and fiats. Governments and countries, currencies and economies across the world infected by the Wall Street's *deregulated,* inflationary, subprime mortgage financial epidemic – rushed helter-skelter to propose and implement damage-control measures. Economic recessions[100] and possible depressions mushroomed everywhere in addicted economies. "Capitalism may be the best economic system man has come up with," wrote Nobel Laureate Stiglitz,[101] "but no one ever said it would create stability. Infact, over the past 30 years, market economies have faced more than 100 crises. That is why I and many economists believe that government regulation and oversight are an essential part of a functioning market economy. Without them, there will continue to be frequent severe economic crises in different parts of the world. *The market on its own is not enough. Government must play a role.*" Stiglitz goes on:

- The troubles we now face were caused largely by the combination of *deregulation* and *low interest rates.* After the collapse of the tech bubble, the economy needed stimulus. But the Bush tax cuts didn't provide much stimulus to the economy. This put the burden of keeping the economy going on the Fed, and it responded by flooding the economy with liquidity. Under

---

(CNY/Fen); Costa Rica: *Costa Rican Colon :* (CRC/Centimo); Czech Rep: *Czech Koruna :* (CZK/Haler); Denmark, Faroe Islands (Faroese Krona): *Danish Krone :* (DKK/Ore); Eritrea: *Eritrean Nokfa :* (ERN/Cent); Estonia: *Esonian Kroon :* (EEK/Sent); Ethiopia: *Ethiopian Birr :* (ETB/Santim); Faroe Islands: Gambia: *Gambian Dalasi :* (GMD/Bulut); Georgia: *Georgian Lar :* (GEL/Tetri); Ghana: *Ghanaian Cedi :* (GHS/Pesewa); Guatemala: *Guatemalan Quetzal :* (GTQ/Centavo); Haiti: *Haitian Gourde :* (HTG/Centime); Honduras: *Honduran Lempira :* (HNL/Centavo); Hungary: *Hungarian Forint :* (HUF/Fill); Iceland: *Islandic Krona :* (ISK/Eyir); Indonesia: *Indonesian Rupiah :* (IDR/Sen); Israel: *Israeli New Shequel :* (ILS/Agora); Japan: *Japanese Yen :* (JPY/Sen); North Korea: *North Korean Won :* (KPW/Chön); South Korea: *South Korean Won :* (KRW/Jeon); Kyrgyzstan: *Kyrgyzstani Som :* (KGS/Tyiyn); Laos: *Lao Kip :* (LAK/Alt); Latvia: *Latvian Lats :* (LVL/Santims); Lesotho: *Lesotho Loti :* (LSL/Sente; ZAR); Lithuania: *Lithuanian Litas :* (LTL/Centas); Macau: *Macanese Pataca :* (MOP/Avo); Madagascar: *Malagasy Ariary :* (MGA/Iraimbilanja); Malawi: *Malawian Kwatcha :* (MWK/Tambala); Malaysia: *Malaysian Ringgit :* (MYR/Sen); Maldives: *Maldivian Rufiyaa :* (MVR/Laari); Mauritania: *Mauritanian Ouguuiya :* (MRO/Khoums); Moldova: *Moldovan Leu :* (MDL/Ban); Mozambique: *Mozambican Metical :* (MZN/Centavo); Myanmar: *Myanmar Kyat :* (MMK/Pya); Nagorno-Karabakh: *Armenian Dram :* (Nagorno-Karabakh Dram); Netherlands Antilles: *Netherlands Antillean Gulden :* (ANG/Cent); Nigeria: *Nigerian Naira :* (NGU/Kobo); Norway: *Norwegian Krone :* (NOK/Ore); Papua New Guinea: *Papua New Guinean Kina :* (PGK/Toea); Paraguay: *Paraguan Guarani :* (PYG/Centimo); Peru: *Peruvian Nuevo Sol :* (PEN/Centivo); Poland: *Polish Zloty :* PLN/Grosz); Turkmenistan: *Turkmenistani Manat :* (TMM/Tennesi); Samoa: *Samoan Tala :* (WST/Sene); Sao Tome/Principe: *Sao Tome/Principe Dobra :* (STD/Centimo); Sierra Leone: *Sierra Leonean Leone :* (SLL/Cent); Swaziland: *Swazi Lilangeni :* (SZI /Cent); Sweden: *Swedish Krona :* (SEK/Ore); Tonga: *Tonga Pa'anga :* (TOP/Seniti); Uzbekistan: *Uzbekistani Som :* (UZS/Tiyin); Vanuatu: *Vanuatu Vatu :* (VUV); Venezuela: *Venezuelan Boliver :* (VEF/Centimo); Vietnam: *Vietnamese Dong* : (VND/Hao); Zambia: *Zambian Kwacha :* (ZMK/Ngwee).

[99] Market Economy & Deregulation: Are the souls of capitalism, the economic system in which the private ownership of property is protected by law; the capitalist political system protects the exchange and distribution of capital between openly competing profit-seeking legal or private persons, and investments, distribution, income, production and pricing of goods and services are predominantly determined; anyone can participate in supply and demand, and form contracts with each other in the operation of a market economy. A free price system or free price mechanism is where prices are set by the interchange of supply and demand, the resulting prices are understood as signals communicated between producers and consumers which serve to guide the production and distribution of resources. In an absolutely free-market economy, all capital goods, services and money flow transfers are unregulated by the governments, *except to stop fraud or collusion* that may take place among market participants. In other words, a free-market economy is free of any government controls; *Deregulation* or decreased state intervention *Laissez-faire* or *Liberalism* or *Neoliberalism* or *Economic Liberalism* or *Reaganomics* or *Thatcherism* - is part-and-parcel of the free-market mechanics. [World Capitalism is based on the works of Scottish philospoher/economist Adam Smith (1723-1790), one of the proponents of the free market theorists on h is "Wealth of Nations", an abbreviation of his "Inquiry into the Nature and Causes of the Wealth of Nations.")

[100] Recession: Is a period of reduced or significant reduced or contracted or significant contracted business activity, spread across the economy, usually within a Calendar Year, that eventually could lead to a depression, (i.e., a sustained, longdrawn recesison.). The Gross Domestic Product (GDP) or Gross Domestic Income Income (GDI) which is based on growth, real personal income, employment, and wholesale-retail sales – is the barometer used to calculate or forecasr a recession, depression or other economic activity parameters. A depression or hyper-recession is characterised by abnormal increases in unemployment, restriction of credit, shrinking output, numerous bankruptcies, reduced amounts of trade and commerce, and volatile relative currency fluctuations, mostly devaöluations.

[101] Joseph Stigltz: *"Nobel Laureate: How to Get Out of the Financial Crisis."* TIME in Partnership with CNN, Saturday, 18 October 2008, 3 pp. Stiglitz is Prof at Columbia University, former World Bank chief economist, and former Clinton Administration Council of Economic Advisers chairman.

normal circumstances, it's fine to have money sloshing around in the system, since that helps the economy to grow. But the economy had already overinvested, and so the extra money wasn't put to productive use. Low interest rates and easy access to funds encouraged reckless lending, the infamous interest-only no-down-payment, no-documentation ("liar") subprime mortgaes. It was clear that if the bubble got inflated even a little, many mortgages would end up under water – with the price less than the value of the mortgage. That has happened – 12 million so far, and more every hour. Not only are the poor losing their homes, but they are also losing their life savings.

- "The climate of deregulation that dominated the Bush-Greenspan years helped the spread of a new banking model. At its core was *securitisation:* mortgage brokers originated mortgages that they sold on to others. Borrowers were told not to worry about paying the ever-mounting debt, because house prices would keep rising and they could refinance, taking out some of the capital gains to buy a car or pay a vacation. Of course, this violated the first law of economics – that *there is no such thing as a free lunch.* The assumption that house prices could continue to go up at a rapid pace looked particularly absurd in an economy in which most Americans were seeing their real incomes declining.

- "The mortgage brokers loved these new products because they ensured an endless stream of fees. They maximised their profits by originating as many mortgages as possible, with frequent refinancing. Their allies in investment banking bought them, sliced and diced the risk and then passed them on – or at least as much as they could. Our bankers forgot that their *job was to prudently manage risk allocation capital.* They became *gambling casinos* – gambling with other people's money, knowing that the taxpaper would step in if the losses were too great. They misallocated capital, with massive amounts going into housing that was ultimately unaffordable. Loose money and light regulation were a toxic mixture. *It exploded.*

- "What made America's recklessness truly dangerous is that we exported it. A few months ago, some talked about *decoupling* – that Europe could carry on even as the U.S., suffered a downturn. I always thought that decoupling was a myth, and events have proven that right. Thanks to globalisation, Wall Street was able to sell off its toxic mortgages around the world. It appears that about half the toxic mortgages were exported. Had they not been, the U.S., would be in even worse shape. Moreover, even as our economy went into a slowdown, exports kept the U.S., going. *But the weakness in America weakened the dollar* and made it more difficult for Europc to scll its goods abroad. Weak exports meant a weak economy, and so the U.S., exported our downturn just as earlier we had exported our toxic mortgages.

- "But now the problems are *richocheting back. The bad mortgages are contributing to forcing many European banks into bankcruptcy.* (We exported not only bad loans but also bad lending and regulatory practices; many of Europe's bad loans are to European borrowers.) And as market participants realised that the fire had spread from America to Europe, there was panic. Part of the concern is *psychological.* But part of it is because our financial and economic systems are closely intertwined. Banks all over the world lend and borrow from one another; they buy and sell complicated financial instruments – which is why bad regulatory practices in one country, leading to bad loans, can infect the global system."

America and the world–at-large, is now facing liquidity, solvency and macroeconomic problems. This needs urgent adjustment to return housing prices to equilibrium levels, get rid of the excess debt, and *bottle the ravaging ghost of crude market-rule into the realm of commonsense.*

Stiglitz provides the following *commonsense measures* to castrate brutal capitalism:

- *Recapitalise Banks:* With all the losses, banks have insufficient equity and will find a hard time raising equity under current circumstances. The government needs to provide or inject equity which will also bailout bondholders; in return, the government should have voting shares in the banks it helps. Right now the market is discounting their bonds, implying there is a high probability of default. There need to be a forced conversion of this debt to equity, and if this is done, the amount of government assistance that will be required will be much reduced. Much of this has happened but it took the Bush administration to figure this out, because the government was bound by the *idea of a free-market solution* that it was unable to accept what economists of all stripes were telling Secretary of the Treasury Paulson: that he needed to recapitalise the banks and provide new money to make up for the losses incurred on their bad loans.

- *"Stem the Tide of Foreclosures:* The original Paulson plan was like massive blood transfusion to a patient with severe internal hemorrhaging. We won't save the patient if we don't do something about the foreclosures...We need to help people stay in their homes, by converting the mortgage-interest and property-tax deductions into cashable tax credits; by reforming bankcruptcy laws to allow expedited restructuring, which will bring down the value of the mortgage when the price of the house is below that of the mortgage; and even government lendings taking advantage of the government's lower cost funds and passing the savings on to poor and middle-income homeowners.

- *"Pass a Stimulus Package that Works:* Helping Wall Street and stopping the foreclosures are only part of the solution. The U.S., economy is headed for a serious recession and needs a big stimulus. We need increased unemployment insurance; if states and localities are not helped, they will have to reduce expenditures as their tax revenues plummet, and their reduced spending will lead to a contraction of the economy. But to kickstart the economy, Washington must make investments in the future. Hurricane Katrina and the collapse of the bridge in Minneapolis were grim reminders of how decrepit our infrastructure has become. Investments in infrastructure and technology will stimulate the economy in the short run and enhance growth in the long run.

- *"Restore Confidence through Regulatory Reform:* Underlying the problems are banks' bad decisions and regulatory failures. These must be addressed if confidence in our financial system is to be restored. Corporate governance structures that lead to flawed incentive structures designed to generously ceos should be changed and so should many of the incentive systems themselves. It is not just the level of the corruption; it is also the form – non-transparent stock options that provide incentives for bad accounting to bloat up reported returns.

- *Create an Effective Multilateral Agency:* As the global economy becomes more interconnected, we need better global oversight. It is unimaginable that American's financial market could function effectively if we had to rely on *50 separate state regulators.* But we are trying to do essentially that at the global level. The recent crisis provides an example of the dangers: as some foreign govenments provided blanket guarantees for their deposits, money started to move to what looked like safe havens. Other countries had to respond. A few European governments have been far more thoughtful than the U.S., in figuring out what needs to be done. Even as the crisis turned global, French President Nicolas Sarkozy, in his address to the U.N. last month, called for a world summit[102] to lay the foundation for more state regulation to

---

[102] The White House, Office of the Secretary: G-20 Saturday Statement After World Summit: *"DECLARATION ON FINANCIAL MARKETS AND THE WORLD ECONOMY."* Preamble: **1.** We, the Leaders of the Group of Twenty, held an initial meeting in Washington on 15 November 2008, amid serious challenges to the world economy and financial markets. We are determined to enhance our cooperation and work together to restore global growth and achieve needed reforms in the world's financial systems; **2.** Over the past months our countries have taken urgent and exceptional measures to support the global economy and stabilise financial markets. These efforts must continue. At the same time, we must lay the foundation for reform to help to ensure that a global crisis, such as this one, does not happen again. Our work will be guided by a shared belief that market principles, open trade and investment regimes, and effectively regulated financial markets foster the dynamism, innovation, and entrepreneurship that are essential for economic growth, employment and poverty reduction; Root Causes of the Current Crisis: **3.** During a period of strong global growth, growing capital flows and prolonged stability earlier this decade, market participants sought higher yields without an adequate appreciation of the risks and failed to exercise proper due diligence. At the same time, weak underwriting standards, unsound risk mannagement practices, increasingly complex and opaque financial products, and consequent *excessive* leverage combined to create vulnerabilities in the system. *Policy-makers, regulators* and *supervisors,* in some advanced countries, did not adequately appreciate and address the risks building up in financial markets, keep pace with financial innovation, or take into account the systemic ramifications of domestic regulatory actions; **4.** Major underlying factors to the current situation, among others, inconsistent and insufficient coordinated macroeconomic policies, inadequate structural reforms, which led to unsustainable global macroeconomic outcomes. These developments, together, contributed to the excesses and ultimately resulted in severe market disruption. Actions Taken and To Be Taken: **5.** We have taken strong and significant actions to date to stimulate our economies, provide liquidity, strengthen the capital of financial institutions, protect savings and deposits, address regulatory inefficiencies, unfreeze creadit markets, and are working to ensure that international financial institutions (IFIs) can provide critical support for the global economy; **6.** But more needs to be done to stabilise financial markets and support economic growth. Economic momentum is slowing substantially in major economies and the global outlook has weakened. Many emerging market economies, which helped sustain the world economy this decade, are still experiencing good growth but increasingly are being adversely impacted by the worldwide slowdown; **7.** Against this background of deteriorating economic conditions worldwide, we agreed that a broader policy response is needed, based on closer macroeconomic cooperation, to restore growth, avoid negative spillovers and support emerging market economies and developing countries. As immediate steps to achieve these objectives, as well as address longer-term challenges, will: * Continue our vigorous efforts and take whatever further actions are necessary to stabilise the financial system; * Recognise the importance of monetary policy support, as deemed appropriate to domestic conditions; * Use fiscal measures to stimulate domestic demand to rapid effects, as appropriate, while maintaining a policy framework conducive to fiscal sustainanability; * Help emerging and developing economies gain access to finance in current difficult financial conditions, including liquidity facilities and programme support. We stress the International Monetary Fund (IMF) important role in crisis response, welcome its new short-term liquidity facility, and urge the ongoing review of its instruments and facilities to ensure flexibility; * Encourage the World Bank and other multinational development banks (MDBs) to use their full capacity in support of their development agenda, and we welcome the recent introduction of new facilities by the World Bank in the areas of infrastructure and trade finance; * Ensure that the IMF, World Bank and other MDBs have sufficient resources to continue playing their role in overcoming the crisis; Common Principles for Reform of Financial Markets: **8.** In addition to the actiuons taken above, we will implement reforms that will strengthen financial markets and regulatory regimes so as *avoid* future crises. Regulation is *first* and *foremost* the responsibility of national regulators who constitute the first line of defence against market instability. However, our financial markets are global in scope, therefore, intensified international cooperation among regulators and strengthening of international standards, where necessary, and their consistent implementation is necessary to protect against adverse *cross-border* regional and global developments affecting international financial stability. Regulators must ensure that their actions support *market discipline,* avoid potentially adverse impacts on other countries, including regulatory arbitrage, and support competition, dynamism and innovation in the marketplace. Financial institutions must also bear their responsibility for the turmoil and should do their part to overcome it including *recognising losses, improving disclosure* and strengthen their governance and risk management practices; **9.** We commit to implementing policies consistent with the following *common principles for reform:* * Strengthening Transparency and Accountability: We will strengthen financial markts transparency, including by enhancing required disclosure on complex financial products and ensuring complete and accurate disclosure by firms of their financial conditions. Incentives should be aligned to avoid excessive risk-taking; * Enhancing Regulation: We pledge to strengthen our regulatory regimes, prudential oversight, and risk management,and ensure that all financial markets, products and participants are regulated or subject to oversight, as appropriate to their circumstances. We will exercise strong oversight over credit rating agencies, consistent with the agreed and strengthened international code of conduct. We will also make regulatory regimes more effective over the economic cycle, while ensuring that regulation is efficient, does not stifle innovation, and encourage expanded trade in financial products and services. We commit to transparent assessment of our national regulatory systems; * Promoting Integrity in Financial Markets: We commit to protect the integrity of the world's financial markets by bolstering investor and consumer protection, avoiding *conflicts of interest,* preventing illegal market manipulation, fraudulent activities and abuse, and protecting against illicit finance risks arising from *non-cooperative* jurisdictions. We will also promote information-sharing, including with respect to jurisdictions that have *yet to commit to international standards* with respect to *bank secrecy* and *transparency;* Reinforcing International Cooperation: We call upon our national and regional regulators to formulate their regulations and other measures in a *consistent manner.* Regulations should ensure their coordination and cooperation across all segments of financial markets, including with respect to *cross-border capital flows.* Regulators and other relevant authorities as a matter of priority should strengthen cooperation on crisis prevention, management, and resolution; Reforming International Financial Institutions: We are committed to advancing the reform of the Bretton Woods Institutions so that they can more adequately reflect changing economic weights in the world economy in order to increase their *legitimacy* and *effectiveness.* In this respect, emerging and developing economies, including the *poorest* countries, should have greater *voice* and *representation.* The Financial Stability Forum (FSF) must expand urgently to a broader membership of emerging economies, and other major standard-setting bodies should *promptly* review their membership. The IMF, in collaboration with the expanded FSF and other bodies, should work to better identify vulnerabilities, anticipate potential stresses, and act swiftly to play a key role in crisis response; Tasking of Ministers and Experts: **10.** We are committed to taking rapid action to implement these principles. We instruct our Finance Ministers, as coordinated by their 2009 G-20 leadership (Brazil, UK, Rep of Korea), to initiate processes and time line to do so. An initial list of specific measures is set forth in the attached Action Plan, including high priority actions to be completed prior to *31 March 2009.* In consultation with other economies and existing bodies, drawing upon the recommendations of such eminent independent experts as they may appoint, we request our Finance Ministers to formulate additional recommendations, including in the folloing specific areas: * Mitigating against pro-cyclicality in regulatory policy; *

Reviewing and aligning global accounting standards, particularly for complex securities in times of stress; * Strengthening the resilience and transparency of *credit derivatives* markets and reducing their systemic risks, including by improving the infrastructure of *over-the-counter* markets; * Review compensation practices as they relate to incentives for risk-taking and innovation; * Reviewing the mandates, governance, and resource requirements of the IFls, and * Defining the scope of systemically important institutions and determining their appropriate regulation or oversignt; **11.** In view of the role of the G-20 in financial systems reform, we will meet again by *30 April 2009,* to review the implementation of the principles and decisions agreed to today; Commitment to an Open Global Economy: **12.** We recognise that these reforms will only be successful if grounded in a commitment to free market priciples, including the *rule of law, respect for private property, open trade and investment, competitivemarkets,* and *efficient, effectively regulated financial systems.* These principles are essential to economic growth and property and have lifted millions out of poverty, and have significantly raised the global standards of living. Recognising the necessity to improve financial sector regulation, we must avoid over-regulation that would hamper economic growth and exacerbate the contraction of capital flows, including to developing countries; **13.** We underscore the critical importance of rejecting *protectionism* and not turning inward in times of financial uncertainty. In this regard, within the next 12 months, we will refrain from raising new barriers to investment or to trade in goods and services, imposing new export restrictions, or implementing World Trade Organisation (WTO) inconsistent measures to stimulate exports. Further, we shall strive to reach agreement this year on modalities that lead to a successful conconclusion to the WTO's Doha Development Agenda with an ambitious and balanced outcome. We instruct our Trade Ministers to achieve this objective and stand ready to assist directly as necessary. We also agree that our countries have the largest stake in the global trading system and therefore must make the positive contribution, necessary to achieve such an outcome; **14.** We are mindful of the impact of the current crisis on developing countries, particularly the most vulnerable. We reaffirm the importance of the Millennium Development Goals, the development assistance commitments we have made, and urge both developed and emerging economies to undertake commitments with their capabilities and roles in the global economy.In this regard, we reaffirm the development principles agreed at the 2002 UN Conference on Financing for Development in Monterrey, Mexico, which emphasised country ownership and mobilising all sources of financing for development; **15.** We remain committed to addressing other critical challenges such as energy security and climate change, food security, the rule of law, and the fight against terrorism, poverty and disease; **16.** As we move forward, we are confident that through continued partnership, cooperation, and multilateralism, we will overcome the challenges before us and restore stability and prosperity to the world economy; Action Plan to Implement Principles for Reform: This Action Plan sets forth a comprehensive work plan to implement the *five* agreed principles for reform. Our finance ministers will work to ensure that the taskings set forth in this Action Plan are fully and rigorously implemented. They are responsible for the development and implementation of these recommendations drawing on the ongoing work of relevant bodies, including the IMF, an expanded FSF, and standard-setting bodies; Strengthening Transparency and Accountability: Immediate Action by 31 March 2009: * The key global accounting standards bodies should work to enhance guidance for valuation of securities, also taking in account the valuation of complex, illiquid products, especially during times of stress; * Accounting standard-setters should enhance work to address weaknesses in accounting and disclosure standards for *off-balance-sheet* vehicles; * Regulators and accounting standard-setters should enhance the required disclosure of complex financial instruments by firms to market participants; * With a view toward presenting financial stability, the governance of international accounting standard-setting body should be further enhanced, including by undertaking a review of its membersip, in particular in order to ensure transparency, accountability and an appropriate relationship between this independent body and the relevant authorities; * Private sector bodies that have already developed *best practices* for private pools of capital and/or hedge funds should bring forward proposals for a set of unified best practices. Finance Ministers should assess these proposals, drawing upon the analysis of regulators, the expanded FSF, and the other relevant bodies; Medium-Term Actions: * The key to local accounting standards bodies should work intensively toward the objective of creating a *single high-quality global standard;* * Regulators, supervisors, and accounting standards-setters, as appropriate, should work with each other and the private sector on an ongoing basis to ensure consistent application and enforcement of high-quality accounting standards; * Financial institutions should provide enhanced risk disclosures in their reporting and disclose all losses on an ongoing basis, consistent with international best practices, as appropriate. Regulators should work to ensure that a financial institution's financial statements include a *complete, accurate,* and *timely picture* of the firm's activities (including *off-balance-sheet* activities) and are reported on a *consistent* and *regular* basis; Enhancing Sound Regulation: Regulating Regimes Immediate Actions by *31 March 2009:* * The IMF, expanded FSF, and other regulators and bodies should develop recommendations to mitigate pro-cyclicality, including the review of how valuation and leverage, bank capital, executive compensation, provisioning practices may exacerbate cyclical trends; Medium-Term Actions: * To the extent countries or regions have already done so, each country or region pledges to review and report on the structure and principles of its regulatory system to ensure it is *compatible* with a modern and increasingly globalised financial system. To this end, all G-20 members commit to undertake a Financial Sector Assessment Programme (FSAP) report and support the transparent assessments of countries' national regulatory systems; * The appropriate bodies should review the differentiated nature of regulation in the banking, securities, and insurance sectors and provide a report outlining the issue and making recommendations on needed improvements. A review of the scope of financial regulation, with a *special emphasis* on instructions, instruments, and markets that are currently *unregulated,* along with ensuring that all systemically-important institutions are appropriately regulated, should also be undertaken; * National and regional authorities should review resolutions regimes and banking laws in light of recent experience to ensure that they permit an orderly wind down of large complex *cross-border* financial institutions; * Definations of capital should be harmonised in order to achieve consistent measures of capital and capital adequacy; Prudential Oversight: immediate Action by *31 March 2009:* * Regulators should take steps to ensure that credit rating agencies meet the highest standards of the international organisation of securities regulators and that they avoid conflicts of interest, provide greater disclosure to investors and to issuers, and differentiate ratings for complex products. This will help ensure that credit rating agencies have the right incentives and appropriate oversight to enable them to perform their important role in providing unbiased information and assessments to markets; * The international organisation of securities regulators should review credit rating agencies' adoption of the standards and mechanisms of monitoring *compliance;* * Authorities should ensure that financial institutions maintain adequate capital in amounts necessary to sustain confidence. International standard-setters should set out strengthened capital requirements for banks' structured credit and *securitisation* activities; * Supervisors and regulators, building on the imminent launch of central counterparty services for credit default swaps (CDS) in some countries, should: speed efforts to reduce the systemic risks of CDS and over-the-counter (OTC) derivatives transactions; insist that market participants support exchange traded or *electronic trading platforms* for CDS contracts; expand OTC derivatives market transparency; and ensure that the infrastructure for OTC derivatives can support growing volumes; Medium-Term Actions: Credit ratings Agencies that provide public ratings should be registered; * Supervisors and central banks should develop robust and internationally consistent approaches for liquidity supervision of, and central bank liquidity operations for, cross-border banks; Risk Management: Immediate Actions by *31 March 2009:* * Regulators should develop enhanced guidance to strengthen banks' risk management practices, in line with international best practices, and should encourage financial firms to reexamine their internal controls and implement strengthened policies for sound risk moment." The post-World War II Bretton Woods system has not adapted well for the new world of globalisation.""

* *President-Elect Obama's Back-On-Tract Agenda:* Michael Grunwald's "First Thing First" recommendations for the new Barack Obama White House Agenda[103] of five items of the Bush Administration which could be *revoked, reversed* or *implemented* by President Obama, are: A New Deal; Repeal Bush; Make Nice with the World; A Sane Farm Policy; and, A First Step on Health Care. *"Repeal Bush"* reads:

* Obama can't undo the last eight years, but he can serve notice on Day One that the Bush Administration is really, really over. He could start by *reversing* Bush's *regulatory efforts to weaken federal oversight* of mining, housing, drilling, finance and other favored industries. He could offer the middle class much needed relief by proposing quickly to *restore* Clinton-era upper-income tax rates and *reduce* the tax burden for everyone else. He could *drop* Bush's

---

management; * Regulators should develop and implement procedures to ensure that financial firms implement policies to better manage liquidity risk; including by creating strong liquidity cushions; * Supervisors should ensure that financial firms develop processes that provide timely and comprehensive measurement of risk concentrations and large counterparty positions across products and geographies; * Firms should reassess their risk management models guard against stress and report to supervisors on their efforts; * The Basel Committee should study the need for and help develop firms' new stress testing models, as appropriate; * Financial institutions should have clear internal incentives to promote stability, and action needs to be taken, through *voluntary* effort or regulatory action, to avoid compensation schemes which reward *excessive* short-term returns or risk taking; * Banks should exercise effective risk management and due diligence over structured products and securitisation; Medium-term Actions: * International standard setting bodies, working with a broad range of economies and other appropriate bodies, should ensure that regulatory policy-makers are aware and able to respond rapidly to evolution and innovation in financial markets and products; * Authorities should monitor substantial changes in asset prices and their implications for the macroeconomy and the financial system; Promoting Integrity in Financial Markets: Immediate Actions by *31 March 2009:* Our national and regional authorities should work together to enhance regulatory cooperation between jurisdictions on a regional and international level; * National and regional authorities should also review business conduct rules to protect markets and investors, especially agsinst *market manipulation* and *fraud* and strengthen their cross-border cooperation to protect the international finance system from *illicit actors.* In case of misconduct, there should be an appropriate sanctions regime; Medium-term actions: * National and regional authorities should implement national and international measures that protect the global financial system from *uncooperative* and *non-transparent* jurisdictions that pose risks of illicit financial activity: * The Financial Action Task Force should continue its important work against *money laundering* and *terrorists financing,* and we support the efforts of the World Bank-UN Stolen Assets Recovery (StAR) Initiative; * Tax authorities, drawing upon the work of relevant bodies such as the Organisation for Economic Cooperation and (OECD), should continue efforts to promote tax information exchange: Lack of transparency and failure to exchange tax information should be vigorously addressed; Reinforcing International Cooperation: Immediate Actions by *31 march 2009:* * Supervisors should collaborate to establish supervisory colleges for all major cross-border financial institutions, as part of efforts to strengthen the surveillance of cross-border firms. Major global banks should meet regularly with their supervisory college for comprehensive discussions of the firm's activities and assessments of the risks it faces; * Regulators should take all steps necessary to strengthen cross-border crises management arrangements including on cooperation and communication with each other and with appropriate authorities, and develop comprehensive contact lists and conduct simulation excercises, as appropriate; Medium-term actions: * Authorities, drawing specifically on the work of regulators, should collect information on areas where convergence in regulatory practices as accounting standards, auditing, and deposit insurance is making progress, is in need of accelerated progress, or where there may be potential for progress; * Authorities should ensure that temporary measures to restore stability and confidence have minimal distortions and are unwound in a timely, well-sequenced and coordinated manner; Reforming International Financial Institutions: Immediate Actions by *31 March 2009:* * The FSF should expand to a broader membership of emerging economies; * The IMF, with its focus on surveillance, and the expanded FSF, with its focus on standard setting, should strengthen their collaboration, enhancing efforts to better integrate regulatory and supervisory responses into the macro-prudential policy framework and conduct early warning excercises; * The IMF, given its universal membership and core macro-financial expertise, should, in close coordination with the FSF and others, take a leading role in drawing lessons from the current crises, consistent with its mandate; * We should review the adequacy of the resources of the IMF, the World Bank Group and the other multilateral development banks and stand ready to increase them where necessary. The Ifls should also continue to review and adapt their lending instruments to adequately meet their members' needs and revise their lending role in the light of the ongoing financial crisis; * We should explore ways torestore emerging and developing countries' access to credit and resume private capital flows which are critical for sustainable growth and development, including ongoing infrastructure investment; * In cases where severe market disruptions have limited access to the necessary financing for counter-cyclical fiscal policies, multilateral development banks must ensure arrangements are in place to support, as needed, those authorities with a good track record and sound policies; Medium-term actions:* We underscored that the Bretton Woods Institutions must be comprehensively reformed so that they can more adequately reflect changing economic weights in the world economy and be more responsive to future challenges. Emerging and developing economies should have greater voice and representation in these institutions; * The IMF should conduct vigorous and evenhanded surveillance reviews of all countries, as well as given greater attention to their financial sectors and better integrating the reviews with the joint IMF/World Bank financial sector assessment programmes. On this basis, the role of the IMF in providing macro-financial policy advice would be strengthened; * Advanced economies, the IMF, and the other international organisations should provide capacity-building programmes for emerging market economies and developing countries on the formulation and the implementation of new major regulations, *consistent* with international standards. [Initial Meeting of the Group of 20 Nations held in Washington on 15 Nov., 2008.]

[103] Ibid., Michael Grunwald: *"Obama's Agenda: Get America Back On Track: First Things First."* 17 November 2008, TIME Commemorative Issue: President-Elect Barack Obama; TIME/CNN; http://www.time.com/.

legal battles to *block* California from enhancing its environmental protections. The *End of the National Nightmare Executive Order* could also include: *No more torture. No more 'threat levels' designed to make people freakout about unnamed dangers. No more 'signing statements' declaring executive prerogative to ignore laws the President doesn't like. No more firing prosecutors for failing to go along with a political agenda. And while he's at it: No more timber lobbyists running the Forest Service, oil lobbyists editing climate reports, Wall Street lobbyists running the SEC or Arabian-horse commissioners running anything....No more T-Ball in the Rose Garden either.* It's cute, but it might bring back memories."

The reform of the wrong intepretation and application of the free-market-economy has become a dire necessity for Americans and the whole world's economic survival. Too many Americans fell prey to the ideology that a free-market "requires nearly complete deregulation of banks and other financial institutions and a government with a hands-off approach to enforcement," wrote New York's former Governor Spitzer.[104] Those who raised red flags about this false image of the free-market were scoffed at for *"failing to understand or even believe" in the 'market'* during stretches of the past 30 years; powerful voices with heavily invested interests accused the 'non-believers' of meddling in the market; *investigative evidence put forward by the New York Justice Department, for example, that warned that some of American International Group's reinsurance transactions were little more than efforts to create the impression of extra capital on the company's balance sheet – were jeered at as 'attacks' on one of the nation's great insurance companies; efforts by the attorneys-general of all 50 U.S., States to investigate some of the subprime lending practices that might be toxic – were blocked by a coalition of the major banks and the Bush Administration. Bush invoked a rarely used statute to preempt the States' ability to probe, under the pretext that the Bush Administration 'had the situation under control' and that an inquiry was unnecessary.*

"Time and time again," Spitzer wrote, "whether at the state level, in Congress or at the Securities and Exchange Commission [SEC] under Bill Donaldson, those who tried to enforce basic principles that would allow the market survive were told that the *'invisible hand'* of the market and *'self-regulation'* could handle the task alone." Spitzer goes on:

- The reality is that unregulated competition drives corporate behavior and risk-taking to unacceptable levels. This is simply one of the ways in which some market participants try to gain a competitive advantage. As one lawyer for my company charged with malfeasance stated in a meeting...(amazingly, this was intended as winning defence): *'You're right about our behavior, but we're not as bad as our competitors.'*

- No major market problem has been resolved through self-regulation, because individual competitive behavior does not concern itself with the larger market. Individual actors care only about performing better than the next guy, doing whatsoever is permitted – or will go undetected. Look at the major bubbles and market crises. *Long-term Capital Management, Enron, the subprime lending scandals*: All are classic demonstrations of the bitter reality that greed, not self-discipline, rules where unfettered behavior is allowed.

---

[104] Eliot Spitzer: *"How to Ground the Street."* Washington Post, 16 November 2008, p. B01; http://www.washingtonpost.com/wp-dyn/content/article/2008/11.

Those who truly understand economics, as did Adam Smith, *do not preach an absence of government participation*. A market doesn't exist in a vacuum. Rather, a market is a product of laws, rules and enforcement. It needs transparency, capital requirements and fidelity to fiduciary duty. The alternative, as we're seeing is anarchy. [One of the great advantages U.S., capital markets have enjoyed over the decades has been the view – held worldwide – that there was an underlying integrity to the representations market participants made, because of the regulatory framework in which they were made was believed to provide *genuine* oversight.] But as we all know, the laws requiring such integrity are meaningless without a government dedicated to enforcing them.

"Second, our corporate governance system has failed. We need to reexamine each of the links in its chain. Boards of directors, compensation and audit committees, the trio of facilitators (lawyers, investment bankers and auditors) whose job it is to create the impression of legal compliance, and shareholders themselves – *all abdicated their responsibilities*. Institutional shareholders, in particular *mutual funds, pension funds,* and *endowments,* must reengage in corporate governance. Over the past decades, arguably the sole challenge to corporate mismanagement and poor corporate strategies has come from private-equity firms or activist *hedge funds.* These firms were among the few shareholders or pools of capital willing to purchase and revamp encrusted corporate machines. So it shouldn't be surprising that the corporate world has taken a sceptical view of them – especially *short-selling hedge funds,* which have often been a rare voice raising the alarm.

"Boards of directors were also missing in action over the past decade: not only did they *not provide answers,* they all too often, *failed even to ask* the appropriate questions. And the roles of compensation committees, of course, must be totally rethought. No longer can Garrison Keillor's observation about our kids – that they are all above average – apply to CEOs and propel failed leaders' paychecks through the roof. Today's monetary public oversight and outrage over executive compensation, while *long overdue*, is no substitute over the long-term for firm standards set by compensation committees and boards of directors.

"...Let's leave aside ideological hesitancy that has hamstrung regulatory agencies. Today's balkernised regulatory framework for financial services no longer matches in any way the needs of a fully integrated global financial system. The divisions of the past – commercial banking vs. hedge funds vs. private equity – have become distinctions without a difference. But these old boxes and formalities still determine how entities are viewed and *regulated*. It should surprise nobody that capital found the crevices in the regulatory framework. That is what capital is paid to do. But we *failed* to respond with a regulatory framework flexible enough to plug the leaks. We do not need *additional* fragmented areas of federal regulation to handle hedge funds, sovereign wealth funds or derivatives. We need a *unified* approach that addresses the underlying issues: *what kinds of leverage we wish to handle; how to measure risk; how much disclosure various trading products should provide.* We cannot survive with the current system: *the SEC; the Office of the Comptroller of the Currency; the Federal Deposit Insurance Corporation; the Fed; the Office of Thrift Supervision,* and so on and on. We must go from the RubeGolberg structure we now have to a sleek iPod design that is cleaner, has better operating software and may even look good.

"Three overriding priorities should guide government actions in the new structure. First, we need *better control* of systemic risk. The currently splintered federal regulatory authority, the continued presence of *off-balance-sheet* transactions for financial entities (even post-Enron) and the failure to subject major players to any government oversight means that *nobody can really understand the full risk financial system;* Second, investors must be protected with adequate, accurate information. Firms must offer *transparency* both to individual investors and to government regulators; and Third, as Eric R. Dinallo, the superintendent of the New York State Insurance Department, has wisely pointed out, we will have to step back from the current environment in which government *has become a guarantor of all major risk.* The so-called *moral hazard* will serve to devalue risk in the market, and this too will have debilitating long-term effect on capital flows. Only if private actors have to bear red risks they incur will the market function properly. We are now perilously close to nationalising risk."

Excessive deregulation breeds excessive inflation[105] with accompyning recession and depression that result in the general rise in the level of prices of goods and services over a period of time. Central banks who are the monetary authorities and control the size of the money supply, can tame inflation or at least keep it low; they do this through the setting of interest rates and banking reserve requirements or required reserve ration, and open market operations, including fixed exchange rates, and wage and price control. The mounting rate of global inflation is troubling especially in developing economies.

Just imagine for a second the mindboggling *monthly* 13,200,000,000,000%, *annual* 516,000,000, 000,000,000,000% on an *inflation index* of 4120,000,000,000,000,000.00 posted by Zimbabwe as of 07 November 2008! By July 2008, Zimbabwe's monthly inflation was *231,000,000%* and still rising. A new ZWD500 million (Zimbabwe dollar) note, worth US$10.00 or €7.50 was introduced on 12 December 2008 by Zimbabwe's central bank, which struggles to print money fast enough to keep pace with price rises several times a day. That was also the same reason why the Zimbabwe central bank on 19 December 2008, issued a ZWD10 billion for Christmas. Cash can be withdrawn from Zimbabwe banks only once a day due to currency shortages and people are allowed to take only ZWD 500 million dollars ($1.00/€7.50) which is not enough to see one through the day; hospitals have no drugs, no equipment and no staff left to treat the cholera epidemic which has now taken more than 1,111 lives and infected 20,580 [as of 19 December 2008], according to the UN. Cholera has spread as sewage and water lines have broken down, contaminating the drinking water supply.

Worldwide inflation rates based on the Consumer Price Index (CPI) provided by the Cato Institute, Bespoke Investment Group, Wikipedia, and MeriNews on Zimbabwe and other countries,[106] include the following:

---

[105] Monetary Inflation: Inflation originally referred to increases in the money supply or monetary inflation, and inflation can also be seen as a decline in the real value of money, (i.e., a loss of purchasing power). When the general level of prices rises, each unit of currency buys fewer goods and services. Price inflation is usually measured by calculating the inflation rate, which is the percentage change in a price index, such as consumer price index (CPI) – a measure of the average price of consumer goods and services purchased by households, which is one of several price indices calculations used by national statistical agencies.

[106] Steve H. Hanke, Prof., Applied Economics, The John Hopkin's University/Senior Fellow, The Cato Institute: *New Hyperinflation Index (HHIZ) Puts Zimbabwe Inflation at 516 Quintillion Percent.."* [Cato Institute, 1000 Massachusetts Ave., N.W., Washington D.C. 2001-5403; http://www.cato.org/zimbabwe.] MeriNews, India's First Journalism Portal: *Inflation in Zimbabwe Breaks all records,"* by C. J. Ravi Kant [http://www.merinews.com/aboutus.jsp]; Bespoke Investment Group, LLC: *"Global Inflation Rates."* 12 Nov., 2008 [http://seekingalpha.com/article782217-global-inflation-rates; 105 Calvert Str., Suite 100, Harrison, NY 10528.] Hanke Hyperinflation for Zimbabwe (HHIZ): 5-Jan-07: **1.00** Index (I), **13.60%** Monthly Inflation Rate (MIR), **0%** Annual Inflation Rate (AIR); 2-Feb-07: 1.78, 77.60 % MIR, 0% AIR; 2-Mar-07: 3.14 I, 76.60% MIR, 0% AIR; 5-Apr-07: 6.90% MIR, 0% AIR; 4-May-07: 6.75 I, -2.13% MIR, 0% AIR; 1-Jun-07: 20.70 I, 207.00% MIR, 0% AIR; 6-Jul-07: 53.00 I, 60.30% MIR, 0% AIR; 3-Aug-07: 49.10 I, -7.25% MIR, 0% AIR; 7-Sep-07: 82.50 I, 70.60% MIR, 0% AIR; 5-Oct-07: 219.00 I, 165.00% MIR, 0% AIR; 2-Nov-07: 6.42 I, 193.00% MIR, 0% AIR; 28-Dec-07: 20,010.00 I, 61.50% MIR, 215,000% AIR; 25-Jan-08: 2,250.00 I, 11.80% MIR, 0% AIR; 29-Feb-08:8,260.00 I, 259.00% MIR, 0% AIR; 28-Mar-08: 17,700.00 I, 114.00% MIR, 0% AIR; 25-Apr-08: 57,700.00 I, 222.00% MIR, 0% AIR; 30-May-08: 442,000.00 I, 497.00% MIR, 0% AIR; 26-Jun-08: 23,700.00 I, 5,260.00% MIR, 41,400,000% AIR; 4-Jun-08: 49,300,000.00 I, 3,740.00% MIR, 93,000,000% AIR; 11-Jun-08: 81,800,000.00 I, 2,080.00% MIR, 167,000,000% AIR; 18-Jul-08: 122,000,000.00 I, 1,030.00% MIR, 250,000,000% AIR; 25-Jul-08: 157,000,000.00 I, 566.00% MIR, 317,000,000% AIR; 29-Jul-08: 66,340,000,000.00 I, 3,190.00% MIR, 9,700,000,000% AIR; 26-Sep-08: 796,000,000,000.00 I, 12,500.00% MIR, 472,000,000,000% AIR; 3-Oct-08: 3,590,000,000,000.00 I, 15,400.00% MIR, 1,640,000,000,000% AIR; 10-Oct-08: 32,400,000,000,000.00 I, 46,000.00% MIR, 11,600,000,000,000% AIR; 17-Oct-08: 1,070,000,000,000,000.00 I, 490,000.00% MIR, 299,000,000,000% AIR; 24-Oct-08: 120,000,000,000,000,000.00 I, 15,100,000.00% MIR, 25,300,000,000,000,000% AIR; 31-Oct-08: 22,600,000,000,000,000.00 I, 630,000,000.00% MIR, 3,520,000,000,000,000,000% AIR; 7-Nov-08: 41260,000,000,000,000.00 I, 13,200,000,000,000% MIR, 516,000,000,000,000,000,000% AIR [Source: Imara Asset Management Zimbabwe/author's calculations.]; MeriNews/Bespoke Global Inflation: Iraq **53.2%**; Venezuela **31.4%**; Ukraine

- *Future Investment Uncertainty:* Uncertainty about future inflation may discourage investment and saving.

- *Variable Income Shifts:* Inflation shifts income from those on fixed incomes to those with variable incomes. Fixed nominal payments, such as rents and wages, are eroded, if they are not inflation-adjusted.

- *Hoarding:* High or excessive inflation causes households to hoard as they buy household consumer durables, such as cars, appliances, business and electronic equipments, and non-durables or urgent goods, such as cosmetics, and foods as stores of wealth.

- *Hyperinflation:* High rates of inflation breed hyperinflation, caused by growth rates of the money supply or money stock, that is, the total amount of money available in an economy at a particular time.

- *Fall in Hard Currency Value:* Inflation causes the face value of coins to fall below the hard currency value of the metals used historically, leading most modern coins being made of base metals, such as: *Cupronickel:* (around 80:20, silver in color), for example: *nickel-brass* (copper 75), nickel (5) and zinc (20), gold in color; *manganese-brass* (copper, zinc, manganese, nickel), bronze or simple-plated steel.

- *Drag on Productivity & Deflation:* Inflation can act as a drag on productivity as companies are forced to shift away from products and services in order to focus on profit and losses from current inflation, as well as increase in hidden taxes. Unemployment and losses occur, when prices are falling relatively to costs of production or deflation.

The wobbling foundations of the market-economy have forced the International Monetary Fund (IMF) and the International Labour Organisation (ILO) to issue warnings, exhortations, and remedies to calm down the economic turbulence and return world economies to normal functioning. In its "World Economic Outlook"[107] report released in October 2008, the IMF said the world's developed economies are headed for the first full-year contraction since World War Two; it urged governments to ramp up spending to support the world economy. The world economy is decelerating quickly as it is still buffeted by an extraordinary financial shock of high energy and commodity prices; many advanced economies are close to or moving into recession. Global growth was to be moderate from 5.0% in 2007, to 3.9% in 2008. This implying that recession in fast growing emerging economies, such as China, for example, be in serious trouble, if its growth fell from 11% to 6%, as projected.[108]

31.10%; Guinea **30.9%**; Sri Lanka **26.20%**; Vietnam **25.20%**; Pakistan **19.27%**; Yemen **20.8%**; Myanmar **20%**; Uzbekistan **19.8%**; Congo DR **18.2%**; Latvia **17.90**; Afghanistan **17%**; Serbia **15.5%**; Russia **15.10%**; Bulgaria **15.00%**; Oman **12.40%**; Iceland **12.32%**; Lithuania **12.00%**; Costa Rica **11.90%**; Estonia **11.30%**; South Africa **11.10%**; Turkey **10.74%**; Saudi Arabia **10.45%**; Indonesia **10.38%**; Philippines **9.60%**; Kuwait **9.53%**; Ecuador **9.29%**; Argentina **9.10%**; Macau **8.95%**; Chile **8.90%**; Romania **8.46%**; China **7.70%**; Thailand **7.60%**; Singapore **7.50%**; Uruguay **7.20%**; Hungary **7.00%**; Czech Rep. **6.80%**; Croatia **6.40%**, Slovenia **6.40%**; Colombia **6.39%**; India **6.02%**; Tunisia **5.83%**; Hong Kong **5.70%**; Brazil **5.58%**; Morocco **5.40%**; Israel **5.40%**; Peru **5.39%**; Belgium **5.21%**; Mexico **4.95%**; Cyprus **4.94%**; Greece **4.90%**; South Korea **4.88%**; Qatar **4.73%**; Ireland **4.70%**; Slovakia **4.60%**; Spain **4.60%**; Poland **4.40%**; Australia **4.20%**; U.S., **4.20%**; Sweden **4.00%**; Malaysia **3.80%**; Taiwan **3.71%**; Eurozone **3.70%**; Italy **3.60%**; New Zealand **3.40%**, Denmark **3.40%**; France **3.30%**; United Kingdom **3.30%**; Luxembourg **3.22%**; Norway **3.10%**; Germany **3.00%**; Switzerland **2.90%**; Portugal **2.80%**; Netherlands **2.30%**; Canada **2.20%**; Japan **0.80%**; Median **5.83%**.

[107] Lesley Wroughton: *"IMF Warns of Deepening Recession in Rich Countries."* Reuters, 8 November 2008.; http://www.reuters.com/businessnews/news.

The ILO expressed worries of exploding global unemployment by an estimated *extra* 20 million[109] with the continued world financial crisis, ILO Director-General Juan Somavia announced. "We need prompt and coordinated government actions to avert social crises that could be severe, long-lasting and global. The number of unemployed could rise from 190 million to 210 million in late 2009...the current crisis would hit hardest such sectors as construction, automotive, finance, services and real estate...This is not simply a crisis on Wall Street; this is a crisis on all streets. We need an economic rescue plan for working families and the real economy, with rules and policies that deliver decent jobs. We must link better productivity to salaries and growth to employment."

World governments, contrary to the vocal opposition of the outgoing U.S., Bush Administration, have called for a new market system that functions with full controls. Bankcruptcies, and failed business enterprises, rising food prices, and rising job layoffs are only inflaming the embers of decreasing job opportunities and rising global unemployment.

1.4.2 Rising Layoffs, Unemployent Prices

THE DEEPENING recessions and encroaching depressions in world economies are sad reminders of what awaits the already 190 million unemployed people around the world. Joblessness through company layoffs, especially in the finance, banking, oil-energy, agriculture, and automobile industries, hit the upward trend everywhere.

For example, during November 2008, various banking and investment institutions planned or carried out mass layoffs of their workers: Citigroup 53,000 in addition to the 23,000 it had already shed off from its payrolls because of massive losses and deteriorating debt tied to bad mortgages; HSBC Holdings PLC announced 500 jobs in Asia were to be cut off due to the economic slump; on 1 December 2008, Credit Suisse, Aston Martin, HSBC and Accessory retailer Halfords, announced they would shed off 2,000 jobs - 650, 600, 500, and 250 each

[108] IMF : "World Outlook Growth Projections, October 2008." World Outlook: **5.0%** (2007), **3.9%** (2008), **3.0%** 2009); Advanced Economies: **2.6%** (2007), **1.5%** (2008), **0.5%** (2009); U.S.: **2.0%** (2007), **1.6%** (2008), **0.1%** (2009); Euro Area: **2.6%** (2007), **1.3%** (2008), **0.2%** (2009); Germany: **2.5%** (2007), **1.8%** (2008), **---%** (2009); France: **2.2%** (2007), **0.8%** (2008), **0.2%** (2009); Italy: **1.5%** (2007), **-0.1%** (2008), **-0.2%** (2009); Spain: **3.7%** (2007), **1.4%** (2008), **-0.2%** (2009); Japan: **2.1%** (2007), **0.7%** (2008), **0.5%** (2009); UK: **3.0%** (2007), **1.0%** (2008), **-0.1%** (2009); Canada: **2.7%** (2007), **0.7%** (2008), **1.2%** (2009); Emerging/Developing Countries: **8.0%** (2007), **6.9%** (2008), **7.7%** (2009); Developing Asia: **10.0%** (2007), **8.4%** (2008), **7.7%** (2009); China: **11.9%** (2007), **9.7%** (2008), **9.3%** (2009); India: **9.3%** (2007), **7.9%** (2008), **6.9%** (2009); Western Hemisphere: **5.6%** (2007), **4.6%** (2008), **3.2%** (2009); Brazil: **3.2%** (2007), **5.2%** (2008), **3.5%** (2009); Mexico: **3.2%** (2007), **2.1%** (2008), **1.8%** (2009). [Note: Real effective exchange rates are assumed to remain constant at the levels prevailing during 18 August – 15 September, 2008.]
[109] International Labour Organisation (ILO): *"Global financial Crisis To Increase Unemployment by 20 Million."* ILO News, ILO/08/45/Press Release, 20 Oct., '08.

respectively; Royal Bank of Scotland (RBS) cut 3,000 jobs; Fidelity Investment eliminated 1,700 jobs; Rolls-Royce announced it would slash 1,500 to 2,000 jobs; Peugeot-Citroen also announced it would shed off 2,700 jobs because of falling demand in Europe; Lehman Brothers bankruptcy left thousands of workers out in the cold. Factories, including automakers, construction companies, especially home builders, retailers, mortgage bankers, securities firms, hotels and motels and educational services – all have cut jobs. Private companies in the U.S., have reportedly cut 263,000 jobs; U.S., bank failures up to 03 October 2009 have been 417 with about $100 billion cost failures, up from 70 billion, according to the U.S., Federal Deposit Insurance (FDIC) Company, compared with only three for all of 2007 and far more than in june 1994. Gordon[110] writes more on this. And it's expected that more banks won't survive the next year of economic tumult because of the pressures of tumbling home prices, rising mortgage foreclosures and tighter credit battering many banks, large and small worldwide. Since May 2008 when the oil-and-food crisis began, the banking industry had lost more than $2.8 trillion.

And job losses keep battering the hemorrhaging world econmy as multinational giants continue shedding off more and more of their work force staffs. On 06 December 2008, the U.S., Labor Department issued the November 2008 unemployment figures as 533,000 at an unemployment rate of 6.8%. In September 2009, this figure had risen to 9.8%.

Smith[111] summarises the 2008-year-end job losses drama, thus:

- The nation's job market, in the final month of a brutal 2008, was dealt a savage blow Thursday [04 December 2008] when major companies – *AT&T Inc., DuPont Co., Viacom Inc., Credit Suisse Group* and *Avis Budget Group* – announced job cuts that total *22,850*. And this is just the latest bit of bad news. On Wednesday, *State Street Corp., Jefferies Group* and *The Carlyle Group* announced job cuts totalling about *3,000*. This follows a brutal November, when U.S., employers announced plans to cut *181,671* jobs, according to Challenge, Gray & Christmas. ADP's month's employment report showed that private sector payrolls fell by *250,000* jobs in November compared to the prior month. Just in the week ended 29 November, *509,000* Americans filed initial jobless claims according to the Labor Department...The Department is expected to announce Friday that the nation hemorrhaged *325,000* [actually *533,000*] jobs in November, according to a consensus of economists surveyed by Briefing.com. This tally will be added to the *1.2 million* jobs that were already lost in the first 10 months of the year. Just in the first 11 months of the year, job losses could match or exceed the 1.5 million job losses that occurred in a 12-month span that overlapped 1990 and 1991 [David Wyss, chief economist for Standards & Poors]...

- "AT&T, a Dallas-based telecom operator, said it would slash 12,000 jobs, totalling 4% of its work force, attributing this to 'economic pressures, a changing business mix and a more streamlined organisational structure' in a news release...would take a charge of $600 million in the fourth quarter to make severance payments, and reduce its 2009 capital expenditures from its 2008 levels.

- "Credit Suisse Group (CS) said it would cut 5,300 staff jobs or 11% of its worldwide work

[110] Marcy Gordon: *"Feds Take Over Two Failed California Banks."* TIME/CNN (Business/Tech), 22 November 2008.
[111] Aaron Smith, CNNMoney.com staff writer: *"Job cuts mount as year-end nears: AT&T, DuPont, Viacom, Credit Suisse and others take another 22,000-plus jobs out of the work force."* CNNMoney.com 04 December 2008, New York. TIME/CNN.

force, as part of a restructuring effort; majority of the cuts were be to investment bank jobs and eliminatng 1,200 contractor positions, said CEO Brady Dougan.

- "DuPont, a chemical company said it would cut 2,500 jobs. DuPont Chief Executive Holliday said his company was making the cuts 'in response to current market challenges' and increase the company's competitiveness in the coming year. DuPont expected a loss of 60 cents-70 cents per share for the fourth quarter, including a 40 cent-per-share charge from the company's restructuring plan. The company expected full year earnings to be $2.25-$2.75 per share in 2009, and full-year earnings of $2.75-$2.85 were expected for 2008, down from the previously announced range of $3.25 - $3.30 per share.

- "Viacom Inc., an entertainment company that includes MTV Networks and Paramount Pictures, was cutting 850 jobs or 7% of its work force. Viacom Chief Executive Phillippe Dauman said restructuring the company allows it to adapt to the challenges presented by the current economic environment. In addition to jobcuts, Viacom was suspending senior level management salary increases throughout 2009. The company expecting the restructuring to result to pre-tax savings of $200-$250 million in 2009, and a pre-tax charge of $400-$450 million in the fourth quarter of 2008, or 42 cents-48 cents per diluted share.

- "Car rental company Avis Budget Group said it has cut more 2,200 jobs and taken other steps to meet its goal of reducing annual costs by $150-$200 million by the middle of 2009. The company will close its claims-processing facility in Orlando, Fla., as well as its customer contact centre in Wichita Fally, Texas. Avis CEO Ronald Nelson said the company will continue its 'relentless focus on cost-containment...'

- "State Street Corp., Jefferies Group, and Carlyle Group announced on Wednesday 03 December 2008 that they were shedding off some of their workers: State Street Corp., 1,600-1,800 workers or 6% of its total work force; Jefferies 358 workers or 18% of its work force throughout 2008 as part of a wide restructuring plan to restore profitability in 2009; Carlyle, a private equity investment firm in Washington 105 of its 1000 work force; State Street's two-thirds of job cuts will occur in North America, with the rest in Europe, Asia and the Pacific region. Jefferies will close its offices in Dubai, Singapore and Tokyo..."

- M Co., a Mapple-based, U.S., manufacturer who had earlier announced laying off 1,000 from its U.S., Europe and other developed nations, on 7 December announced it was cutting 1,800 jobs in its fourth quarter. It ordered some workers to take vacation or unpaid time off for the last two weeks of 2008.

- J. P. Morgan said on 02 December, it would make 9,000 workers redundant, with Glasgow-based Bowie Castle Bank Group shedding off 817 jobs. On 09 December 2008, Sony, Japan's electronics giant, and Barclays, the British banking giant – both announced they were slashing jobs: Sony by 8,000 of its global electronics work force by the end of 2010, an action aimed at cost-cutting by $11 billion a year as global downturn battles profits; and Barclays by 109 affecting its outsource department work that deals with accounts when a client dies. The next day, 10 December, Rio Tinto, the world's largest mining company announced it would lay off 14,000 workers in an effort to reduce its $10 billion debt; and Yahoo, Inc., also put out a statement it was to lay off 1,500 workers. The century-old British retail business Woolworths on 10 December labored under pain of bankcruptcy as it sought to strike off more than 25,000 jobs from its 813-strong worldwide retail stores. Woolworhs has been seating under its mounting debts and losses and was looking vainfully for a buyer. Bank of America with 308,000 employees, on 11 December 2008 shocked with its dreary announcement to strike 30,000-35,000 jobs over three years, as it faced a deteriorating economic environment and tried to absorb Merrill Lynch.

The Eurozone was on 14 November 2008, formally declared as being immersed in recession, based on the dismal growth rates projected by the IMF; on 1

December 2008 the U.S., National Bureau of Economic Research (NBER) said the U.S., formally was in recession, adding to the 2007 U.S. informal recession. Some 12 months ago as of November 2008, European banks scented and caught the American sub-prime bank flue and the credit crunch, write-downs, write-offs ritual as the epidemic swelled on globally. French, German and Dutch banks confessed to being hit by their exposure to soured U.S., sub-prime mortgages, as Smith[112] partly explains:

- "For banks across Europe, as for their U.S., counterparts, 2008 is proving painfully difficult. Globally, banks could write-down as much as $450 billion more over the next three to four years, according to research from Deutsche Bank. Lenders, it says, are short of funds equivalent to 4% of their balance sheets, with those in Ireland, Spain and Britain finding fund-raising particularly tricky. As the U.S., sputtered over the past year, Europe's economies initially drew praise for motoring on. But housing markets in Ireland, Spain and the U.K., have turned down fast in the past few months and food and fuel bills have soared. Europe, it seems, has finally caught America's cold.

- "Take Britain. Just as it did in Ireland and Spain, consumer confidence in the U.K., swelled in recent years on the back of rising housing prices. But in all three countries, red-hot housing markets have suddenly gone cold. With jittery banks slashing the range of available mortgages, and rocketing gas prices nudging inflation to 3.8% - well above the Bank of England's 2% target – demand in Britain's housing market has been choked. House prices fell 1.7% last month, according to Halifax, a major mortgage lender, and a total of 8.8% over the past year. That hit Britain's construction business hard. Shares in Taylor Wimpey, the U.K.'s largest house builder, have fallen more than 80% over the past 12 months. The construction downturn slowed growth in Britain to an anemic 0.2% in the second quarter of this year, the lowest quarter-on-quarter rate for three years..."

Unemployment and welfare[113] rolls (where these exist, mostly in developed countries) rose dramatically. Unemployment or joblessness[114] results from turmoil caused by inefficiencies in the labor market that lead to consumers consuming less of the goods and services provided by the industries. Low or fixed workers' wages and more consumer taxes tend to encourage *less consumption* and *less production*. Higher workers' wages tend to entice *higher consumption* and thus stimulate *higher production*. Lower producer's taxes tend to lead to *more production* and *more investment*. High oil and energy prices tend to lead to *high production costs* and *lower consumption rates*. When banks speculate on commodities futures and lose money or invest in bad ventures, such as the U.S., subprime lending fiasco, they tend to *lose money* and offer *less credit* to producers and service industries and companies. When consumers are burdened with too many taxes, fixed wages, higher commodity and mortgage prices and dwindling efforts to save – *consumption and production suffer* and come to a

---

[112] Adam Smith: *"The Credit Crisis Spreads to Europe."*, TIME/CNN, (Business/Tech) 11 August 2008.

[113] Unemployment Welfare: Industrialised countries (known as welfare states) and some emerging economies provide unemployment benefits to the unemployed, aimed at alleviating hardship and more time to search for jobs. Unemployment benefits include insurance, and monthly welfare payments for rents and food, and unemployment compensation benefits. Subsidies and retraining programmes are also provided.

[114] Unempoyment Classified: *Cyclical* or *Structural* Unemployment: Caused by inefficiencies in the labor market, and largely involuntary in nature; *Classical* Unemployment: Caused by rigidities imposed on the labor market such as minimum wage laws, taxes, and other regulations that may discourage the hiring of workers; *Voluntary* or *Involuntary* Unemployment: Voluntary unemployment is attributed to an individual's decision; involuntary unemployment exists because of the lack of the socio-economic environment, including market structure, government intervention, level of aggregate demand by individuals; *Open* Unemployment: Open Unemployment occurs where employers don't hire workers when wages tend to be higher than production costs, or when workers make higher wage demands believed by employers to be jeopardising to further investments. If employers hire workers, they pass the resulting costs on to consumers, which in turn results in inflationary instability; *Frictional* Unemployment: these are people in the midst of transiting between jobs, in search of new, permanent jobs or full employment.

virtual standtill. When producers and investors are blessed with generous tax cuts, less regulation, and more lending, they tend generally to *over-invest, over-speculate, under-produce* and *under-pay* for production and services.

Betwixt these inefficiencies in the labor market, the UN International Labor Organisation's (ILO) Global Employment Trends issued its report. According to a Reuter News Agency[115] dispatch the ILO said more people will be out of work in 2008 as a result of global economic cooling; and any major slowdown could cause disruption and further hike unemployment. The ILO estimates 3 billion people aged 15 and older had jobs in 2007, up nearly 2% from the year before and more than 17% highter than in 2007. There were 190 million unemployed in 2007. Among those employed, about 487 million people did not earn enough to lift themselves and their families above $2.00 a day poverty line; and 1.3 billion earned less than $2.00 a day. In addition, looming economic trouble and rapid technological advances would present a major challenge for workers in the years ahead, particularly in rich markets such as the U.S., Europe and Japan; in these rich countries, jobs are increasingly being moved to poorer countries with cheaper labor costs.

The unemployment rate[116] is used as a country's macroeconomic basis that deals with performance, structure and behavior of a national or regional economy as a

---

[115] Laura McInnis, Reuters: *"Global Unemployment Rate to Climb in 2008."* Reuters, 24 January 2008. http://uk.reuters.com/article/businessNews/.

[116] The U.S., Central Intelligence (CIA) World Factbook: *"Rank Order – Unemployment Rate"* covering most of the countries of the world, updated 06 November 2008: Andorra **0.00%** 1996 estimates (est); Monaco **0.00%** 2005 est; Qatar **0.70%** 2007 est; Uzbekistan **0.80%** 2007 est; Guernsey **0.90%** Mar 2006 est; Azerbaijan **1.00%** est; Iceland **1.00%** 2007 est; Liechtenstein **1.30%** Sep 2007 est; Thailand **1.40%** 2007 est; Isle of Man **1.5%** Dec 2006 est; Belarus **1.60%** 2005; Vanuatu **1.70%** 1999; Cuba **1.80%** 2007 est; Papua New Guinea **1.90%** 2004; Kiribati **2.00%** 1992 est; Seychelles **2.00%** 2006 est; Bermuda **2.10%** 2004 est; Singapore **2.10%** 2007 est; Moldova **2.10%** 2007 est; Faroe Islands **2.10%** 2006; Jersey **2.20%** 2006 est; Kuwait **2.20%** 2004 est; Ukraine **2.30%** 2007 est; United Arab Emirates (UAE) **2.40%** 2001; Laos **2.40%** 2005 est; Tajikistan **2.40%** 2007 est; Bangladesh **2.50%** 2007 est; Cambodia **2.50%** 2000 est; Bhutan **2.50%** 2004; Norway **2.50%** 2007 est; Denmark **2.80%** 2007 est; Switzerland **2.80%** 2007 est; Gibraltar **3.00%** 2005 est; Mongolia **3.00%** 2007; Macau **3.10** 2006; Guatemala **3.20%** 2006; Malaysia **3.20%** 2007 est; South Korea **3.30%** 2007 est; Lithuania **3.50%** 2007 est; New Zealand **3.60%** 2007 est; British Virgin Islands **3.60%** 1997; Mexico **3.70%** 2007 est; Japan **3.80%** 2007 est; San Marino **3.80%** 2004; Northern Mariana Islands **3.90%** 2001; Cyprus **3.90%** 2007 est; Taiwan **3.90%** 2007 est; Brunei **4.00%** 2006; China **4.00%** 2007 est; Hong Kong **4.00%** 2007 est; Romania **4.10%** 2007 est; Palau **4.20%** 2005 est; Vietnam **4.30%** 2007 est; Austria **4.40%** 2007 est; Australia **4.40%** 2007 est; Cayman Islands **4.40%** 2007 est; Luxembourg **4.40%** 2004; Saint Kitts/Nevis **4.50%** 1997; Trinidad/Tobago **4.50%** 2007 est; Costa Rica **4.60%** 2007 est; Ireland **4.60%** 2007 est; Netherlands **4.60%** 2007 est, U.S. **4.60%** 2007 est; Estonia **4.70%** 2007 est; Nigeria **4.90%** 2007 est; Nicaragua **4.90%** 2007 est; Burma **5.20%** 2007 est; Namibia **5.20%** 2007 est; U.K. **5.30%** 2007 est; Paraguay **5.60%** 2007 est; Pakistan **5.60%** 2007 est; Latvia **5.70%** 2007 est; Canada **6.00%** 2007 est; Montserrat **6.00%** 1998 est; Sri Lanka **6.00%** 2007 est; Sweden **6.10%** 2007 est; El Salvador **6.20%** 2007 est; Italy **6.20%** 2007 est; Virgin Islands **6.20%** 2004; Russia **6.20%** 2007 est; Malta **6.40%** 2007 est; Panama **6.4%** 2007 est; Czech Rep **6.60%** 2007 est; Aruba **6.90%** 2005 est; Peru **6.90%** 2007 est; Finland **6.90%** 2007 est; Chile **7.00%** 2007 est, Armenia **7.10%** 2007 est; India **7.20%** 2007 est; Hungary **7.30%** 2007 est; Israel **7.30%** 2007 est; Kazakhstan **7.30%** 2007 est; Philippines **7.30%** 2007 est; Botswana **7.50%** 2007 est; Belgium **7.50%** 2007 est; Bolivia **7.50%** 2007 est; Bahamas **7.60%** 2006 est; Fiji **7.60%** 1999; Bulgaria **7.70%** 2007 est; Slovenia **7.70%** 2007 est; France **7.90%** 2007 est; Anguilla **8.00%** 2002; Portugal **8.00%** 2007 est; Central African Rep **8.00%** 2001 est; Greece **8.30%** 2007 est; Spain **8.30%** 2007 est; Slovakia **8.40%** 2007 est; Argentina **8.50%** 2007 est; Venezuela **8.50%** 2007 est; European Union (EU) **8.50%** 2007 est; Ecuador **8.80%** 2007 est; Mauritius **8.80%** 2007 est; Germany **9.00%** 2007 est; Syria **9.00%** 2007 est; Egypt **9.10%** 2007 est; Indonesia **9.10%** 2007 est; Uruguay **9.20%** 2007 est; Brazil **9.30%** 2007 est; Greenland **9.30%** 2005 est; Belize **9.40%** 2006; Suriname **9.50%** 2004; Morocco **9.80%** 2007 est; Jamaica **9.90%** 2007 est; Turkey **9.90%** 2007 est; Turks/Caicos Islands **10.00%** 1997 est; Saint Pierre/Miquelon **10.30%** 1997 est; Barbados **10.70%** 2003 est; Antigua/Barbuda **11.00%** 2001 est; Ghana **11.00%** 2000 est; Colombia **11.20%** 2007 est; Guam **11.40%** 2002 est; French Polynesia **11.70%** 2005; Algeria **11.80%** 2007 est; Croatia **11.80%** 2007 est; Iran **12.00%** 2007 est; Niue **12.00%** 2001; Puerto Rico **12.00%** 2002; Grenada **12.50%** 2000; Poland **12.80%** 2007 est; Saudi Arabia **13.00%** 2004 est; Tonga **13.00%** FY03/04 est; Cook Islands **13.10%** 2005; Albania **13.20%** 2007 est; Jordan **13.50%** 2007 est; Georgia **13.60%** 2006 est; Saint Helena **14.00%** 1998 est; Tunisia **14.10%** 2007 est; Montenegro **14.70%** 2007 est; Bharain **15.00%** 2005 est; Saint Vincent/Grenadines **15.00%** 2001 est; Oman **15.00%** 2004 est; Wallis/Futunna **15.20%** 2003; Dominican Rep **15.60%** 2007 est; Netherlands Antilles **17.00%** 2002 est; New Caledonia **17.10%** 2004; Iraq **18.00%** 2006 est; Kyrgyzstan **18.00%** 2004 est; West Bank **18.60%** 2006 est; Sudan **18.70%** 2002 est; Serbia **18.50%** 2007 est; Comoros **20.00%** 1996 est; Lebanon **20.00%** 2006 est; Mauritania **20.00%** 2004 est; Saint Lucia **20.00%** 2003 est; Cape Verde **21.00%** 2000 est; Gabon **21.00%** 1997 est; Mozambique **21.00%** 1997 est; Micronesia **22.00%** 2000 est; Dominica **23.00%** 2000 est; South Africa **24.30%** 2007 est; Mayotte **25.40%** 2005; Honduras **27.80%** 2007 est; American Samoa **29.80%** 2005; Cameroon **30.00%** 2000 est; Libya **30.00%** 2004 est; Equatorial guinea **30.00%** 1998 est; Mali **30.00%** 2004 est; World **30.00%** 2007 est; Marshall Islands **30.90%** 2000 est; Gaza Strip **34.80%** 2006;

whole. The microeconomics of a country deals on how individuals, households and firms and some nations make decisions to allocate limited resources available in the labor market. Employment and unemployment are important factors of any economy, but in opposite interdependent directions. *Permanent pools of unemployed enable employers to hire and fire workers and thus hold down costs;* the natural rate of unemployment is mainly determined by the economy's supply side, and hence production possibilities and economic institutions. The demand side of an economy which includes unemployment features in:

- *Inability to Earn Money:* Unemployed individuals are unable to earn money to meet their financial obligations, such as paying their monthly rents or monthly mortgage costs. Failure of this happening leads to homelessness, foreclosure or eviction. Susceptibility to malnutrition, illness, mental stress, loss of self-esteem, and even depression and suicides, are caused by states of unemployment. Unemployed are blacklisted in the employers' credit of debtors and never eligible for a loan.

- *Low-Income Job vs. Welfare:* In many circumstances, many low-income jobs are no real better options to unemploment with welfare, because it is difficult, if not impossible, to get unemployment insurance benefits without having worked in the past. Low-paying jobs and unemployment are more complementary rather than substitutes as low-paying jobs are short-term held either by students or by those trying to gain experience; unemployment insurance keeps an available supply of workers for the low-paying jobs, because the existence of the insurance is formulated with employers' choice of management techniques in mind: low wages and benefits, and few chances for advancement. And: the combination of low wages and benefits and unemployment insurance *feeds* and promotes the existence of *frictional* unemployment, (i.e., job-seekers transiting between jobs for better jobs or full-time jobs.)

World economies appear to be situated between the *Rising Prices Devil* and the *Escalating Poverty Deep Sea:* bulging unemployment rolls, poverty, and humiliating, inhuman indignities bred by joblessness; a shrinking world economy bred by *Financial Ramboism*; speculative energy and food price hikes and shortfalls in food commoditites; hightened environmental pollution, Climate Change and Global Warming; prolonged wars and conflicts; and high living- and death-prices. The devil side also entails the *continued* en masse production and use of fossil fuels and petrochemicals, the root cause of the world's dilemma. The deep sea side also entails the world's seeming indifference *not to discontinue* the en masse production and use of fossil fuels and petrochemicals, and *embrace* renewable energies production and consumption en masse.

1.5 Rising Environmental Pollution Costs

Macedonia **34.90%** 2007 est; Yemen **35.00%** 2003 est; Afghanistan **40.00%** 2005 est; Kenya **40.00%** 2001 est; Swaziland **40.00%** 2006 est; Nepal **42.00%** 2004 est; Kosovo **43.00%** 2001 est; Lesotho **45.00%** 2002; Bosnia/Herzegovina **45.50%** 31 Dec 2004 est; Senegal **48.00%** 2007 est; Timor-Leste **50.00%** 2001 est; Zambia **50.00%** 2002 est; Djibouti **59.00%** 2007 est; Cocos (Keeling) Islands **60.00%** 2000 est; Turkmenistan **60.00%** 2004 est; Burkina Faso **77.00%** 2004; Zimbabwe **80.00%** 2005 est; Liberia **85.00%** 2003 est; Nauru **90.00%** 2004 est. [Countries for which no information is available are not included. Office of Public Affairs (OPA), Washington, D.C. 20505.]

ENVIRONMENTAL POLLUTION and its costs are rising and escalating, not the other way round. The scar imprimatur on the world by en masse use and production of fossil fuels will forever haunt and hunt humanity and this civilisation. Much of the world's industries and the global economy itself are fossil-fuel-based. European Union (EU) leaders meeting under France's presidency on 12 December 2008, agreed on a plan for its 27-member states to cut carbon emissions 20% by 2020. The EU plan Emission Trading Scheme (ETS) includes concessions meant to ensure that hefty cost curbing pollution will not impede economic growth, particularly in the Eastern European countries. By 28 January 2009, the EU made a global call for a global carbon trading market as part of a plan to tackle climate change, that could link other carbon trading systems with its ETS by 2015. The goal is to include emerging economies by 2020.

But the costs of environmental pollution continue to rise stiffly and steeply. They influence some of the largest current global health burdens which include: approx.: *800,000 annual deaths from ambient urban air pollution; 1.2 million from road-traffic accidents; 1.9 million from physical inactivity and 1.5 million per year from indoor air pollution,* according to the World Health Organisation (WHO).[117] In their Abstract which features climate change, cities, energy, equity, health and transport, Campbell-Lendrum and Corvalan also note that "climate change is an emerging threat to global public health. It is also a highly inequitable, as the greatest risks are to the poorest populations, who have contributed least to greenhouse gas (GHG) emissions." The authors add:

- ...The rapid economic development and the concurrent urbanisation of poorer countries mean that developing cities will be both vulnerable to health hazards from climate change and, simultaneously, an increasing contributor to the problem...Common vulnerability factors include coastal locations, exposure to the urban heat-island effect, high levels outdoor and indoor air pollution, high population density, and poor sanitation. There are clear opportunities for simultaneously improving health and cutting emissions most obviously through policies related to transport systems, urban planning, building regulations and household energy supply...GHG emissions and health protection in developing country-cities are likely to become increasingly prominent in policy development..." [in view of heatwaves, floods and storms, communicable diseases, and air pollution].

- *"Heatwaves:* These result from lowered evaporative cooling, increased heat storage and sensible heat flux caused by lowered vegetation cover, increased impervious cover and complex surfaces, and possibly from heat trapping by elevated levels of locally produced carbon dioxide...

- *"Floods and Storms:* Many of the world's largest and fastest growing cities are located on the coast, and therefore vulnerable to sea-level rise. They are also exposed to the more frequent severe windstorms and floods that some studies are already linking to past and future climate change...construction patterns in many developing cities have resulted in a combination of degradation of natural protection (e.g., through deforestation and building of floodplains), poor-quality housing construction on exposed slopes, and extensive ground coverage of concrete without adequate drainage. Heavy rains, therefore, often result in intense, sometimes lethal, flash floods, such as those that occurred in and around Caracas, Venezuela in 1999 and Mumbai, India in July 2001...

---

[117] Diarmid Campbell-Lendrum/Carlos Corvalan:*"Climate Change and Developing-Country Cities: Implications For Environmental Health and Equity."* Department of Public Health & Environment, WHO, Geneva   27, Switzerland; campbellendrumd@who.int.

- *"Communicable Diseases:* Many water-borne and vector-borne infections and diseases are strongly influenced by climate conditions, and several are common within cities. The clearest example is *dengue,* the most important vector-borne virus globally. Dengue transmission has increased dramatically in tropical developing regions in the past few decades due to the weakening of vertical control programmes in many regions; this is coupled with rapid unplanned urbanization, producing breeding sites for *Aedes* mosquitoes, and high human population densities, supplying a large pool of susceptible individuals. Increasing travel has helped to spread the four different stereotypes of dengue around the world, heightening exposure multiple strains, which in turn increases the severity of clinical diseases...the distribution of *Aedes* and dengue are favored by high absolute humidity, which increases with high temperatures and rainfall...

- *"Air Pollution:* Levels of many pollutants, such as ozone, are affected by atmospheric conditions and tend to go higher on warmer days. Epidemiological evidence from developing-country cities is weak, but inferences from developed countries, suggest significant risks associated with increasing temperatures..."

In an environmental pollution costs' study by researchers Boyd and Genius in Canada, Salvadore[118] points out that most environmental pollution-caused diseases result from a combination of lifestyle, socio-economic status, environmental exposure, cultural, and genetic factors. They estimated the environmental burden of disease (EBD) in Canada, based on environmental contributions of respiratory disease, cardiovascular disease, cancer, and congenital affliction at: 10,000-25,000 deaths; 78,000-194,000 hospitalisations; 600,000-1.5 million days spent in hospital; 1.1 million – 1.8 million restricted activity days for asthma sufferers; 8,000-24,000 new cases of cancer; and 500-2,500 low birth weight babies. Air pollution was cited to cause cancer of the lungs, trachea, and bronchitis, primarily from *fine particulate matter* (PM), *benzene,* and other chemicals. The cost of adverse environmental exposure in Canada was estimated at between CAD$3.6 billion – CAD$9.1 billion each year, primarily due to respiratory disease, cardiovascular illness, cancer, and congenital affliction.

The U.S., California State, plagued by smoggy skies and rising asthma rates, on 12 December 2008, adopted the nation's toughest diesel emission standards for the trucks and buses that crowd its highways; the new diesel emission standard was aimed to reduce the state's GHG and would cost businesses, school districts and transit agencies an estimated $5.5 billion over 16 years. Nearly a million vehicles will have to be replaced or retrofitted with smog traps, filters or cleaner-burning technology beginning 2011. By 2014, all trucks must have soot filters, so that by the time the rule is fully implemented in 2023, no trucks or bus in California will be allowed to be older than 13 years unless it has equipment to cut nitrogen oxide emissions. The compliance cost is to be outweighed by an estimated $48-$69 billion in health benefits for illnesses caused by breathing diesel fumes.

Air, water, soil, noise, light, visual, and thermal pollution of the Earth and our environments caused by man-made or *anthropogenic* substances that ooze out of *chemicals, products* and *systems. Carbon dioxide* (CO_2) was the primary GHG emitted by human activities, which in the U.S., represents 85% largely from

---

[118] Lourdes Salvadore: *"Environmental Pollution Costs Billions in Illnesses Each Year."* American Chronicle, 28 November 2008.

combustion of fossil fuels. Declining *methane* emissions since 1990, were primarily sourced from *decomposition of wastes in landfills, natural gas systems and activities associated with domestic livestock.* Agricultural management and *mobile-source fossil fuel combustion* were the major sources of *nitrous oxide* emissions. *Hydrocarbons* which are substitutes for *ozone-depleting* substances are the primary component of *fluorinated gas* emissions. Emissions from *electricity* generation and *transportation* account for the second largest, *industry* for the third largest, and *residential* and *commercial* sectors, the largest contributors to carbon dioxide emissions. The other important GHG gases include: *Carbon monoxide (CO); Sulphur dioxide (SO); Chlorofluorocarbons (CFCs); Nitrogen oxides (N²Os); Photochemical ozone and Smog, Particulate matter (PM); Persistent organic pollutants (POPs* – sub-classified: PBTs, TOMPs, PAHs, BFRs, TBTs); *Volatile organic compounds (VOCs);*[119] *Herbicides, Pesticides; Radioactive wastes; Traffic over-illumination; Power lines; Mining, Solid wastes; Oil refineries, Oil rigs, Oil spills; Petrochemical plants, Pvc, Metals, Plastic factories; Heavy industries; Chlorinated hydrocarbons; Batteries; Aviation fuel; Gasoline, Trucks, Automobiles; Acid rain; Ocean acidification; Biomagnification; Cruise, Marine, Tanker, Ship, Marine debris; Genetically-engineered organisms (GMOs); and Heat.*

Global anthropogenic or man-generated greenhouse gases (GHGs) are broken down into eight different sectors for the year 2000[120] in the production of the three chief GHGs *Carbon dioxide, Methane,* and *Nitrous oxide,* thus:

- ower Stations 21.3%; *Industrial Processes* 16.8%; *Transportation Fuels* 14.0%; *Agricultural By-products* 12.5%; *Fossil Fuel retrieval, processing and distribution* 11.3%; *Residential, commercial, and other sources* 10.3%; *Land Use and biomass burning* 10.0%; and *Waste disposal and treatment* 3.4%.

- *Carbon Dioxide* is produced by: Fuel & Cement 55%; Power Stations 29.5%; Industrial Processes 20.6%; Transportation Fuels 19.2% ; Land Use Change & Forestry 19% (EPA); Residential, Commercial, and other sources 12.9%; Land Use and biomasss burning 9.1%; Fossil Fuel retrieval processing and distribution 8.4%. [Carbon dioxide enters the atmosphere through the burning of fossil fuels (oil, natural gas, coal), solid waste, trees and wood products, and also as a result of other chemical reactions (e.g., manufacture of cement).]

- *Methane* is produced by: Agricultural By-products 40.0%; Transportation Fuels 29.6%; Waste Disposal & Treatment 18.1%; Land Use and biomass burning 6.6%; Residential, Commercial, and other sources 4.8%. [Methane is emitted during the

---

[119] Volatile organic compounds (VOCs): Are organic chemicals with high vapor pressures, such as the wide range of carbon-based molecules as ketones, aldehydes and other light hydrocarbons; methane gas from wetlands; ruminants, such as cows, pigs and cattle; energy use; rice agriculture; landfills; burning biomass such as wood/natural gas composed mainly of methane; trees; a common source of VOCs is methane gas from wetlands, ruminants such as cows, energy use, rice fields, landfills, burning of biomass such as wood or trees or natural gas composed mainly of methane; Common VOC Products include: Paint thinners; dry cleaning solvents; semiconductor cleaners; constituents of petroleum fuels such as gasoline and natural gas; crude oil tanking releases before and after onloading and offloading crude oil in tankers; odors from oxidizing photocopiers, or carpets and furnishings and hundreds of office components when they oxidize formaldehyde present in laminated furniture, shelving, wallcover, and evaporates from painsts, varnishes and chemicals used for sealing/finishing walls; tobacco smoke contributes to high levels of VOCs; halogenide/sulfide emission through human respiration, and formaldehyde emitted at a lower rate from the surface of the human body; land previously used for industrial purposes.

[120] U.S. Department of Commerce/National Oceanic & Atmospheric Administration (NOAA) Research, Earth System Research Laboratory, Global Monitoring Division (GMD): *"Annual GHGs by Sector."*; U.S. Environmental Protection Agency (EPA); International Energy Agency (IEA); Carbon Dioxide Information Analysis Center (CDIAC) at Oak Ridge National laboratory; World Resources Institute (WRI) – provide CO² emission reports as well.

production and transport of coal, natural gas, oil; from livestock and other agricultural practices, and by decaying organic waste in municipal waste landfills.]

- *Nitrous Oxide* is produced by: Agricultural By-products 62.0%; Land Use and biomass burning 5.9%; Industrial Processes 26.0%; Waste Disposal & Treatment 2.3%; Residential, Commercial and other sources 1.5%; Transportation Fuels 1.1%. Nitrogen oxides include: Nitric oxide (NO), Nitrogen (II) oxide; Nitrogen (III) oxide; Nitrogen dioxide (NO^2), Nitrogen (IV) oxide; Nitrous oxide (N^2O), Nitrogen oxide (I); Dinitrogen trioxide (N^2O^3), nitrogen (IV) oxide; Dinitrogen tretoxide (N^2O4), nitrogen (IV) oxide; Dinitrogen pentoxide (N^2O5), nitrogen (V) oxide. [Nitrous oxide is emitted during agricultural and industrial activities, as well as during combustion of fossil fuels and solid waste.]

- *Sulphur Oxides Pollutants* Include: Lower sulphur oxides (SnO, $S7O^2$ and $S6O^2$); Sulphur monoxide (SO); Sulphur dioxide (SO^2); Sulphur trioxide (SO^3); Higher sulphur oxides (SO^3+x).

- *Particulate Matter, Particulates, Fine Particles Pollutants:* Are tiny particles or liquid suspended in a gas produced by burning of fossil fuels in vehicles, power plants, and various industrial processes for aerosols; wind blown dust from construction sites, other land areas devoid of water or vegetation, mineral dust, etc.; Aerosol is a combination of the gas and particles.

- *Fluorinated Gases: Hydrocarbons, perfluorocarbons,* and *sulphur hexafluoride* (SF6) are synthetic, powerful greenhouse gases that are emitted from a variety of industrial processes, according to the U.S., Environmental Protection Agency (EPA)[121]. Fluorinated gases sometimes used as *substitutes* for ozone-depleting substances (i.e., FCs, HCFCs, and halons) – are released as by-products of industrial operations and are extremely potential GHGs with very long atmospheric lifetimes. These gases are typically emitted in smaller quantities but because they are potent greenhouse gases, they are sometimes referred to as *High Global Warming Potential* gases (High GWP gases).

- *Sources of High Global Warming Potential* (GWP) *Gases:* The sources for these gases include: HCFC-22 production; electric utilities; magnesium; aluminum; and semiconductor manufacturing. Other sources are: substitutes for ozone-depleting substances (ODSs) used for refrigeration and air-conditioning, solvents, foams, aerosols, and fire extinguishing. Cost-effective opportunities for reducing or mitigating emissions of these include reusing or recycling the gas, reducing use of the gas, or using the gas more effectively. The EPA has since February 2002, launched a Voluntary Programme[122] partnership between it and industry aimed at

---

[121] U.S. Environmental Protection Agency (EPA): *"EPA Greenhouse Gas Emissions: 2008 Inventory of Greenhouse Gas Emissions & Sinks Report."* Prepared annually by the EPA since 1990 through 2005, which also discusses the methods and data usded to calculate the emission estimates.

[122] EPA Voluntary Programme: A set of voluntary partnership between the EPA and industry for reduction of high global warming potential (high GWP) gases. U.S. Companies participating in the Voluntary Programme Pleadging to Reduce GHG Emissions nationally and globally, include those with *Reduction Targets;* and *Without Reduction Targets as yet.* With GHG Emission Reduction Targets: *3 Degrees* / Marketing / CA / to achieve net zero by 2007 and maintain that level through 2012; *3M* / Manufacturing / MN/ achieved its initial total goal by 60% from 2002-2007; *Abbot* / Health Services / IL / to reduce total of 2% from 2006-2011; *ACE Group of Companies* / Insurance / PA / to reduce 8% per employee from 2006-2012; *Advanced Micro Devices* (AMD) Inc. / to reduce 33% per manufacturing index from 2006-2010; achieved its initial 53% goal reduction per manufacturing index from 2002-2006; *Agilent Technologies* / Hardware Manufacturing / CA / to reduce total global GHG emissions by 10% from 2006-2011; Charter Partner *Alcoa Inc* / Aluminum Manufacturing / PA / to reduce by 45 from 2008-2013; *American Power* / Utilities / OH / to reduce by 6% from 2001-2010; achieved its initial total U.S. goal by 4% from 2001-2006;

reducing the release of GWPs and its extremely potent GHG by-products with very

American Package Companies, Inc. / Manufacturing / MO / to reduce total U.S. emissions by 5% from 2005-2010; *Applied Materials, Inc.* / CA / to reduce total GHD emissions by 20% from 2006-2012; *Anheuser-Busch Companies, Inc.* / to reduce 5% from 2005-2010; [Charter Partner] *Ball Corporation* / Manufacturing / CO / to reduce by 16% per production index from 2002-2012; *Baltimore Aircoil Company* / Manufacturing / MD / to reduce by 15% per ton of steel processed from 2004-2009; *Bank of America Corporation* / Financial Services / NC / to reduce by 9% from 2004-2009; [Charter Partner] *Baxter International Inc.* / Health Services / IL / to reduce 5% from 2005-2012; achieved its initial goal of 2.7% per unit production value from 2000-2005; *Best Buy Co., Inc.* / Retail / t / MN / to reduce by 8% per square ft from 2005-2012; *Boise Paper* / Manufacturing / ID / to reduce by 10% from 2004-2014; *Burt's Bees, Inc.* / Manufacturing / NC / to reduce by 35% dollar sales from 2006-2011; *Calpine* / Utilities / CA / to reduce by 4% per megawatt hour from 2003-2008; *CalPortland Company* / Cement / CA / to reduce by 9% from2003-2012; *Campbell Soup Company* / Manufacturing / NJ / to reduce by 12% per adjusted case of product from 2005-2010; [Charter Partner] *Casella WasteSystems, Inc.* / Waste Management / VT / to reduce by 10% from 2005-2012; *Caterpillar Inc.* / Manufacturing / IL / to reduce by 3% from 2006-2015; achieved its initial goal to reduce by 28% per dollar from 2002-2006; *Cherokee Investment Partners* / Investment Services / NC / to reduce net zero by 2007 and maintain that level through 2011; *Cisco Systems, Inc.* / Telecommunications / CA / to reduce by 25% from 2007-2012; *CitigroupInc.* / Financial Services / NY / to reduce by 10% from2005-2011; *Codding Enterprises* / Real Estate / CA / to reduce by 50% per sq ft from 2005-2010; *Conservation Services Group* / Consulting / MA / to achieve net zero by 2006 and maintain that level through 2010; *Coors Brewing Company* / Manufacturing / CO / to reduce by 12% per production index from 2005-2010; *Cummin Inc.* / Manufacturing / IN / to reduce by 25% per dollar revenue from 2005-2010; *Deere & Company* / Manufacturing / IL / to reduce by 25% per dollar revenue from 2005-2014; *Dell Inc.* / Manufacturing / TX / to reduce by 15% per dollar revenue from 2007-2012; achieved its initial net zero global GHG emissions by 2008, and maintains that level through 2012; *DPR Construction, In.* / Construction / CA / to reduce by 25% per employee from 2007-2015; *DuPont Company* / Chemical / DE / to reduce by 15% from 2004-2015; *EarthColor* / Printing / NJ / to reduce by 40% per dollar sales from 2006-2012; Charter Partner *Eastman Kodak Company* / Manufacturing / NY / to reduce by 10% from 2002-2008; *EcoPrint* / Printing / MD / to achieve net zero by 2006, and maintain that level through 2010; *Entergy Corporation* / Utilities / LA / to reduce by 20% from 2000-2010; *Exelon Corporation* / Utilities / IL / to reduce by 8% from 2001-2008; *Fairchild Semiconductor* / Semiconductor / ME / to reduce by 30% permanent ring index from 2003-2010; *First Environment, Inc.* / Consulting / NJ / to achieve net zero by 2008; [Charter Partner] *FPL Group, Inc.* / Utilities / FL / to reduce by 21% per kWh from 2001-2007; *Frito-Lay, Inc.* / Food Services / TX / to reduce by 14% per pound of production from 2002-2010; *Gap Inc.* / Retail / CA / to reduce by 11% per sq ft from 2003-2008; *General Electric Company* / Manufacturing / CT / to reduce by 1% from 2004-2012; [Charter Partner] *General Motors Corporation* / Automative / MI / to reduce by 40% from 2000-2010; achieved its initial goal by reducing total North American GHG emissions by 23% from 2000-2005; *Green Mountain Energy Company* / Marketing / TX / to reduce by net zero by 2005, and maintain that level through 2009; [Charter Partner] *Hasbor, Inc.* / Manufacturing / RI / to reduce by 43% from 2004-2007; *Haworth, Inc.* / Manufacturing / MI / to reduce by 20% per dollar sales from 2004-2009; [Charter Partner] *Holcim (US)* / Cement Manufacturing / MI / to reduce by 12% per ton of cement from 2000-2008; *HSBC-North America* / Financial Services / IL / to reduce by 10% from 2005-2010; [Charter Partner] *IBM Corportation* / Hardware manufacturing / NY / to reduce by 7% from 2005-2010; achieved its initial goal by reducing total global energy-related GHG emissions by an average of 6% per year and PFC emissions by 58% from 2000-2005; *Intel Corporation* / Semiconductor / CA / to reduce by 30% per production unit from 2004-2010; [Charter Partner] *Interface, Inc.* / Manufacturing / GA / to reduce by 15% per unit of production from 2001-2010; [Charter Partner] *International Paper* / manufacturing / CT / to reduce by 15% from 2000-2010; [Charter Partner] *Johnson & Johnson* / Health Services / NJ / to reduce by 14% from 2001-2010; *Lincus, Incorporated* / Consulting / AZ / to reduce by 30% per sq ft from 2006-2010; [Charter Partner] *Lockheed Martin Corp.* / Engineering / MD / to reduce by 30% per dollar revenue from 2001-2010; *LSI Corporation* / Information / CA / to reduce by 15% from 2007-2012; *Mack Trucks, Inc.(MTI)* / Automative / PA / to reduce 12% per unit produced from 2007-2012; achieved its initial goal by 32% per unit produced from 2003-2007; *Marriot International, Inc.* / Hotel Services / DC / to reduce by 6% per available room from 2004-2010; *Melaver, Inc.* / Real Estate / GA / to achieve net zero by 2006 and maintain that level through 2009; *Merck & Co., Inc.* / Pharmaceutical / NJ / to reduce by 12% from 2004-2012; [Charter Partner] *Miller Brewing Company* / Manufacturing / WI / to reduce by 18% per barrel of production from 2001-2006; *Millipore Corporation* / Pharmaceutical 7 MA / to reduce by 20% from 2006-2011; [Charter Partner] *National Renewable Energy Laboratory* (NREL) / Federal Gov't / to reduce by 75% from 2005-2009; achieved its initial goal of 10% per sq ft from 2000-2005; *NVIDIA Corporation* / Semiconductor / CA / to reduce by 9% per sq ft on non-data centre space, and 9% in RUE for data centres from 2007-20012; *North Bay Construction* / Construction / / CA / to reduce by 20% from 2005-2010; *Oracle Corporation* / Software Manufacturing / CA / to reduce by 65 per sq ft from 2003-2010 for all non-data centre space, and to purchase 5% green power for data centres; *Owens Corning* / Manufacturing / OH / to reduce by 25% per unit of production from 2006-2012; *PepsiCo* / Manufacturing / NY / to reduce by 25% per ton of production from 2006-2015; *Petaluma Poultry* / Food Services / CA / to reduce by 20% from 2006-2011; [Charter Partner] *Pfizer Inc.* / Pharmaceutical / NY / to reduce by 20% from 2007-2012; achieved its initial goal of 43% per million dollar of revenue from 2000-2007; *PPG Industries, Inc.* / Manufacturing / PA / to reduce by 10% from 2006-2011; [Charter Partner] *Public Service Enterprise Group* (PSEG) / Utilities / NJ / to reduce by 18% per kWh from2000-2008; *Quad/Ggraphics Inc.* / Printing / WI / to reduce by 25% per page printed from 2003-2013; [Charter Partner] *Raytheon Company* / Engineering / MA / to reduce 33% per dollar revenue from 2002-2009; *Roche Group U.S. Affiliates* / Health Services / Switzerland / to reduce by 15% from 2001-2010; achieved its initial goal by 11% from 2001-2006; *Sandy Alexander* / Printing / / NJ / to reduce by 11% per dollar revenue from 2006-2012; [Charter Partner] *SC Johnson* / Manufacturing / WI / to reduce by 8% from 2005-2010; achieved its initial goal of 17% from 2000-2005; [Charter Partner] *Shaklee Corporation* / Consumer Products / CA / to reduce by net zero from 2006-2009; *Sonoma Wine Company* / Food Processing / CA / to reduce by 15% from 2005-2010; *Sprint* / Telecommunications / VA / to reduce by 15% from 2007-2017; [Charter Partner] *St-Lawrence Cement* / Cement / Quebec / to reduce by 2% per ton of cementitious product from 2000-2012; achieved its initial goal of 16% per ton of cementitious product from 2000-2006; [Charter Partner] *Staples,Inc.* / Retail / MA / to reduce by 7% from 2001-2010; *Steelcase, Inc.* / Manufacturing / MI / to reduce by 40% per dollar sales from 2004-2009; *Sterling Planet, Inc.* / Manufacturing / GA / to reduce to net zero GHG emissions by 2006 and maintain that level through 2010; [Charter Partner] *STMicroelectronics* / Semiconductor / TX / to reduce by 50% per manufacturing unit from 2000-2010; [Charter Partner] *Sun Microsystems, Inc.* / Software Manufacturing / CA / to reduce by 20% from 2007-2012; achieved its initial goal of 23% from 2002-2007; *Tetra Tech EM inc.* / Consulting / CA / to reduce by 20% from 2006-2011; *The Boeing Company* / Aerospace / IL / to reduce by 1% from 2007-2012; *The Collins Companies* / Manufacturing / OR / to reduce by 18% from 2000-2010; *The Tower Companies* / Real estate / MD / to reduce by net zero by 2008 and maintain that level through 2012; *The World Bank* / Other / DC / to reduce by 7% per employee from 2006-2011; *Thomas Rutherfoorf, Inc.* / Insurance / VA / to reduce by 7% per employee from 2006-2012; *Travelers Companies* / Insurance / MN / to reduce by 7% from 2006-2011; *Turner Construction Company* / Construction / NY / to reduce by 5% from 2006-2011; [Charter Partner] *Unilever* / Manufacturing / NJ / to reduce by 25% per ton of production from 2004-2012; *United Technologies Corporation* / Manufacturing / CT / to reduce by 12% from 2006-2010; achieved its initial goal of GHG emission reduction of 43% per dollar revenue from 2001-

long atmospheric lifetimes. The following are places in the world known to be the most polluted.[123]

2006; *Volvo Trucks North America, inc.* (VTNA) / Automative / NC / to reduce 20% per truck produced from 2003-2010; *Xeros Corporation* / Manufacturing / CT / to reduce by 20% from 2002-2012; achieved its initial goal by reducing total global GHG emissions by 18% from 2002-2006; <u>GHG Emission Reduction Targets Under Development:</u> *Aggregate Industries* / Construction / MD; *Air Products & Chemical, Inc.* / Chemical / PA; Charter Partner *Alcan Aluminum Corporation* / Aluminum Manufacturing / OH; *Acticor Inc.* / Manufacturing / MI; *American Packaging Corporation* / manufacturing / NY; *American Water* / Utilities / NJ; *Ash Grove Cement Co.* / Cement / KS; *Autodesk* / Software manufacturing / CA; *Belkin* / Manufacturing / CA; *Benziger Family Winery* / Manufacturing / CA; *Berry Plastics Corp.* / Manufacturing / IN; *Best Worldwide Chaufeured Transportation* / Transportation / CA; *Bluebonnet Electric Cooperative* / Utilities / TX; *Boise Cascade* / Manufacturing / ID; *California Limousine* / Transportion / CA; *Capital One Financial Corporation* / Financial Services / CA; *Carlisle Construction Materials* / Manufacturing / ME; *Classique Limousines* / Transportation / CA; *Clements Environmental* / Consulting / CA; *Coca-Cola Enterprises* / Manufacturing / GA; *Company Car 6 Limousine* / Transportation / OH; *ConAgra Foods* / Food Services / NE; *Conestoga-Rovers & Associates* / Construction / NY; *Continuum* / Consulting / MA; *CSX Transportation, Inc.* / Transportation / FL; *Cytec Industries, Inc.* / Chemical / NJ; *DuBois Chemicals Inc.* / Chemical / OH; *EcoLab Inc.* / Chemical / MN; *EMC Corporation* / Software manufacturing / NJ; *EmpireCLC* / Transportation / NJ; *Evelyn Hill Inc.* / NY; *Fetter Printing Co.* / Printing / KY; [Charter Partner] *Tetzer Vineyards* / Agriculture / CA; *Fleet Transportation, LLC* / Transportation / VA; *Freescale Semiconductor, Inc.* / Semiconductor / TX; *FXFOWLE Architects, PC* / architectural Services / NY; *Genesis Microchip Inc.* / Semiconductor / CA; *Greenworth Financial* / Financial Services / VA; *Genzyme Corporation* / Biotechnology / MA; *Grand Canyon North Rim, LLC* / Travel–Tourism /AZ; *Greenstar North America* / Waste Management / TX; *GXS* / Consulting / MD; *Harbec Plastics, Inc.* / Manufacturing / CA; *Harrah's Entertainment, Inc.* / Entertainment / NY; *Honeywell Inc.* / Manufacturing / NJ; *Hormel Foods Corp.* / Food Processing / MN; *Invitrogen Corporation* / Biotechnology / CA; *Jackson National Life Insurance Company* / Financial Services / MI; *Kellog Company* / Manufacturing / MI; *Kimberly-Clark Corporation* / Health Services / TX; *Kohl's Department Stores* / Retail / WI; *Kroenke Sports Enterprises* / Entertainment / CO; *L.L. Bean, Inc.* / Retail / ME; *Lafarge North America, Inc.* / Cement / CA; *Lexmark International, Inc.* / Printing / KY; *Limited Brands, Inc.* / Retail / OH; *Lucent Technologies, Inc.* / Communications / NJ; *Luxury Limousines of Sacramento* / Transportation / CA; *Mantria Corporation* / Real estate / PA; *Mahawk Fine Papers Inc.* / Manufacturing / NY; *Mouadnock Paper Mills, Inc.* / Paper Goods/ NH; *MOSAIC* / Printing / MD; *MTC Limousine & Corporate Coach, Inc.* / Transportation / NY; *MWH Global, Inc.* / Engineering / CO; *National Geographic Society* / NGO / DC; *Navistar, Inc.* / Manufacturing / IL; *NCR Corporation* / Manufacturing / OH; *News Corporation* / Communications / NY; *Niagara Conservation* / Manufacturing / NJ; *Nicholas Earth Printing, LLC* / Printing / TX; *NiSource Inc.* / Utilities / IN; *Nortel* / Communications / NC; *Norvartis Corporation* / Pharmaceutical / NJ; *Novelis Corporation* / Aluminum Manufacturing / OH; *Office Depot* / Retail / FL; *One Boston Place LLC* / Real estate / MA; *OSRAM SYLVANA* / Manufacturing / MA; *Pall Corporation* / Manufacturing / MI; *Partner's Executive Transportation* / Transportation / NY; *PHH Arval* / trucking / MD; *Pizza Fusion* / Food Services / FL; *Potomac-Hudson Engineering, Inc.* / Engineering / ME; *PrintFast, LLC* / Printing / NJ; *Progressive Environmental & Safety* / Consulting / KS; *Publix Super Markets,Inc.* / Retail / FL; *Pure & Gentle Soap* / Consumer Products / TX; *Pure Luxury Transportation* / Transportation (CA; *Puronics, Incorporated* / Manufacturing / CA; *Ram Offset* / Printing / OR; *Random House, Inc.* / Publishing / NY; *Rizco Design* / Marketing / NJ; *Rockwell Automation* / Manufacturing / WI; *Rockwell Collins, Inc.* / Engineering / IA; *Royal Coachman Worldwide* / Transportation / CT; *Schering-Plough Corp.* / none / NJ; *Scout real Estate Capital, LLC* / Real estate / MA; *Sid Richardson Carbon & Energy Company* / Manufacturing / TX; *SKF USA Inc.* / Manufacturing / PA; *Smithfield Foods, Inc.* / Food Services / VA; *STERIS Corporation* / Health Services / OH; *Stora Enso North America Corp* / Manufacturing / WI; *Syngenta* / Agriculture / Switzerland; [Charter Partner] *Target Corporation* / Retail / MN; *Tate Access Floors* / Manufacturing / MD; *Teneco* / Automative / IL; *Teradata Corporation* / Manufacturing / OH; *The Clorox Company* / Manufacturing / CA; *The Dow Chemical Company* / Chemical / MI; *The Estee Lauder Companies, Inc.* / Manufacturing / NY; *The Hartford* / Insurance / CT; *The Inter-American Development Bank* / Other / DC; *The Mosaic Company* / Chemical 7 FL; *The Sherwin-Williams Co.* / Manufacturing / OH; *The Tidewater Group* / Consulting / ME; *Tiffany & Co.* / Retail / NY; *Trane* / manufacturing / NJ; *True manufacturing Co., Inc.* / Manufacturing / MO; *Tua Latin Valley Water District* / Utilities / OR; *Tyson Foods, Inc.* / Food Services / AR; *U.S. Forest Service* / Federal Gov't 7 DC; [Charter Partner] *U.S. Steel Corporation* / Manufacturing / PA; *Union bank of California* / Banking / CA; *University Corporation for Atmospheric Research* / NGO / CO; *UPS,Inc.* / Shipping / GA; *VF Outdoor, Inc.* / Retail / CA; *Virgin America* / Air Transportation / CA; *Wafertech L.L.C.* / Semiconductor / NA; *Wells Fargo* / Banking / CA; *Western States Envelope Co.* / manufacturing / WI; *WaterWave Foods Co.* / manufacturing / CO; *Whole Foods Market* / Food Services / TX; *Wilton Armetale* / manufacturing / PA; *Yahoo Inc.* / Networking / CA.

[123] Bryan Walsh:*"The World's Most Polluted Places."* 12 September 2007, TIME/CNN: <u>Linfen; China:</u> No. of people potetially affected – *3,000,000*; Type of pollutant – *Coal and Particulates*; Source of Pollution – *Automobile and Industrial Emissions*; This soot-blackened city in China's inland Shanxi province makes Dickensian London look as pristine as a nature park. Shanxi is the heart of China's coal belt, and the hills around Linfen are dotted with burning coal. Don't bother hanging your laundry – it'll turn black before it dries. China's State Environmental Protection Agency says that Linfen has the worst air in the country, which is saying something, considering that the World Bank has reported that 16 of the 20 most polluted cities in the world are Chinese. One Linfen native summed up the city's plight to a TIME reporter last year [2007]: "This place of ours is not good." <u>Tianying, China:</u> No. of eople potentially affected – *140,000*; Type of pollutant – *Lead and other Heavy Metals*; Source of pollution – *Mining and Processing;* An industrial city – though China doesn't really have any other kind – in the country's northeastern belt, Tianying accounts for over half if China's lead production. Thanks to poor technology and worse regulation, much of that toxic metal ends up in Tianying's soil and water, and then in the bloodstream of its children, where it can cause lowered IQ. Wheat has been found to contain lead levels up to 24 times Chinese standards, which are even more stringent than U.S., restrictions on lead. "China has a commitment to environmental protection, but it also has a commitment to industry," says Fuller.It's a constant push that's mostly won by industry."; <u>Sukinda, India:</u> No. of people potentially affected – *2,600,000*; Type of pollutant – *Hexavalent chromiumand other metals*; Source of pollution – *Chromite Mines and Processing*; If you watched Erin Brockovich, then you know what hexavalent chromium is: a nasty heavy metal used for stainless steel production and leather tanning that is *carcinogenic* if inhaled or ingested. In Sukinda, which contains one of the largest open cast chromium ore mines in the world, 60% of the drinking water contains hexavalent chromium at levels more than double international standards. An Indian health group estimates that 84.75% of deaths in the mining areas – where regulations are nonexistent – are due to chromite-related diseases. There has been virtually no attempt to clean up the contamination; <u>Vapi, India:</u> No. of people potentially affected – *71,000*; Type of pollutant – *Chemicals and heavy Metals*; Source of pollution – *Industrial Estates*; If India's environment is on the whole healthier than its giant neighbor's China, that's because India is developing much mire slowly. But that's changing, starting in towns like Vapi, which sits at the southern end of a 400-km-long belt of industrial estates. For the citizens of Vapi, the cost of

• *Pharmaceutical Pollution:* Pharmaceutical contamination in rivers, lakes and streams is an emerging concern worldwide, that could be harmful and lead to the *proliferation* of *drug-resitant* bacteria. When researchers analysed vials of treated wastewater from a plant in Patancheru, India, where about *90 Indian drug factories* dump their residues, they were shocked. Enough of a single, powerful antibiotic, *Ciprofloxacin,* was being spewed into one stream each day to treat every person in a city of 90,000.[124] Ciprofloxacin was once considered a *powerful* antibiotic of last resort, used to treat especially tenacious infections, but in recent years, many bacteria *have developed resistance to the drug* – leaving it *insignificantly* less effective. Antibiotic Ciprofloxacin and the popular *antihistamine cetirizine,* with the highest levels in the wells of six Indian villages tested, were confirmed in the study conducted by Joakim Larsson, an environmental scientist at the University of Gothenburg in Sweden. Larsson presented his study results at a U.S., scientific conference in November 2008. Not only ciprofloxacin was detected, but also a "soup of 21 different active pharmaceutical ingredients, used in generics for treatment of hypertension, heart disease, chronic liver ailments, depression, gonorrhea, ulcers and other ailments. Half of the drugs measured at the highest levels of pharmaceuticals ever detected in the environment, researchers say," wrote Mason. Associated Press (AP) series of articles in 2008, documenting the commonplace presence of *minute concentrations of pharmaceuticals* being almost ubiquitous in rivers, lakes and streams shows that: the medicines taken by patients are excreted without being fully metabolised; hospitals and long-term care facilities annually flush millions of pounds of unused pills down the drain; human cells fail to grow normally in the laboratory

growth has been severe: levels of mercury in the city's groundwater are reportedly 96 times higher than WHO safety levels, and heavy metals are present in the air and the local produce. "It's a disaster," says Fuller.; La Oroya, Peru: No. of people potentially affected – *35,000*; Type of pollutant – *Lead, Copper, Zinc and Sulphur Dioxide*; Source of pollution – *Heavy Metal Mining and Processing*; Lead is the contaminant that shows up most frequently on Blacksmith's list because the toll it takes on children can be so devastating. In La Oroya, a mining town in the Peruvian Andes, 99% of children have blood levels that exceed acceptable limits, thanks to an American-owned smelter that has been pilluting the city since 1922. The average lead level, according to a 1999 survey, was triple the WHO limit. Even after active emissions fromsmelter are reduced, the extended lead will remain in La Oroya for centuries – and there's currently no plan to clean it up; Dzerzhinsk, Russia: No. of people potentially affected – *300,000*; Type of pollutants – *Chemicals and Toxic Byproducts, incl. Sarin and VX Gas*; Source of pollution – *Cold War-era Chemical Weapons*; The legacy of Cold War weapons programmes has left environmental blackspots throughout the former Soviet Union, but Dzerzhinsk is by far the worst. The city's own environmental agency estimates that almost *300,000 tons of chemical waste* – including some of the most dangerous neurotoxins known to man – were improperly dumped in Dzerzhinsk between 1930 and 1998.Part of the city's water are infected with dioxins and phenol at levels that are reportedly 17 million times the safe limit. The Guinness Book of World Records named Dzerzhinsk the most chemically polluted city on Earth, and in 2003 its death rate exceeded its birth rate by 260%; Norilsk, Russia: No. of people potentially affected – *134,000*; Type of pollutant – *Air Pollution – Particulates, Sulphur dioxide, Heavy Metals, Phenols*; Source of pollution – *Major nickel and Metal Mining and Processing*; Norilsk was founded in 1935 as a Siberian slave labor camp, and life there has pretty much gone downhill since. Home to the world's largest heavy metal smelting complex, more than 4 million tons of cadmium, copper, lead, nickel, arsenic, selenium, and zinc are released into the air every year. Air samples exceed themaximum allowance for both copper and nickel, and mortality from respiratory diseases is much higher than in Russia as a whole. "Within 30 miles (48 km) of the nickel smelter there's not a single living tree," says Fuller."It's just a wasteland."; Chernobyl, Ukraine: No. of people potentially affected – *5.5 million*, currently disputed; Type of pollutant – *Radiation*; Source of pollution – *Nuclear Meltdown*; When Chernobyl melted down on 26 April 1986, the ruined plant released 100 times more radiation into the air than the fallout from the nuclear bombs at Hiroshima and Nagasaki. Today the 19-miles (30-km) exclusion zone around the planet remains uninhabited, and between 1922 and 2002 more than 4,000 cases of *thyroid cancer* cases were diagnosed among Russian, Ukrainian and Belarussian children living in the fallout zone. "It's the largest industrialaccident in the world," says Fuller. "It'll be contaminated for tens of thousands of years." Fortunately, work is being done to prevent further radiation spill from the ruined sarcophagus of the nuclear plant; Sumgayit, Aserbaijan: No. of people potentially affected – *275,000*; Type of pollutant – *Organic Chemicals, Oil and Heavy Metals*; Source of pollution – *Petrochemical and Industrial Complexes*; Another legacy of the Soviet Union's utter disregard for the environment – Stalin once boasted that he could correct nature's mistakes – Sumgayit's many factories, while they were operational, released as much as 120,000 tons of harmful emissions, including mercury, into the air every year. Most of the factories have been shutdown, but the pollutants remain – and no one is stepping up to take responsibility for them. "It's a huge, abandoned industrial wasteland," says Fuller; Kabwe, Zambia: No. of people potentially affected – *255,000*; Type of pollutant – *Lead and Cadmium*; Source of pollution – *Lead Mining and Processing*; When rich deposits of lead were discovered near Kabwe in 1902, Zambia was a British colony called Northern Rhodesia, and little concern was given for the impact that the toxic metal might have on native Zambians. Sadly, there's been almost no improvement in the decades since, and though the mines and smelter are no longer operating, lead concentrations in children are five to 10 times the permissible U.S., EPA levels, and can even be high enough to kill." We did blood tests on some of these kids, and they were literally broke our machines," says Fuller. "There is a long, nasty history here." But there's also a bit of hope: the World Bank has recently allocated US$40 million for a clean-up project.

[124] Margie Mason, Associated Press (AP) Reporter: *"World's Highest Drug Level Found in India Stream."* Sunday, 25 January 2009, quoted by TIME/CNN: World..

when exposed to trace concentrations of certain pharmaceuticals; some water-borne drugs also promote antibiotic-resistant germs, especially when they are mixed with bacteria in human sewage; extremely diluted concentrations of drug residues harm the reproductive systems of fish, frogs and other aquatic species in the wild; the discharge from wastewater treatment facility is spawning drug resistance, and the entire biological food web could be affected; the more bacteria is exposed to a drug, the more likely that bacteria will mutate in a way that renders the drug *ineffective*, and such resistant bacteria can then possibly infect others who spread the bugs as they travel; high drug concentrations in lakes upstream from the treatment plant in Patancheru, indicating potential *illegal dumping*.

The participants of the voluntary programme, known as Climate Leaders Partners (CLP), pledge to commit to develop a corporate-wide GHG inventory including all emission sources of the six major GHGs – CO^2, N^2O, HFCs, PFCs, SF6 – using the CL HG Inventory Management Plan; Set an aggressive corporate-wide GHG emissions reduction goal to be achieved over the next 5-10 years; Annually report inventory data and document progress towards their reduction goal; Publicise their participation, reduction goal, and accomplishments achieved through the programme.

The high GWP partnership programme involves several industries, including *HCFC-22 producers, primary aluminum smelters, semiconductor manufacturers, electric power companies,* and *magnesium smelters* and *die-casters.* Benefits of the Programme include: Reduces impacts on the global environment; Develops a credible GHG inventory, an Inventory Management Plan, and sets an aggressive, corporate-wide long-term GHG reduction goal; Better manages GHG emissions and associated risks; Realises cost-savings through energy efficiency; Receives expert EPA technical assistance on inventories and reporting; Partners participate in national public recognition campaigns, and engage with other partner companies demonstrating climate leadership; Accesses the latest GHG tools, technologies, and protocols; Integrates climate change strategies with state, regional, and international GHG accounting schemes.

The voluntary programmes which has so far been carried through *Electric Power, Aluminum* (VAIP), *Magnesium,* and *Semiconductor,* seem to have had a significant difference of improvement:

- *mission Reduction Partnership for Electric Power Systems* (ERPEPS): The SF6 Emission Reduction Partnership for Electric Power Systems (ERPEPS) is a collaborative effort between the EPA and the electric power industry to identify and implement cost-cutting solutions to reduce *sulphur hexafluoride* (SF6) emissions. SF6 is a highly potent GHG used in the industry for insulation and current interruption in electric transmission and distribution equipment. Currently 81 utilities participate in this voluntary programme. Accomplishments include: starting with 49 Charter Partners in 1999, the Partnership now totals 77 Partners, representing 45% of the industry. Since the number of reporting Parties varies from year to year, the SF6 emission rate provides a valuable assessment of Partnership trends because it normalises SF6 emission relative to the total amount of SF6 containing electrical equipment (calculated by dividing total emissions by total nameplate capacity). SF6 emission rate has dropped from 17% in 1999 to 6.5% in 2006, and is projected to drop to 0% by 2010.

- *Voluntary Aluminum Industrial Partnership* (VAIP): This is an innovative pollution prevention programme developed jointly by the EPA and the primary aluminum industry. Participating companies work with EPA to improve aluminum production efficiency, while reducing *perfluorocarbon* (PFC) emissions, potent GHGs that remain in the atmosphere for thousands of years. Accomplishments include: In 2004, emissions were 2.18 million metric tons carbon equivalent (MMT CE) less than in 1990. This reduction is equivalent to eliminating emissions from over *1.5 million cars* (assuming an average of 11,450 lbs CO^2/car/year) in 2004 alone, *plus* the environment receives this benefit annually for over 10,000 year atmospheric lifetimes of PFCs.

- *Emission Reduction Partnership for the Magnesium Industry* (ERPMI)*:* The SF6 Emission Reduction Partnership for the Magnesium Industry (ERPMI) is a cooperative effort between EPA and the U.S., magnesium industry to better understand and reduce emissions of SF6, a potent GHG from magnesium production and casting processes. In February 2003, EPA's Partners and the International Magnesium Association (IMA), committed to eliminate SF6 emissions by year-end 2010. This is facilitating remarkable progress towards eliminating SF6 emissions by identifying, evaluating and implementing cost-effective climate protection strategies and technologies, including alternative cover gases. Accomplishments include: The Partners continue to improve their operational efficiencies and environmental performance by optimising SF6 cover gas concentrations, flow rates, and delivery mechanisms, as well as actively seeking and repairing leaks in the gas distribution systems. Growing demand for lightweight magnesium automative parts and hand-held electronic devices (e.g., 3 C's: computers, communications, cameras) supports strong growth in the magnesium casting sector. The growing demand for primary magnesium alloys is expected to be met by U.S., Magnesium as well as international suppliers.

- *PFC Reduction/Climate Partnership for the Semiconductor Industry* (PFC R/CPSI)*:* PFC Reduction/Climate Partnership for the Semiconductor Industry (PFC R/CPSI): Supports the industry's voluntary efforts to reduce high global warming potential (GWP) GHG-emissions by following a pollution prevention strategy. The GHG emissions of primary concern are *perfluorocarbons, trifluoromethane* (CHF^3), *nitrogen trifluoride* (NF^3), and *sulphur hexafluoride* (SF6) - collectively termed *perfluorocompounds* (PFCs). EPA's partners have committed to reduce PFC emissions by 10% below their 1995 baseline by 2010. Accomplishments include: the semiconductor's impressive growth pattern is historically cyclical. While production slowed and declined in 2001 and 2002, rising demand for mobile consumer products (e.g., iPods, cell phones) and computers in 2005 is driving recovery and continued growth, and analysts predict growth approaching 11% annually through 2007. The difference between the actual, projected emissions and the "Business-as-usual" (BAU) emissions representing the partnership's environmental benefits, is expressed in millions metric tons of carbon equivalent (MMTCE). The partnership interim report in December 2003, showed a 10% PFC reduction, and the goal to be achieved by 2010.

Still, inspite of the *voluntary* GWP GHG-emission partnership programme, the carbon trading market and carbon standard perimeter aimed at mitigating anthropogenic carbon dioxide, are in a flux. Moves to *carbon-neutralise* or *zero-carbon economise* are millions of light years away. Natural reservoirs or carbon sinks predicate Nature and are part of Nature; but carbon dioxide (CO^2), the chief greenhouse gas, and the rebel leader of man-generated greenhouse gases (GHGs) contained in these carbon reservoirs or sinks, is contained by Nature through absorption by forests and oceans. Plants and algae use this carbon dioxide during photosynthesis – the process which converts carbon dioxide into carbohydrates by chlorophyll under influence of light.

The EPA maintains a greenhouse gas (GHG) inventory, which is an accounting of the amount of greenhouse gases emitted to, or removed from the atmosphere over a specific period of time. A GHG inventory provides information on the activities that cause emissions and removals, as well as background on the methods used to make the calculations. Policy-makers use GHG inventories to track emission trends, develop strategies and policies to assess progress; scientists use GHG inventories as inputs to atmospheric models, and can be tracked at the global, state, regional, local, companies' and individual levels. The U.S., GHG inventory also presents emissions by more commonly used categories: *agriculture, commerce, electricity generation, industry, residential & transportation.* The EPA uses global warming potentials (GWPs) to compare and combine emissions of different GHGs into a national total. GWPs compare the *radiative forcing* or ability to trap heat of one metric ton of a GHG to a metric ton of CO^2.

Carbon sequestration can also serve the purpose: injecting carbon dioxide into underground geological formations; trapping carbon dioxide in the form of solid carbonate in minerals; absorption of carbon dioxide onto various amine-based solvents; using pressure swing absorption (PSA), temperature swing absorption (TSA), gas separation membranes (GSM), industrial distillation (cryogenics), flue capture, conversion to baking soda; or use of algae for conversion to fuel or feed. But *Regenerative Agriculture*[125] in extension agricultural works can both prevent and absorb carbon dioxide.

Two notable attempts to successfully sequester carbon dioxide emissions are Masdar City in Abu Dhabi[126] and Kuzumaki Town[127] in Japan. Masdar City, still in construction, is the most ambitious sustainable development that will be the world's first *zero-carbon, zero-waste, car-free* city, powered entirely by renewable energy sources. Kuzumaki is a "manual of energy self-sufficiency," writes Masters, "whose signs of...comprehensive focus on environmental sustainability are visible from its mountaintops to the pens of the dairy cows that once were the bedrock of local commerce."

The *Masdar Initiative* is a long-term strategic endeavour by the Abu Dhabi Government to accelerate the development and deployment of clean future energy solutions; Masdar City is a free-zone cleantech cluster, which is already attracting the world's best in all areas of sustainability: from *renewable energy* to *biomass*, including *innovators, incubators, research* and *develoment, pioneers* and *solution providers* who will be part of the contribution to create, work and live in it; by fall 2009, phase one of Masdar City – building the *Masdar Institute of Science & Technology* to house 100 students and faculty - will be complete. The city will be built over seven years at an estimated investment in excess of USD20 billion. With this model of a self-sustained, self-efficient, carbon dioxide and other greenhouse gases emissions-free, renewable energy-based, Masdar City, Abu Dhabi is achieving two objectives: display its position as a world-class research

---

[125] Regenerative Agriculture: Is traced to agricultural extension works and researches done by U.S., Dr Washington Carver in the early 20th century in the Tuskegee University, later popularised by Dr Whatley of the same University.
[126] Abu Dhabi: *"The Masdar Initiative."* The Government of Abu Dhabi in the United Arab Emirates (UAE) is inviting tenders from international companies and individuals to participate in the reaslisation of Masdar City. Careers in renewable energies can also be made at Masdar City. Abu Dhabi has also planned a World Future Energy Summit (WFES) from 19-21 January 2009. The WFES will feature The Zayed Future Energy Prize of $1.5 million (linked to the vision of the late Ruler of Abu Dhabi and Founding Father of the UAE, HH Sheikh Zayed bin Sultan Al Nahyan) on 19 January. Two finalists will receive $350,000 each. The Zayed Future Energy Prize is aimed to inspire the next generation of global energy innovators to create solutions.
[127] Coco Masters: *"A Japanese Town That Kicked the Oil Habit."* 22 December 2008, TIME/CNN.

and development hub for new energy technologies, effectively balancing its strong position in an evolving world energy market; and drives commercialisation and adoption of these and other techniques in sustainable energy, carbon management and water conservation. Masdar will also play a decisive role in Abu Dhabi's transition from technology-*consumer* to technology-*producer*. This will lead to the establishment of a new economic sector in Abu Dhabi around these new industries, expected to assist in economic diversification and the development of knowledge-based industries.

Kuzumaki Town is a successful example of resource-poor Japan's struggle to wiggle out of its 90% import fuel dependence. Kuzumaki has attained energy independence through the use of wind, and biomass energy, to produce sufficient energy for electricity and heat and still has sufficient to sell to neighboring towns and villages: the town's Kamisodegawa Mountain runs 12 wind turbines, each (305 ft 93m tall) that generate 21,000 kw of electricity; its Highland Farm with 200 dairy cows processes manure to fertiliser and methane gas, where the methane gas is used as fuel for an electrical generator at the town's biomass facility; wood chips are used to create gas to power the production of farm milk and cheese operations, and barks from trees are converted into pellets for heating stoves. Kuzumaki's total clean energy generated 161% of its electricity in 2007.

1.6 Surviving Escalating Rising Prices

THE TIMES are tough and getting tougher. The high cost-of-living and high-cost-of-dying-for-everything is unprecedented. It is upward-trending amidst a bubbling sea of increasingly polarizing multi-coloring conflicts: social, religious, cultural; fiscal, monetary, economic; oil-and-food-, energy-and-agriculture-, transportation-price-hikes; and, high- and-low-profile prices of wars. Faced with these worst of prospects for survival, the survival instinct urges many to undertake appropriate survival methods. Some of these survival kits may seem "primitive" but healthy technology practices: raising crops, food and animals free of chemicals; using herbs for health, hygiene and beauty; adopting and adapting regenerative agriculture; taping energy from less costly and inexhaustible natural sources; erecting natural buildings; even using alternative currencies[128] side-by-side the conventional ones; and practical altruism.[129]

---

[128] Alternative Currency: Currency used as an alternative to the dominant national or multinational currency systems – usually referred to as *fiat money*. Alternative currencies can be created by an individual, corporation, or organisation, national, state or local governments, or they can arise nationally as people begin to use certain commodity as a currency. *Mutual Credit* is a form of alternative currency, and thus any form of lending that does not go through the banking system can be considred a form of alternative currency. When used in *combination* with or when designed to work in combination with national or multinational fiat currencies, they can be referred to as *complementary currency*. If the use of an alternative currency is limited to a certain region, it is called a *local currency;* Often, there are issues related to paying tax. Some alternative currencies are considered tax-exempt, but most of them are fully taxed as if they were national currency. The legality and tax-status of alternative currencies varies individually from country to country, some systems in use in some countries would be illegal in others. [Source: Wikipedia, the free encyclopediai]

[129] Altruism: Is a selfless concern for the welfare of others, and a traditional virtue in many cultures, and a core aspect of various religious traditions, such as Judaism, Christianity, Islam, Buddhism, Sikhism, Hinduism and many others. Altruism is also a key aspect of many humanitarian and philanthropic causes, exemplified in leaders such as *Dr. Martin King, Jr., Mahatma Ghandi*, and *Mother Teresa.* Selfishness is the opposite of altruism. [Source: Wikipedia, the free encyclopedia]

Specifically approriated survival "primitive" technology practices[130] include: supplies and equipment, and ancestral skills; raising chickens, ducks, geese, rabbits, goats, sheep, cows; greenhousing for growing flowers, food, vegetables, fruits, and grain; building aquaculture ponds for small-scale fish farming salmon, catfish, trout, and tilapia; growing farm worms to replace chemical fertilisers and for manure; raising honey-bees for honey; engaging hydroponic gardening without using soils; using healthy storage of food *sans* chemicals; using herbs, medicine plants and plants for traditional healing, bathing, and easing labor pains; and learning forgotten arts and skills of past civilisations. "From deep within the caverns of time, like-minded primitive technologists have felt an irresistible urge to come together around the camp fire and share ideas. This urge to share has reverberated throughout the centuries," writes Westcott.[131] Living in modern society, Westcott qualifies, we have become disassociated from the earth, from the essence of ourselves, and the need is awakened in us to return to the wilderness – physically and emotionally. We long to feel a sense of connection with our ancient roots. This urge is what has prompted man's fascination with primitive skills: producing objects from natural materials and using methods similar to prehistoric cultures.

Thus, campers, hunters, anglers, nature lovers, take refuge deep in the woods or right in the backyard. They enjoy the natural world in their "own rustic retreats" of low-cost, sturdy, beautiful outdoor structures including garden pavillions, grape arbors, hillside hunts, sauna huts, water gazebos, triangular tree houses, log cabins, wigwams, river rafts, yurts, add Stiles[132] and Roy:[133] "Instead of marring a grassy knoll or field with the construction of a conventional house, design and build a home in around nature – one that's *underground*. Whether you choose to create an *earth-sheltered* house that blends into the world around it or one that is truly *underground*...you will be able to bring your environmentally friendly dream house to fruition without spending a lot of money...sceptical about living in a non-traditional home? Listen to people who live in underground houses sing their praises: Far from their being dingy caves, you'll find natural light streaming in from above and the beauty of Mother Earth all around you. And you'll be helping to conserve the planet."

Self-sufficiency, self-reliance, self-accommodation, self-help, and self-food-production modes - are in high demand during tough economic times; these aspects are buoyed on by price hikes that breed more poverty, such as the world is passing through right now. You learn and engage to preserve and save seeds for the next sowing season, and perishables: fruits, vegetables, wild and delicious edible herbs, vegetables, leaves, plants and nuts - without the use of chemical substances and preservatives; make ice cream, cheese, bake bread; harvest eggs and chickens from your poultry, meat from your cattle, and wildlife; accumulate and use natural manure from your barnyard animals' pool; breed and harvest fish in ponds; weave or spin wool for clothing.

---

[130] Primitive Technology: Used in this context means: Simple, natural, old-fashioned, crude, wild, practices of food and shelter production involving no chemicals in food and diets and natural energy.
[131] David Westcott: *"Primitive Technology."* Gibbs-Smith Publisher. Amazon.com
[132] David Stiles: *"Rustic Retreats: A Build-It-Yourself-Guide."* Storey Publishing, Amazon.com; J. Wayne Fears: *"How To Build Your Dream Cabin in the Woods: The Ultimate Guide to Building and Maintaining a Backcountry Gateway."*
[133] Rob Roy: *"The Complete Book of Underground Houses: How To Build a Low-Cost Home."* Sterling Publishing Co., Inc., New York; Amazon.com.

Aubert[134] and Madison[135] show you how to preserve food without nutrient loss. Coleman/Damrosch,[136] Ashworth[137] and Seymour/Headon[138] will initiate you into the lore of how to battle, survive and win the soaring costs of living and dying, thus:

- .... *Seed Saving & Preservation:* Complete and seed-saving guide for 160 vegetable crops, with detailed information on each vegetable: botanical classification, flower structure and pollination method, isolation distances, caging and hand pollination techniques, and proper methods for harvesting, drying, clearing and storing the seeds. Beginning or experienced gardeners can easily learn how to save all of their own seeds, resulting in substantial annual savings and the satisfaction that comes from a garden which is truly self-perpetuating.

- *Organic Vegetables:* At the end of the summer keep harvesting! If you love the joys of home-garden vegetables but always thought those joys had to stop at the end of summer,..successfully use the sun to raise a wide variety of traditional winter vegetables in backyard cold frames and plastic-covered tunnel greenhouse without supplementally heat...Apply Farm-generated Fertility: how to meet your soil fertility needs from the resources of your own land, even if manure is not available; The Moveable Feast: how to construct home-garden and commercial-scale greenhouses that can be easily moved to benefit plants and avoid insect and disease buildup; Winter Garden: how to plant, harvest, and sell hardy salad crops all winter long from unheated or minimally heated greenhouses; Pests: how to find "plant-positive" rather than "pest-negative" solutions by growing, healthy, naturally resistant plants; The Information Resource: how and where to learn what you need to know to grow delicious organic vegetables, no matter where you live.

- *Spin Wool, Make Cheese:* Live off the land and reap the harvest, respect the land, stay healthy, and waste nothing; bale hay, breed rabbits using 150 artworks and classic illutrations on ethical living, organic gardening, creating an urban garden, generating power from wind, water, and the sun, making the break and changing your way of life; Clearing land, planting, harvesting, feeding, clothing, and sheltering yourself; Make bricks, cure bacon, engage in crafts and skills, gathering food from garden, animals or from the wild, self-sufficiency in the kitchen. Raise your own livestock:[139] chickens, ducks, geese, rabbits, sheep and cows. Barnyard basics include: before you put down money on that hen house, turn a single goat loose in the yard, or fence in a couple of acres for frisky calves...study the spotlight on single animals with simple, clear, instructions on what it takes to keep your livestock healthy and happy. Learn what type of housing and how much land your animals need, what to feed them, how to breed them, and how to handle routine health care.

- *Practical Skills:* Learn to make ice cream, or sharpen an axe, clean a chimney or grow raspberries, milk a cow, tap a maple tree, clean a fish, lead a horse, build the best chicken coop, heat your house with wood or by the sun, or de-skunk a country dog, or identify edible and medicinal plants[140] you need to do this in a treasury of time-honored country wisdom.

---

[134] Elaine Aubert: *"Keeping Food Fresh: Old World Techniques & Recipes."* Centre Terre Vivante ISBN 1-890 132-10. Amazon.com.

[135] Deborah Madison: *"Preparing Food Without Greezing or Canning: Traditional Techniques Using Salt, Oil, Sugar, Alcohol, Vinegar, Drying, Cold Storage & Lactic Fermentation."* Gardeners and Farmers of Centre Terre Vivante. Amazon.com.

[136] Eliot Coleman/Barbara Damrosch: *"Four Season Harvest Organic Vegetables from Your Home Garden All Year Long,"* Amazon.com; Eliot Coleman: *"The New Organic Grower: A Master's Manual of Tools & Techniques for Home and Market Gardener* (A Gardener's Supply Book)." Amazon.com.

[137] Suzanne Ashworth: *"Seed to Seed: Seed Saving & Growing Technique."* Amazon.com.

[138] John Seymour/Deidre Headon: *"Self-sufficient Life and How to Live It.."* Amazon.com; John/Martha Storey: *"Storey's Basic Country Skills: A Practical Guide to Self-Reliance."* Amazon.com.

[139] Gail Damerow (editor): *"Barnyard in Your Backyard: A Beginner's Guide to Raising Chickens, Ducks, Rabbits, Goats, Sheep, and Cows."* Amazon.com.

[140] Steve Brill: *"Identifying And Harvesting Edible and Medicinal Plants In Wild* (And Not So Wild) *Places."*Amazon.com; Bradford Angier: *"Field Guide to Edible Wild Plants."* Stackpole Books. Amazon.com

There are hundreds of fascinating, delicious wild vegetables, fruits, nuts and seeds, and herbs growing in our neighborhoods, backyards, parks, and forests that we overlook: you can find and prepare more than 500 different plants for nutrition and better health, including such common plants as *mullein* (a tea made from the leaves and flowers that suppresses a cough), *stinging nettle* (steam the leaves and you have a tasty dish rich in iron), *cattail* (cooked stalks taste similar to corn and are rich in protein), and *wild apricots* (an infusion made with the leaves is good for stomachaches and digestive disorders). More than 260 detailed line drawings help identify a wide range of plants – many of which are suited for cooking by following the more than 30 recipes included. There are literally hundreds of plants available *underfoot* waiting to be harvested and used either as *food* or a *potential therapeutic;* or *acorns* or *quercus* - also known as Red-, Black-, Blackjack-, Yellow-, Turkey-, Georgia-, Northern Pin-, Chinquapin-, Laurel-, Bluejack-, Live-, Oglethorpe-, Durand-, Shingle-, Scrub-, Water-, Bur-, Post-, Valley-, Gambel-, Low-. Swamp-White-, Basket-, Mossycup-, Spanish-, Oregon-, California-Oaks - topped wildfoods relied on by the Indians for soup or mush which was probably the main food of more than three-fourths of the native Californians.

Survive we must at all cost. But even as the oil price hikes appeared to backdown in late December 2008, high food prices held fast to their highs, and self-vigilance refused to backdown. Moreover, Carbon Dioxide Information Analysis Center's (CDIAC)[141] data collected for the UN only for burning fossil fuels (excluding deforestation or other sources) showed huge growth in Asia. This is not even the start of the begin of divesting our world of the reign of terror of greenhouse gases.

1.7 Conclusion

---

[141] Carbon Dioxide Information Analysis Center (CDIAC): *"CDIAC 2004 Data Collected for the UN on CO^2 Emissions by Sovereign States."*: World *27,245,758* annual CO^2 emissions metric tons – 100.0% percentage of total, of which the following countries emitted: U.S., 6,049,435 or 22.2%; China/Taiwan 5,010,170 or 18.4%; European Union 4,001,224 or 14.7%; Russia 1,524,993 or 5.6%; India 1,342,962 or 4.9%; Japan 1,257,963 or 4.6%; Germany 860,522 or 3.1%; Canada 639,403 or 2.3%; U.K. 587,261 or 2.2%; South Korea 465,643 / Italy 449,948 - or 1.7%; Mexico 438,022 / South Africa 437,032 / Iran 433,571 - or 1.6%; Indonesia 378,250 / France 373,693 - or 1.4%; Brazil 331,795 / Spain 330,497 / Ukraine 330,039 / Australia 326,756 - or 1.2%; Saudi Arabia 308,397 / Poland 307,238 – or 1.1%; Thailand 268,082 or 1.0%; Turkey 226,125 or 0.8%; Kazakhstan 200,278 /Algeria 194,001 / Malaysia 177,84 – or 0.7%; Venezuela 172,623 / Egypt 158,237 - or 0.6%; UnitedArab Emirates 149,188 / Argentina 141,786 / Uzbekistan 137,907 / Pakistan 125,669 - or 0.5%; Czech Rep 116,991 / Nigeria 114,025 / Belgium 100,716 / Kuwait 99,364 / Vietnam 98,663 / Greece 96,695 - or 0.4%; Romania 90,425 / Iraq 81,652 / Philippines 80,512 / North Korea 79,111 / Israel 71,247 / Austria 69,846 / Syria 68,420 - or 0.3%; Finland 65,799 / Belarus 64,890 / Chile 62,418 / Libya 59,914 / Portugal 58,906 / Hungary 57,183 / Colombia 53,634 / Serbia/Montenegro 53,322 / Sweden 53,033 / Denmark 52,956 / Qatar 52,904 / Singapore 52,252 / Norway 43,149 / Bulgaria 42,558 / Ireland 42,353 / Turkmenistan 41,726 / Morocco 41,169 / Switzerland 40,457 - or 0.2%; Hong Kong 37,411 / Bangladesh 37,165 / Slovakia 36,289 / Trinisdad/Tobago 32,557 / New Zealand 31,570 / Peru 31,493 / Azerbaijan 31,365 / Oman 30,899 / Ecuador 29,268 / Cuba 25,818 / Croatia 23,501 / Tunisia 22,885 / Yemen 21,114 / Dominican Rep 19,640 / Estonia 18,944 / Bahrain 16,949 / Jordan 16,465 / Lebanon 16,263 / Slovenia 16,212 / Bosnia/Herzegovina 15,596 / Lithuania 13,309 / Guatemala 12,220 / Sri Lanka 11,534 / Luxembourg 11,277 / Jamaica 10,592 / Kenya 10,588 / Zimbabwe 10,559 / Macedonia 10,420 / Sudan 10,372 / Myanmar 9,760 / Brunei 8,810 / Mongolia 8553 / Ethiopia 7,982 / Angola 7,897 / Moldovia 7,685 / Honduras / Ghana 7,190 / Latvia 7,098 / Bolivia 6,973 / Cyprus 6,750 / Costa Rica 6,405 / El Salvador 6,167 / Kyrgyzstan 5,727 / Panama 5,661 / Uruguay 5,477 / Equatorial Guinea 5,426 / Cote d'Ivoire 5,162 / Tajikistan 5,004 / Senegal 4,993 / Tanzania 4,352 / Botswana 4,301 / Paraguay 4,180 / Netherlands Antilles 4,088 / Nicaragua 4,007 / Goergia 3,912 / Cameroon 3,839 / Albania 3,674 / Armenia 3,648 / Rep Congo 3,542 / Mauritius 3,197 / Nepal 3,043 / Madagascar 2,731 / New Caledonia 2,577 / Mauritania 2,555 / Namibia 2,471 / Malta 2,453 / Papua New Guinea 2,449 / Benin 2,387 / Togo 2,310 / Zambia 2,288 / Suriname 2,284 / Reunion 2,277 / Iceland 2,229 / Macau 2,207 / Mozambique 2,167 / Aruba 2,156 / DR Congo 2,104 / Bahama 2,009 / Uganda 1,826 / Haiti 1,756 / Guadeloupe 1,734 / Guyana 1,445 / Gabon 1,371 / Guinea 1,338 / Martinique 1,291 / Laos 1,280 / Barbados 1,269 / Niger 1,214 / Burkina Faso 1,096 / Fiji 1,071 / Malawi 1,045 / French Giana 1,005 / Sierra leone 994 7 Swaziland 957 / Belize 792 7 Eritrea 755 / Maldives 726 / Afghanuistan 693 / French Polynesia 671 / Faroe Islands 660 / Palestine Authority 649 / Greenland 572 / Rwanda 572 / Mali 565 / Bermuda 550 / Seychelles 546 7 Cambodia 535 / Liberia 470 / Bhutan 414 / Antigua/Barbuda 414 / Gibraltar 374 / Saint Lucia 367 / Djibouti 367 / Cayman islands 312 / Gambia 286 / Cape Verde 275 / Guinea-Bissau 271 / Central African Republic 253 / Western Sahara (SADR) 238 / Palau 238 / Burundi 220 / Grenada 216 / Saint Vincent/Grenadines 196 / Timor-Leste 176 / Solomon Islands 176 / Samoa 150 / Nauru 143 / Saint Kitts/Nevis 125 / Chad 125 / Tonga 117 / Dominica 106 / Sao Tome/Principe 92 / Vanuatu 88 / Comoros 88 / British Virgin Islands 84 / Saint Pierre/Miquelon 62 / Montserrat 62 / Falkland Islands 44 / Kiribati 29 / Cook Islands 29 / Saint Helena 11 / Niue 4 – or 0.1%.

SO LONG as fossil fuels continue to call the shots of living and dying, and predicate the lifestyle of living things and economics of our globe, further disasters are all but programmed to happen. We assume that we will continue to bury our heads ostrich-like in the sands, to let the fossil fuels-dominated hurricanes pass away; but we will be shocked to realise that this won't happen after we lift off our heads from the sand; that only renewable energies are capable to ensure better and lasting sustainable agriculture, transportation and energy management, and the ultimate destruction of greenhouse gases.

Chapter 2

RENEWABLE ENERGIES AND FOOD PRODUCTION

2.1. Introduction: Renewable Energies, Increasing Populations

INCREASING WORLD populations, increasing poverty, increasing diseases resistant to antibiotics,   increasing malnutrition, and advanced, almost irreversible climate change and global warming - obligate the world economic and industrial powers, to embrace renewable energies hook line and sinker. Conventional food production methods must give way to multicultural food production. Regenerative and organic farming must replace inorganic, chemical fertilisers, herbicides and, factoring farming agrobusiness practices. The common denominator for all of these inefficient, unsustainable, unhealthful and wasteful systems, is *fossil fuels, petrochemicals* and their *myriad derivatives*.

2.2 Conventional Food Production, Multiculture Food Production

MONOCULTURE CONVENTIONAL food production methods invariably mean the use of fossil fuel in one form or another. These methods entail the generation and production of more greenhouse gases and more pollution of air, water, soil through the uses of more fertilisers, herbicides, pesticides, and insecticides, hormones and growth agents, and the involved health risks and effects. Food quality and sustainability are not improved by the addition of additives, flavourings, enhancers, preservatives, as well as genetic modified (GM) foods. Multiculture food production methods from seed preservation, sowing, harvesitng, and consumption, on the other hand, significantly reduce, mitigate or wipe out the lion share portions of monocultural conventional food production negatives, through the use of renewable energies.

2.2.1 Regenerative, Organic, Cooperative Farming

THE APPLICATION and use of renewable energy sources for food production and industrial purposes *conserves* nature and ecosystems, energy, agriculture, and transport costs; it results in good quality fresh food and near-to-zero greenhouse emissions. It embraces many tried-and-tested methods, among others, regenerative, organic and cooperative farming. Regenerative farming is a method of sustainable agriculture that integrates three main goals of environmental responsibility, farm profitability and prosperous farming communities. Organic farming is a form of agriculture that rellies on crop rotation, green manure, compost, biological pest control, and mechanical cultivation to maintain soil productivity and strictly limiting the use of synthetic fertilisers, synthetic pesticides and plant growth regulators. In cooperative farming, several farmers democratically organise collectively in a joint venture in farming activities, and pool their resources. Cooperative or collective farming provides social and economic self-esteem and social responsibility to its members.

Conventional agriculture and industrial food production methods that are prevalent today and provide us with most of the meat, dairy, eggs, fruits and vegetables available in supermarkets – are significantly different from regenerative and organic farming methods. Conventional agriculture or agribusiness *depends* heavily on the use of fossil fuels, agrichemicals or agrochemicals such as insecticides, herbicides, fungicides, fertilisers, hormones and plant growth agents,[142] and concentrated stores of animal manure. Many of

---

[142] Hormones & Plant Growth Agents: <u>Hormones:</u> Are chemicals released by plant and animal cells in other parts of the organism which are required to alter metabolism; in animals and humans, glands in organisms synthetise the release of hormones into the bloodstream or cavities; in plants, the five major and other hormones activate and stimulate various functions in the plant from budding to harvest; <u>Plant Growth Agents/Phytohormones:</u> Plants lack glands, so hormones in plants are known as plant growth agents or phytohormones, produced within the plant in extremely low concentrations, and regulate cellular processes in targeted cells locally and when moved to other locations of the plant.

the agrochemicals are *toxic* and bulk storage may pose high environmental and health risks, particulaly in the event of accidental spills. Multicultural methods of agriculture in which regenerative, organic, and cooperative farming methods are prevalent, directly or indirectly conform to the use and necessity of renewable energies.

Case-studies show that 3.5 billion tillable acres could sequester up to 40% of current carbon dioxide emissions, if regenerative and organic agriculture could be practised worldwide; even when using biologically-based regenerative practices, this dramatic benefit can be accomplished with no decrease in harvest or farmer profits. Organically managed soils can convert carbon dioxide from a greenhouse gas into a *food-producing asset*. According to the U.S., EPA, in 2006, U.S., carbon dioxide emissions from fossil fuel combustion were estimated at nearly 6.5 billion tons. If a 2,000 lb/ac/year sequestration rate was achieved on all 434,000,000 acres (1,700,000 sq km) of cropland in the U.S., *nearly 1.6 billion tons of carbon dioxide* would be sequestered per year, mitigating close to one-quarter of the country's total fossil fuel emissions. This is the emissions cutting equivalent of *taking one car off the road for every two acres* under 21st Century regenerative agricultural management – based on a vehicle acreage of 15,000 miles per year at 23 mpg.

Regenerative agriculture and farming can be traced back to the works and scientific contributions regarding the nitrogen cycle and the biological regeneration of soils in southern U.S., by Carver[143] at the Tuskegee University, Alabama, U.S., in the early 20th century. Carver introduced crop rotation methods in combination with the planting of nitrogen-fixing legumes, such as peanuts, peas, and soybeans, now known as *agricultural extension:* application of scientific research and new knowledge to agricultural practices through farmer education; and, field of extension.[144] Field of extension now encompases a wider range of communication and learning activities organised for rural people by professionals from different disciplines, including agriculture, health, and business studies. Carver's works were elaborated upon and popularised by Whatley[145] in the 1970s at the same university: Whatley campaigned and championed for smaller and smarter farms as a successful strategy for small farmers, rather than competing for the same market as large farmers and going broke in the process. His *Regenerative Farming System* for small farms involved four core components:

- *Creating a Diversified Pick-Your-Own* (PYO)/*U-Pick) Farm:* The farm should measure between 10-100 acres (0.81 sq km); producing at least 10 diverse products (agricultural and/or artisanal) on a year-round basis that are supported through a Clientele Membership Club (CSA)- consisting of a community of individuals who pledge support to a farm operation so that the farmland becomes either legally or spiritually, the community farm, with the growers

---

[143] Dr. George Washington Carver (1864-Jan 5, 1943: *"How the Farmer can save his Sweet Potatoes and Ways of Preparing them for the Table."* Edited by Geo W. Carver, M.S., Agric., D.Sc., Director of Experiment Station, Tuskegee Institute. 4th ed., revised and updated.

[144] Field Extension Practitioners: These are found working throughout the world, usually for government agencies, and represented by several professional organisations such as: Australian-Pacific Extension Networks (APEN); Agricultural Research and Extension Network (AGREN); Journal of Extension; Agricultural Extension agencies in developing countries which receive large amounts of support from international development organisations, such as the World Bank and UN Food & Agricultural Organisation (FAO).

[145] Dr. Booker T. Whatley (1915-3 Sept 2005): *"Handbook on how to Make $100,000 Farming 25 acres (100,000 m²): With Special Plans for Prospering on 10 to 200 acres (0.81 sq km)."* Edited by George DeVault, Emmaus, PA Regenerative Agriculture Association. Distributed by Rodale Press, 1987 xi, 180 p.

and consumers providing mutual support and sharing the risks and benefits of food production.

- *Clientele Membership Club (CMC/CSA):* Focuses usually on a system of weekly delivery or pickups of vegetables and fruits, sometimes dairy products and meat. CSA and modern-day CSA, the Clientele Membership Club (CMC) similar to the USDA-supported Community Supported Agriculture (CSA) - operate country-wide areas having a population centre of about 50,000 residents, marketing to CMC members for 40% of supermarket pricing and yielding profits.

- *Small Farm Sufficiency Law:* Dr Whatley's Small-Farm Sufficiency Law or the so-called '10 Commandments' are that a small farm *shalt:* provide year-round daily cash flow; and year-round, full-time employment; be a pick-your-own operation; have a guaranteed market with a CMC; be located on a hard-surface road within a radius of 40 miles of a population centre of at least 50,000, with well-drained soil and an excellent source of water; produce only what clients demand – and nothing else! Shunning middlemen and middle women like a plague, for they are a curse upon thee; consist of compatible, complementary crop components that earn a minimum of $3,000 per acre annually; be 'weatherproof', at least as far as possible with both drip and sprinkler irrigation; and be covered by a minimum of $250,000-$1 million worth of liability insurance.

- *Essentials of Regenerative Agriculture:* Essentials of regenerative agriculture about which Dr Whatley wrote included: U-pick operations; community-supported agriculture (CSA); drip irrigation; rabbit production; farm-owned hunting preserves; kiwi vines; shiitake mushrooms; veneer-grade hardwood stands; on-the-farmbed and breakfasts; direct marketing; organic gardening and great cheese production; practical, positive entrepreneurial options for small farmer operations that included production diversification, organic farming practices, farm value-added products and innovative, direct marketing schemes. These sustainable alternatives have grown, taken root and flourished worldwide, and adopted by the U.S., Agricultural Department (USDA) and several U.S., states. Whatley created several plant varieties: Foxy Lottie grave cultivar (named after Lottie, his wife of 56 years); Five sweet potato varieties, including the popular yellow-meated Carver sweet potato; and 15 varieties of muscadine grapes.

When agriculture is regenerative, soils, water, vegetation and productivity continually *improve* rather than staying the same or slowly getting worse, states Jones.[146] Regenerative agriculture is productive and profitable, and instils a deep sense of personal satisfaction in farmers, rural communities and observers. Revitalizing the natural resource base rekindles our sense of self and our sense of place in the environment. But many of what we term 'sustainable' agricultural practices represent only small improvements in current methodology, and at best, they impart a fleeting tinge of green to a deteriorating landscape, criticises Jones. She contributes citing Hill (1998): "Regenerative practices embody fundamental redesign. They utilise natural ecological services to replenish and reactivate the resource base." Agricultural practices can be productive, profitable and restorative *provided* they: *regenerate* rather than merely *'sustain'* the natural base; *enhance,* rather than *replace* natural ecosystem processes; and *stimulate* the formation, rather than *attempt to reduce* the loss of topsoil.

---

[146] Dr. Christine E. Jones, Australian scientist, Founder: 'Amazing Carbon'; Organizer: 'Managing the Carbon Cycle Forum'; "Christine Jones – short CV"; *"Regenerative Land Management: a whole of landscape approach to the restoration of water balance and water quality.'Beyond Streambark Fencing'"*; Landcore Forum, Gloucester, NSW, Dec., 2001; *"Regenerative grassland management"*. Balala-Bushgrove Landcore Group Pinrush and Grazing management Trials Summary Technical Report 1991-1998, 2nd ed. Armidale NSW; 13 Laurence Avenue, Armidale NSW 2350, Australia; Christinejones22@aol.com; www.amazingcarbon.com.

Jones adds with emphasis that:

- The traditional approach to land management has been one of 'simplistic replacement' (with exotic species and unbalanced chemical fertilisers, rather than a *multi-level, multi-species* approach. In recent times, there have been concerted attempts to make over-simplified ecosystems 'sustainable.' It is a battle which cannot be won. Until a *preventive,* rather than a *curative* approach to land management is adopted, agricultural 'problems' such as soil *compaction, low fertility, weeds, pests*and *disease,* and their 'treatments', will continue *indefinitely.*

- "The more compounds in an ecosystem, the greater the synergy between the parts. To improve the diversity and health of agricultural landscapes, we need to think *creatively* and *MANAGE* for change, rather than embarking on the broadscale replacement of natural biological processes with expensive technology. It is difficult to step off the replacement treadmill, because nutrient acquisition and distribution no longer occur naturally in dysfunctional soils. However, the costs of production continue to increase for as long as the replacement philosophy endures.

- "Biologically, active, self-renewing topsoil is the cornerstone for a productive agricultural sector and a robust environment. It is essential for the health of plants, humans and other animals. The appropriate management of soil biology in agriculture, horticulture, forestry, amenity, wildlife and conservation areas is the most vital, and most *neglected,* of the natural resource issues facing Australia. Most of our grasslands and croplands aren't as healthy as we'd like them to be. They are often characterised by areas of bare ground, sheet and gully erosion, the presence of weeds and the lack of desirable plant species. It is easy to assume that removing the weeds and replanting some 'better' species will solve the problems. Decades of experience have demonstrated that this simplistic approach *rarely* works.

- "The interactions between animals, plants and soil biota remain out of balance because the overriding importance of soil management has not been addressed. The resulting shortfalls in ecosystem services, such as nutrient availability, need to be supplemented at our cost – and it's usually quite an expense. Landscapes are not degraded because they *lack* desirable species. Rather, desirable species will not *flourish* when landscapes are *degraded.* In the agricultural context, grazing and cropping account for the major portion of the land area. If the primary focus of natural resource management is to be the maintenance of high levels of humic materials to rebuild topsoil, then *radical departures* from conventional methods of production will be required."

Radical conventional agricultural production methods are provided by the International Assessment of Agricultural Knowledge, Science and Technology (IAASTD)[147] which focuses on agriculture as the provider of food, nutrition, health, environmental services, and economic growth that is both *sustainable* and *socially equitable.* IAASTD assessment recognises the diversity of agricultural ecosystems and of local social and cultural conditions, and a fundamental rethink of the role of agricultural knowledge, science and technology in achieving

---

[147] International Assessment of Agricultural Knowledge, Science and Technology (IAASTD): *"Agriculrual Questions & Answers."* Secretariat: Robert T. Watson, Chief scientist, World Bank, and the IAASTD Chairs – Judi Wahhungu, Exco Dir., African Centre for Technology Studies, and Hans R. Herren, Pres. Millennium Institute – are responsible for the intellectual leadership of the project. The IAASTD assessment process was initiated by the World Bank in open partnership with a multistakeholder group of organisations, incl., the FAO, GEF, UNDP, UNEP, WHO,UNESCO and representatives of governments, civil society, private sector and scientific institutions. It uses a strong consultative 'bottom-up' process in its assessments that recognises the different needs of different regions and communities. The IAASTD was formed in 2002 in Joahennesburg, South Africa.

equitable development and *sustainability.* The focus is on small farmers in diverse ecosystems and to areas with the greatest needs to improve rural livelihoods, empowering marginalised stakeholders, sustaining natural resources, enhancing multiple benefits provided by ecosystems, considering diverse forms of knowledge and providing fair market access for farm products. The IAASTD "Q&A" reads:

- *What Challenges Does Agriculture Face Today?* For decades, agricultural science has focused in boosting production through the development of new technologies. It has achieved enormous gains as well as lower costs for large-scale farming. *But this success has come at a high environmental cost. Further, it has not solved the social and economic problems of the poor in developing countries, which have generally benefited the least from this boost in production.* Today, the world is a place of uneven development, unsustainable use of natural resources, worsening impact of climate change, and continued poverty and malnutrition. Poor food quality and diets are partly responsible for increase of chronic diseases like obesity and heart disease. Agriculture is closely linked to these concerns, including the loss of biodiversity, global warming and water availability.

- *What Are The Pros And Cons Of Bioenergy?* Bioenergy is heat, electricity or transport fuel produced from plant or animal materials. Millions of people still depend on traditional bioenergy like wood or charcoal for cooking and heating, which can be unsustainable and pose health risks. In many developed countries, the rising costs of fossil fuels, as well as concerns about energy security and climate change, are generating new interest in *other forms of bioenergy.* For example, new liquid biofuels are made from *crops* or from agricultural and forestry *residues.* However, energy is needed to grow, transport and process bioenergy crops, causing considerable debate about their net benefit in terms of greenhouse gas reduction. Another major concern is that using crop land to produce fuel could raise food prices, drive small-scale farmers off their land and prolong hunger in the world. Electricity and heat can also be obtained from plant *residues* and animal *wastes,* either by burning them directly or by first producing biogas then burning it. These *renewable* energy sources usually produce less greenhouse gas emissions than other fuels. They can be effective, for instance, in places not connected to the electric grid. Decision-makers should compare all forms of bioenergy to other sustainable energy options and carefully weigh full social, environmental and economic costs against *realistically achievable* benefits. Decisions in this context are heavily influenced by local conditions.

- *Can Biotechnology Help Meet The Growing Demand For Food?* Biotechnologies are techniques that use living organisms to make or modify a product. Some *conventional* biotechnologies are well-accepted, such as *fermentation* for bread or alcohol production. Another [well-accepted] example is plant and animal *breeding* to create varieties with better characteristics or increased yields. *Modern* biotechnologies change the *genetic code* of living organisms using a technology called *"genetic modification".* These technologies have been widely adopted in industrial applications such as *enzyme* production. Other applications remain contentious, such as the use of *genetically modified* (GM) crops created by inserting genes from other organisms. Some GM crops can bring yields in some places and declines in others. Because new techniques are rapidly being developed, longer-term assessments of environmental and health risks and benefits tend to lag behind discoveries. This increases speculation and uncertainty. The possibility of *patenting* genetic modifications can attract investment in agricultural research. But it tends to *concentrate ownership of resources, drive up costs, inhibit independent research,* and *undermine local farming practices* such as *seed-saving* that are especially important in developing countries. It could mean new *liabilities,* for example, if a genetically modified plant spreads to nearby farms. Many problems could be solved, if biotechnologies would focus on local priorities identified through transparent processes involving the full spectrum of stakeholders.

- *How Is Climate Change Threatening Agriculture?* Agriculture has contributed to climate change in many ways, for instance, through the conversion of forests to farmland and the release of greenhouse gases. Conversely, climate change now threatens to *irreversibly* damage natural resources on which agriculture depends. The effects of global warming are *already visible* in much of the world. In some areas, moderate warming can slightly increase crop yields. But overall, *negative impacts* will increasingly dominate. Floods and droughts become more frequent and severe, which is likely to seriously affect farm productivity and the likelihood of rural communities, and increase the risk of conflicts over land and water. Also, climate change encourages the *spread of pests and invasive species* and may increase geographical range of some diseases. Some land use management approaches can help mitigate global warming, which include: *planting trees, restoring degraded land, conserving natural habitats, improving* soil and and *fertility management.* Policy options include: financial incentives to grow trees, reduce deforestation and develop *renewable energy* sources. Agriculture and other rural activities must be integrated in future international policy agreements on climate change. However, since some changes in climate are now *inevitable,* adaptation of new measures are also imperative.

- *How Is Food Production Affecting Health?* Although food production has increased in recent decades, many people are *undernourished,* a problem accounting for about 15% global disease. Many population groups still lack *protein, micronutrient* and suffer *vitamin deficiency.* Meanwhile, *obesity* and chronic diseases are increasing across the world because of people eating *too much of the wrong foods.* Agricultural research and policies should be devised to increase *dietary diversity,* improve food quality and promote better food processing, preservation and distribution. Global trade and growing consumer awareness have increased the need for *proactive* food safety systems. Health concerns include the presence of *pesticide residues, heavy metals, hormones, antibiotics* and *additives* in food systems, as well as risks related to *large-scale livestock farming.* Worldwide, agriculture accounts for at least 170,000 work-related deaths each year. Accidents with farm equipments like tractors and harvesters cause many of these deaths. Other important health hazards for agricultural workers include *noise, transmissible* animal diseases, and exposure to *toxic* substances, such as pesticides. Agriculture can contribute to the emergence and spread of infectious diseases. Therefore, robust surveillance, detection and response programmes are critical across the food chain.

- *How Can Agriculture Make Better Use Of Natural Resources?* Historically, agricultural development was geared towards increasing productivity and exploiting natural resources, but *ignored* complex interactions between agricultural activities and, local ecosystems, and society. These interactions must be considered to enable sustainable use of resources as water, soil, biodiversity and fossil fuels. Much of the agricultural knowledge, science and technology needed to resolve today's challenges are available and well understood, but putting them into *practice* requires creative efforts from all stakeholders. Existing agricultural science and technology can tackle some of the underlying causes of declining productivity. But further development based on a *multidisciplinary* approach are needed, starting with more monitoring of how natural resources are used. Other options for action include research into how to use natural resources *responsibly* and efforts to foster public awareness of their importance.

- *Why Haven't Small Farmers Benefited From Global Trade?* Small farmers and rural communities in developing countries have *not benefited* from opportunities that agricultural trade can offer. Opening farm markets *prematurely* to international competition can further weaken the agricultural sector of a developing country, causing more poverty, hunger and harm to the environment in the long-term. Trade reforms could make relations *more equitable.* Developing countries would benefit from key changes such as *removing trade barriers* on products for which they have a competitive advantage, *lowering tariffs* on imports of processed commodities; and improving their *access* to export markets. The capacity of developing countries to analyse and negotiate trade agreements needs to be strengthened and more *transparent* decisions concerning the agricultural sector. The environmental footprint of agriculture could be reduced by adapting market and trade policies, for instance, by *removing detrimental subsidies, changing taxation policies,* and *improving property laws.*

- *Can Traditional Knowledge Contribute To Agriculture?* Many effective innovations are generated locally based on the knowledge and expertise of indigenous and local communities, rather than on formal scientific research. Traditional farmers embody ways of life beneficial to the conservation of biodiversity and to sustainable rural development. Local and traditional knowledge has been successfully built into several areas of agriculture, for example, in the domestication of *wild trees,* in *plant breeding,* and in *soil and water management.* Scientists should work more closely with local communities and traditional practices should have a higher profile in science education. Efforts should be made to archive and evaluate the knowledge of local people and to protect it under *fairer* international intellectual property legislation.

- *What Is The Role Of Women In Agriculture And Development?* Current trends in *globalisation* and *rising environmental* and *sustainability concerns* are redefining the relationship of women to agriculture and development. The proportion of women involved in agricultural activities ranges from 20%-70%, a number which is climbing in many developing countries, especially where agriculture is geared towards export. Although some progress has been made, women continue to struggle with *low income, limited access to education, credit* and *land, job insecurity,* and *deteriorating* work conditions. Growing competition has fuelled demand for *cheap, flexible labor,* and conflicts over access to natural resources have added to the pressure. Poor rural households are increasingly threatened by natural disaster, environmental change and health and safety risks – this at a time of *diminishing* government support.

- *What Are The Options For Action?* Small-scale farmers would benefit from greater access to knowledge, technology, and credit, and, *critically,* from more political power and better infrastructure. They need laws that secure access to land and natural resources as well as fair intellectual property rights. Ensuring food security is not merely a matter of producing enough to eat: food must also be *available* to those who need it. Possible policy actions that can embrace access to food include *reducing transaction costs* for small-scale producers, strengthening local markets and improving food safety and quality. Global systems are needed to watch out for *sudden price changes* and extreme weather events that could lead to food shortages and *price-induced hunger.* Agricultural sustainability means *maintaining productivity while protecting natural resources base.* Possible actions include: improving low impact practices such as *organic agriculture* and providing incentives for sustainable management of water, livestock, forests, and fisheries. Science and technology should focus on ensuring that agriculture not only provides food but also *fulfils* environmental, social and economic functions, such as mitigating climate change and preserving biodiversity. Policy-makers could provide end subsidies encouraging *unsustainable* practices and provide incentives for *sustainable* resource management. Human health can be improved through efforts to diversify diets and enhance their nutritional value, through advances in technologies for *processing, preserving* and *distributing* food, and through better health policies and systems. Food safety can be increased by investing in *infrastructure, public health* and *veterinary* capacity, and by developing legal frameworks for controlling biological and chemical hazards. Work-related health risks can be reduced by enforcing health and safety regulations. The spread of infectious disease like *bird flu* can be limited through better coordination across the food chain. Achieving *greater equity* in agriculture requires investment to bring technology and education to rural areas.

Meanwhile, an IAASTD Global Report[148] concluded that regenerative farming practices, local knowledge and regionally appropriate technology were *favored* over biotech and industrial agriculture. "Regenerative agriculture is by definition, sustainable, using easily available and affordable techniques *instead* of expensive *petroleum-based chemical fertilisers* and *herbicides...*" said one of the report's lead co-authors, Amadou Makhtar Diop, who is international projects

---

[148] Dan Sullivan: International Assessment of Agricultural Knowledge, Science and Technology Development (IAASTD): *"Global Report."* Johannesburg, South Africa, 7 April 2008. Lead/co-authors of the IAASTD Report: Amadou Makhtar Diop, Rodale Institute International Programme Dir; Marcia Ishii-Eiteman, PhD., senior scientist at the Pesticide Action Network; Achim Steiner, UNEP Executive Dir.; Benny Haerlin, Greenpeace Germany, as well as more than 400 scientists.

coordinator for the Rodale Institute. The 57-nation signed report action-plan set a bold new course for developing nations to feed themselves while also addressing pressing environmental concerns. The report's key findings were:

- Development and sustainability must go together. Policy and market incentives must encourage sustainable choices that appeal beyond personal benefit. Business as usual is no longer an option. Solutions moving forward must consider modern science and technology as well as local and traditional knowledge. Science and technology have increased production while failing to address social and environmental consequences. New priorities in science, technology, institutions, development and investment must recognise and address the multiplicity of agriculture within diverse social and ecological contexts. Choices made at this juncture in history will determine how we protect our planet and secure our future.

- Agriculture is as complex and diverse as the various cultures and landscapes in which it takes place. Agriculture impacts biodiversity and ecosystem services, climate change and water resources. Agricultural knowledge, science and technology must be retooled to address the needs of the poor and small-scale farmers in diverse ecosystems. The new bottom line must take into account relationships among productions, social and environmental systems. Agricultural practices and policy must empower marginalised stakeholders to sustain the diversity of agriculture and food systems, including cultural dimensions.

- The adverse consequences of the new global economy have had the most significant negative impact on the poorest and most vulnerable. Climate change may have the most adverse consequences where the potential to improve productivity is lowest. The mounting crisis in food security is like nothing we've seen before. Food, fibre and fuel must be produced in a manner that enhances environmental and cultural services. It is critical to assess the potential environmental, health and social impacts of any technology.

- Farming communities, farm households and farmers are producers and managers of ecosystems. Institutional changes should benefit those who have historically been served the least by agricultural knowledge, science and technology and must improve their access to food, seeds, germplasm and improved technologies. Appropriate technology can help rehabilitate degraded land, reduce environmental and health risks associated with food production and consumption and sustainably increase production. Organic, fair-trade and other value-added mechanisms should be encouraged locally and provided markets for locally produced food and for export.

- Success will require reprioritised, redirected public and private investment in agricultural, science and technology, supporting policies and institutions, acknowledgement and interdisciplinary, systems-based holistic approach to knowledge-gathering and sharing. Knowledge systems combined with human ingenuity and a shift to no hierarchical developments can meet the challenge of increasing productivity while protecting the environment and preserving human dignity.

Organic farming or organic agriculture and biodiversity are key factors for resisting climate change to lessen its food-production impact; methods such as using cover crops, crop rotation and composting play a lead role in helping to transform poor soil into productive farms. According to the Rodale Institute,[149] research shows that organically-managed soils can store (sequester) more than 1,000 pounds of carbon per acre, while non-organic systems can cause carbon

---

[149] Rodale Institute: *"Leaders in organic Solutions since 1947."* The institute runs "New Farm" which has inspired and informed farmers worldwide for more than 29 years, explaining how to make regenerative farming profitable and build supportive communities; 611 Siegfriedale Road, Kutztown, PA 1930-9320, U.S. http://www.rodaleinstitute.org/contact.

loss. For consumers, this means that the simple act of buying organic products can help to reduce global warming. Organic farming excludes or *strictly* limits the use of synthetic fertilisers and pesticides, plant growth hormones, regulators, additives and genetically modified organisms (GMOs) used to feed livestock or plants to induce rapid growth. Since 1990, the market for organic products has grown at a rapid pace, averaging 20-25% per year to reach $33 billion in 2005. A similar increase in organically managed farmland grew worldwide to approx. 306,000 sq km (3.6 million hectares) worldwide, representing about 23% of total as of 2005, and organic wild products are farmed on approx. 62 million hectares.

Organic wild products are collected from wildland. As of 2005, organic wild products are famed on approx. 62 million hectares, according to the International Federation of Organic Agricultural Movements (IFOAM).[150] Thirty-six percent of these organic wild plants are *bamboo shoots,* 21% *fruits* and *berries,* 19% *nuts.* IFOAM's report summarises its objectives partly in its "Executive Summary":

- *Reorienting and Redesigning Agriculture In an Organic Direction:* Requires several different functions, actions and strategies that complement and reinforce each other – a solid foundation of common values among key actors; knowledge and experience; and engaged and dynamic individuals and organisations; possibilities for consumers to identify organic products in the evermore anonymous food market; political lobbying; and interested and bold market actors. Deep and broad cooperation and dialogue among the stakeholders in the whole food sector, from consumers to decision-makers, from farmers to scientists, is essential, and their participation in strategic deciscions is fundamental for success.

- *Consumer Interest and Willingness:* Consumer interest and willingness to buy organic food is the foundation for market development. Consumer awareness is built with availability of good quality products and positive promotion, and a common logo and standard is an efficient tool for promotion. The media play an immportant role in spreading the values of organics, informing about the logo and presenting good examples. Market information is an important tool for all market actors, not least the public sector and the farmers. To organise the farmers and producers for marketing is important for the supply and quality improvement. The initial markcting efforts should be oriented towards simple chains and direct marketing, but for long-term growth of the organic sector, development of a diversity of market channels is essential. A combination of market and demand measures is the most effective strategy. Export often plays a big role in the initial stage, and exporters need to consider the special demands of the organic markets.

- *Certification:* Is a strong market tool that serves to build trust in organic agriculture and products. One organic standard that is applied by all organic producers, certified or not, helps to build energy and joint activities in the sector. Stakeholder involvement is critical in standard development, especially in the early stage. Third-party certification is by far the most common, but there is a growing interest in *alternatives* like PGS, and it is important that governments do *not inhibit* this development, as formal certification may not be what is demanded in the domestic market. The initial standard should be developed with local market development in mind, and a locally based certification body often plays a big role in this. The introduction of an organic regulation means an official recognition of organic that will strengthen the sector and make it *visible* and *credible* in both the public and private sectors. However, a *mandatory* regulation is not the only way for a government to accomplish this.

---

[150] International Federation of Organic Agricultural Movements (IFOAM):*"Building Sustainable Organic Sectors Report."* IFOAM headquartered in Bonn, Germany, was formed in 1972, and has 710 members. Its mission is: leading, uniting and assisting the organic movement in its full diversity, its goal being the worldwide adoption of ecologically, socially and economically sound systems that are based on the principles of Organic Agriculture; www.ifoam.org.

- *Government Support:* Governments support organic farming for a mix of reasons: *reduction* of imports of agrochemicals; *income* generation through exports; *environmental* protection; and *low-cost,* environmentally-friendly farming accessible to small-scale farmers. National strategies or action plans have contributed to organic development in many countries. They are most efficient when they relate to goals or consist of a combination of specific measures, including direct income support through the agro-environment, rural development programmes; marketing and processing support; certification support, producer information initiatives (research, training and advice); consumer education; and infrastructure support. Bread stakeholder involvement facilitates this process and gives the best results. Without this kind of dialogue, the organic sector will be less successful and development will be slower. There are also other incentives that can have importance for the farmers, e.g., a *reduced* credit rate for entrepreneurs in the organic sector, access to crop insurance and agricultural disaster programmes. Involvement of local or regional governments and authorities with the organic sector often leads to constructive and relevant developments; and churches and international institutions play a supportive role in organic agriculture policy development. Information and statistics on the sector activities (size and expansion of production and markets, policies, etc.) and active organisations are a major resource for strategy building.

- *Conversion to Organic:* This means a mental conversion of the whole food sector, and positive attitudes must be created with adequate information. Education, extension and research are therefore essential in all organic development. A research programme descrbing the most urgent research needs is a help in prioritising research projects, and all relevant stakeholders should be involved in its elaboration. A new approach has been developed where dialogue, participation, and exchange of experience inspire both farmers and researchers, and where traditional knowledge is appreciated and integrated. Extension services need to consider all aspects of the farmer's situation, from production to marketing; economy and social situation. Cooperation and linkages among farmers, advisors and scientists are important for relevance and efficiency, and the farmers and advisors can be generators of creative and feasible research projects. In the healthy development of an organic sector, a wide range of relevant stakeholders are invited to cooperate and contribute. It is a winning concept to have a dialogue not only with those who from the beginning are *positive* towards organic, but also with conventional farmers' organisations, authorities, market actors, etc. For strategic decisions an ongoing analysis about the development measures is vital. Unification on the national level creating common concepts and messages also is a great strength, while development of local organisations and activities is an important life nerve of the organic movement. In a young organic sector, a good strategy to win respect and allies is to focus on the positive contribution of organic and common points of interest instead of criticising the current policies of institutions and organisations. When the sector grows, different perspectives of organic will develop and one challenge is to find new forms of communication. How to keep the integrity of organic agriculture while allowing growth and expansion is a main discussion issue for the organic sector.

To *strictly* limit the use of agrochemicals in agriculture that suit organic farming, biological control of pests, insects, mites, weeds and plant diseases relies on *predation, parasitism, herbivory,* or other natural mechanisms. These natural mechanisms can be an important component of integrated pest management (IPM) programmes. IPM's strategy uses an array of contemporary methods: mechanical and physical devices which involve genetic, biological, legal, and cultural methods, accomplished in three stages: *prevention, observation,* and *intervention.* This ecological approach has the main goal of significantly *reducing* or *eliminating* the use of pesticides.

Cooperative farming or collective farming, social or worker cooperative unions, savings, banking or consumer, agricultural cooperatives, are forms of farm enterprises in which the farmers pool their farm resources together and democratically run them. These enterprises help the poor and marginalised as well as generally use regenerative farming and organic farming methods. Various forms of cooperative federation farming organisations exist throughout the world. The International Cooperatives Alliance's (ICA)[151] top 300 farming organisations are described as follows:

---

[151] International Cooperatives Alliance (ICA): *"ICA Global 300 Ranking by Turnover USD (2000 Data)."*: The 300 Cooperatives include: *Zen-Noh*/*SOK Corporation* / (National Federation of Agricultural Cooperatives) / Food&Agri / Japan 1948 / www.zenoh.or.jp; *Zenkyoren /* Insurance/Japan 1951/ www.ja-keyosai.or.jp; *Credit Agricole Group /* Div. Financials / France 1897 / www.credit-agricole.fr; *National Agricultural Cooperative Federation* (NACF) / Div. Financials / Korea 1961 / www.nonghyup.com; *China National Agricultural Means of Production Group* Corporation/Food&Agrii//China 1982/www.sino-agric.com; *Nationwide Mutual Insurance* Company / Insurance /U.S., 1952 / www.nationwide.com; *Edeka Zentrale AG /* Retailing/Germany 1898/www.edeka.de; *Groupama* / France 1899 / www.groupama.com; *Eureko /* Insurance / Netherlands 1992 / www.eureko.net; *Mondragon Corporation /* Materials/Spain 1956 / www.moc.es; *Migros* / Retailing / Switzerland 1925 / www.migros.ch; *The Norinchukin Bank /* Banks / Japan 1923 / www.nochubank.or.jp; *Groupe Caisse D'Espargne /* Banks / France 1983 / www.groupe.caisse-esparge.fr; *CHS Inc /* Food&Agri / Retailing / UK 1863 / www.chsinc.com; *The Cooperative Group /* Retailing / UK 1863 / www.co-op.co.uk; *Confederation Nationale du Credit Mutuel /* Banks / France / 1880/http://www.creditmutuel.com; *Robobank Group /* Div. Financials / Netherlands 1898 / www.rabobank.com; *Debeka Group /* Insurance/Germany 1905 / www.debeka.de; *John Lewis Partnership PLC* / Retailing / UK.../ www.johnlewispartnership.co.uk; *Coop Norden* / Retailing / Sweden 2002 S&N&D / www.coopnorden.com; *Metsäliitto* / Food&Agri / Finland 1934 / www.metsaliito.com; *Coop Swiss* / Retailing / Switzerland 1890 / www.coop.ch; *Conad* (Consorzio Nationale Dettaglianti) / Retailing / Italy 1962 / www.conadi.it; *R+V Versicherung AG* / Insurance / Germany 1922 / www.ruv.de; *UNIPOL* / Div. Financials / Italy 1963 / www.unipolonline.it; *Group Banques Populaires* / Div. Financials / France 1999 / www.banquespopulaires.fr; *Edeka Minden eG* / Retailing / Germany 1920 / www.edeka.de; *BayWa Group* / Food&Agri / Germany 1923 / www.baywa.de; *SOK Corporation* / Retailing / Finland 1904 / www.s-kanava.net; *VGZ-IZA-Tria* / Health / Netherlands 1992 / www.vgz.nl; *Fonterra Cooperative Group* / Food&Agri / New Zealand 2001 7 www.fonterra.com; *Danish Crown* / Food&Agri / Denmark 1887 / www.danishcrown.dk; *Desjardins Group* / Div. Financials / Canada 1900 / www.desjardins.com; *Arla Foods* / Food&Agri / Denmark 1975 / www.arlafoods.com; *Dairy Farmers of America* / Food&Agri / U.S., 1998 7 www.dfamilk.com; *Wakefern Food Corp* / Retailing / U.S., 1946 / www.shoprite.com; *Edeka Südwest eG* / Retaing / Germany 1950 / www.edeka.de; *Land O'Lakes* / Food&Agri / U.S., 1921 / www.landolakesinc.com; *La Mondiale Groupe* / Insurance / France 1905 / www.lamondialpartenaire.co.fr; *Huk Coburg* / Insurance / Germany 1933 / www.huk.de; *MACIF* / Insurance / France 1960 / www.macif.fr; *Azur GMF* / Insurance / France 1934 / www.gmf.fr; *MMA* / Insurance / France 1828 / www.mma.fr; *Suedzucker* / Food&Agri / Germany 1837 / www.suedzucker.de; *Royal Friesland Foods* / Food&Agri / Netherlands 1879 / www.frieslandfoods.com; *Signal Iduna* / Insurance / Germany 1907 / www.signal-iduna.com; *RZB* / Banks / Austria 1927 / www.rzb.at; *DZ Bank* / Banks / Germany 2001 / www.dg.dzbank.de; *CZ* / Health / Netherlands 1930 / www.cz.nl; *Catholica Assicurazioni* / Insurance / Italy 1896 / www.catholicassicurazioni.it; *Zonrosai* / Insurance / Japan 1957 / www.zonrosai.org.jp; *Ethias* / Insurance / Belgium 1919 / www.europe.ethias.be; *AGRAVIS Raiffeisen AG* / Food&Agri / Germany 1875 / www.agravis.de; *Associated Wholesale Grocers* / Retailing / U.S., 1926 / www.awginc.com; *Federated Cooperatives Limited* / Retailing / Canada 1955 / www.fel.ca; *Campina* / Food&Agri / Netherlands 1979 / www.campina.com; *Lantmännen* / Food&Agri / Sweden 1905 / www.lantmannen.com; *SNS REAAL* / Div. Financials / Netherlands 1997 / www.snsreaal.nl; *Gothaer* / Insurance / Germany 1820 / www.gothaer.de; *MAAF* / Insurance / France 1950 / www.maaf.fr; *WAZ Bank* / Banks / Germany / www.wgz-bank.de; *Cooperative d'Exploitation et de Repartition Pharmacieutiques de Rouens* / Retailing / France / www.cerp-rouen.fr; *Community Credit Cooperative* (cc) / Div. Financials / Korea 1963 / www.kfcc.co.kr; *Cooperatif Terrena Group* / Food&Financials / France 2003 / www.terrena.fr; *MAIF Group* / Insurance / France 1934 / www.maif.fr; *Ace Hardware* / Retailing / U.S., 1924 / www.acehardware.com; *Fenaco* / Food&Agri / Switzerland 1925 / www.fenaco.ch; *Länsförsäkringar* / Insurance / Sweden 1936 / www.lansforsakringar.se; *JCCU* / Retailing / Japan 1951 / www.jccu.coop; *Sodra Skogsagama* / Food&Agri / Sweden 1938 / www.sodra.com; *United Cooperatives Ltd* / Retailing / UK 2002 / www.united.coop; *AGIS* / Health / Netherlands 1827 / www.agisweb.nl; *Societa Reale Mutua di assicurazioni* / Insurance / Italy / www.realemutua.it; *Invivo* / Food&Agri / France 2001 / www.invivo-group.com; *Growmark Inc.* / Food&Agri / U.S., 1927 / www.growmark.com, *Humana Milch union* / Food&Agri / Germany 1934 / www.humana.de; *KF Group* (The Swedish Cooperative Union) / Retailing / Sweden 1899 / www.kf.se; *Nationwide Building Society* / Banks / UK 1848 / www.nationwide.co.uk; *Britannia Building Society* / Banks / UK 1848 / www.britannia.co.uk; *Cheques Cooperatifs Dejeuners-Societe Cooperative de Production* / Consumer Services / Franc / www.chequesdejeuner.fr; *Mutual of Omaha* / Insurance / U.S., 1909 / www.mutualofomaha.com; *OP Bank Group* / Banks / Finland 1902 / http://www.oko.fi; *DL Group* / Food&Agri / Denmark 1955 / www.dlg.dk; *Unicoop Firenze* / Retailing / Italy 1854 7 www.coopfirenze.it; *Do-it-Best Corp* / Retailing / U.S., 1945 / www.doitbestcorp.com; *Shinkin Central Bank* / Banks / Japan 1950 / www.shinkin-central-bank.jp; *Unified Western Grocers Inc* / Retailing / U.S., 1925 / www.uwgrocers.com; *SMABTP* / Insurance / France 1859 / www.smabtp.fr; *TINE BA* / Food&Agri / Norway 1881 /www.tine.no; *Tereos* / Food&Agri / France 2004 / www.tereos.com; *Noweda eG Apothekergennossenschaft* / Retailing / Germany / www.noweda.de; *Cooperative Federee du Quebec* / Food&Agri / Canada 1939 / www.coopfed.qc.ca; *E. Leclerc* (S.C. Galec) / Retailing / France 1962 / www.e-leclerc.com, *FloraHolland* / Food&Agri / Netherlands 1931 / www.floraholland.nl; *Irish Dairy Board Cooperative Ltd* / Food&Agri / Ireland 1961 / www.idb.ie; *AgriBank, FCB* / Div.financials / U.S., 1992 / www.agribank.com; *Sodiaal Union* / Food&Agri / France 1990 / www.sodiaal.fr; *The Greenery* / Food&Agri / Netherlands 1996 / www.thegreenery.com; *Nordmilch* / Food&Agri / Germany 1993 / www.nordmilch.de; *Socopa* / Food&Agri / France 1998 / www.socopa.fr; *Health Partners* / Health / U.S., 1957 / www.healthpartners.com; *ReWe Dortmund AG* / Retailing / Germany 1913 / www.rewe.de; *Coop Adriatica* / Retailing / Italy 1995 / www.adriatica.e-coop.it; *KLP Insurance* / Insurance / Norway 1949 / www.klp.no; *Glanbia* / Food&Agri / Ireland 1964 / www.glanbia.com; *Coop Kobe* / Retailing / Japan 1921 / www.kobe.coop.or.jp; *Gilde* / Food&Agri / Norway 1964 / www.gilde.no; *Ag Processing Inc.* / Food&Agri / U.S., 1983 / www.AGP.com; *Folksam* / Insurance / Sweden 1908 / www.folsam.se; *California Dairies Inc.* / Food&Agri / U.S., 1990 / www.californiadairies.com; *Group Health Cooperative Puget Sound* / Health / U.S., 1947 / www.ghc.org; *Copersucar* / Eood&Agri / Brazil 1959 / www.copersucar.com.br; *Bloemenveiling Aalsmeer* (VBA) / Food&Agri / Netherlands 1911 / www.vba.nl; *Cobank* / Banks / U.S., 1916 / www.cobank.com; *Mobiliar* / Insurance / Switzerland 1826 / www.mobi.ch; *Kyoei Fire & Marine Insurance Co Ltd* / Insurance / Japan 1942 / www.kyoeikasai.co.jp; *Gjensidige Forsikring* / Insurance / Norway 1921 7 www.gjensidigenor.no; *The Cooperatives* / Insurance / Canada 1949 / www.cooperators,ca; *RWZ Rhein-Main* / Food&Agri / Germany 1890 / www.zwz.de; *Valio Group* 7 Food&Agri /

Finland 1905 / www.val.fi; *Coop Expansion Repartion Pharmacie Lorraine* / Retailing / France / www.cerp-lor.com; *PAC.2000 A* / Retaiking / Italy 1972 / www.PAC2000a.it; *True Value Cooperation* / Retailing / U.S., 1948 7 www.truevalue company.com; *Agropur* / Food&Agri / Canada 1938 / www.agrapur.com; *Agrial* / Food&Agri / France 2000 / www.agrial.fr; *CUNA Mutual Group* / Insurance / U.S., 1935 / www.cunamutual.com; *Groupe MACSF* / Insurance / France 1935 / www.macsf.fr; *U.S. Central Credit Union* / Div. Financials / U.S., 1974 / www.uscentral.org; *COSUN* / Food&Agri / Netherlands 1970 / www.cosun.com; *Navy Federal Credit Union* (NFCU) / Div. Financials / U.S., 1933 / www.navyfeu.org; *Coopagri Bretagne* / Food&Agri / France 1911 / www.coopagri-bretagne.com; *Emmi* / Food&Agri / Switzerland 1925 / www.emmi.ch; *Portman Building Society* / Div. Financials / UK 1846 / www.portman.co.uk; *Coop Sapporo* / Retailing / Japan 1965 / www.coop.sapporo.or.jp; *Foodstuffs (Auckland) Limited* / Retailing / New Zealand 1925 / www.foodstuffs.co.nz; *CERP Rhim Rhone Mediterranee* / Retailing / France / www.cerp-rrm.com; *Unicopa* / Food&Agri / France 1960 / www.unicopa.fr; *Coop Schleswig-Holstein* / Retailing / Germany 1899 / www.coop.de; *Swiss Union of Raiffeisen Banks* / Banks / Switzerland 1902 / www.raiffeisen.ch; *Taiwan Cooperative Bank* / Banks / Taiwan 1946 / www.tcb-bank.com.tw; *West Fleisch* / Food&Agri / Germany 1928 / www.westfleisch.de; *Tradeka* / Retailing / Finland 1917 / www.tradeka.fi; *Sperwer* / Retailing / Netherlands 1928 / www.sperwer.nl; *CECAB Group* / Food&Agri / France 1968 / www.cecab.com; *Ag First farm Credit Bank* / Div. Financials / U.S., 1916 / www.agfirst.com; *Alte Leipziger Hallesche* / Insurance / Germany 1819 / www.schole.al-h.de; *Norske Felleskjop* / Retailing / Norway 1896 / www.fk.no; *Royal London Mutual Insurance* / Insurance / UK 1861 / www.royallondongroup.com; *Groupe MGEN* / Insurance / France / www.mgen.fr; *NFU Mutual* / Insurance / UK 1908 / www.nfumutual.co.uk; *Coop Estense* / Retailing / Italy 1989; *National Cable Television Cooperative, Inc.* / Utilities / U.S., 1984 / www.cabletvcoop.org; *HOK Elanto* / Retailing / Finland 2003 / www.hok-elanto.fi; *Southern States Cooperative* / Food&Agri / U.S., 1923 / www.southernstates.com; *Landgard* / Food&Agri / Germany 1932 / www.landgard.de; *HAGe Kiel (RHG Nord)* / Retailing / Germany / www.hage-kiel.de; *Murray Goulburn Cooperative Co. Ltd* / Food&Agri / Australia 1950 / www.mgc.com.au; *P&V* / Insurance / Belgium 1907 / www.pv.be; *Unicoop Tirreno* / Retailing / Italy 1945; *Swedish Meats* / Food&Agri / Sweden / www.swedishmeats.com; *Atria Group* / Food&Agri / Finland 1990 / www.atria.fi; *Darigold* / Food&Agri / U.S., 1918 / www.darigold.com, *Coop Lombardia* / retailing / Italy 1945 / www.cooplombadia.it; *SACMI* / Materials / Italy 1919 / www.sacmi.com; *Associated Food Stores Inc* / Retailing / U.S., 1940 / www.afstores.com; *National Agricultural Cooperative Marketing Federation of India* (NAFED) / Food&Agri / India 1958 / www.nafed-india.com, *Conad del Tirreno* / Retailing / Italy 1997 / www.conad.it; *Hochwald* / Food&Agri / Germany 1932 / www.hochwald.de; *Champagne Cereales* / Food&Agri / France 1927 / www.champagne-cereales.co.fr; *Limagrain* / Food&Agri / France 1942 / www.limagrain.fr; *Cristal Union* / Food&Agri / France 1998 / www.cristal-union.fr; *United Farms of Alberta* / Retailing / Canada 1909 / www.ufa.net; *Staplcotn, Inc.* / Food&Agri / U.S., 1921 / www.staplcotn.com; *Consum* / Retailing / Spain 1959 / www.consum.es; *Coop Tokyo* / Retailing / Japan 1957 / www.coopnet.or.jp; *Western Coroorate Federal Credit Union* (WestCorp) 7 Div. Financials / U.S., 1969 / www.westcorp.org; *Prairie Farms Dairy, Incorporated* / Food&Agri / U.S., 1938 / www.prairiefarmsdairy.com, *Foodstuffs (Wellington) Cooperative Society Ltd* / Retailing / New Zealand 1925 / www.foodstuffs.co.nz; *Midlands Cooperative Society* / Retailing / UK 1925 / www.midlandsco-op.com, *NTUC Income* / Insurance / Singapore 1970 7 www.income.coop; *Cooperl Hunaudaye* / Food&Agri / France 1996 / www.cooperl.hunaudaye.fr; *PPCS Limited* / Food&Agri / New Zealand 1948 / www.ppcs.co.nz; *Coop Atlantique* 7 retailing / France 1912 / www.coop-atlantique.fr, *Intres* / Retailing / Netherlands / www.intres.nl; *Associated Wholesalers, Inc* / Retailing / U.S., 1962 / www.awweb.com; *Febelco* / Retailing / Belgium 7 / www.febelco.be; *Liverpool Victoria* / Insurance / UK 1843 / www.liverpoolvictoria.co.uk; *Plains Cotton Coop Association* 7 Food&Agri / U.S., 1953 / www.pcca.com/; *Foremost Farms* 7 Food&Agri / U.S., 1995 / www.foremostfarms.com; *Coop Consumatori Nordest* / Retailing / Italy 1905 / www.e-coop.it; *Foodstuffs Ltd.* (South Island) / Retailing / New Zealand 1925 / www.foodstuffs.co.nz; *HK Ruokatalo* / Food&Agri / Finland 1975 / yrity.hk.ruokatalo.fi; *Epis* Centre / Food&Agri / France 1937 / www.epicentre.com; *Coop* Kanagawa / Retailing / Japan 1946 / www.kanagawa-coop.or.jp; GranLatte / Food&Agri / Italy 1959 / www.granlatte.it; *Bäko-Zentrale* Süddeutschland / Food&Agri / Germany 1917 / www.baeko-zentrale-sued.de; *Ocean* Spray / Food&Agri / India 1969 / www.oceanspray.com; **IFFCO-Indian Farmers Fertiliser Cooperative** Limited / Food&Agri / India 1967 / www.iffco.nic.in; **SSQ Financial** Group / Div. Financials / Canada 1944 / www.ssq.ca; Novacoop / Retailing / Italy 1960 / wwwnovacoop.it; **Recreational Equipment** Inc / Retailing / U.S., 1938 / www.rei.com/; **Groupe** Even / Food& Agri / France; **Seminole Electric Corporation** Inc / Utilities / U.S., 1948 / www.seminole-electric.com; **Groupo de Alimentation** Coren / Food&Agri / Spain 1959 / www.coren.es; **Associated Milk Producers Inc.** / Food&Agri / U.S., 1969 / www.ampi.com; **Affiliated Foods Incorporated** Company / Retailing / U.S., 1946 / www.afiama.com; **COAMO-Agroindustrial** cooperativa / Food&Agri / Brazil 1968 / www.coamo.com.br; **Ogelthorpe Power** Corporation / Energy / U.S., 1974 / www.opc.com, **Consorzio Cooperative** Construzioni / Capital Goods / Italy; Les *Cooperateurs de Normandie-Picardie* / Retailing / France / www.coop-cnp.coop; *Topiola* / Div. Financials / Finland 1858 / www.tapiola.fi; *Central Grocers Cooperative Inc* / Retailing / U.S., 1918 / www.central-grocers.com; *Nordiconad* / Retailing / Italy 1993 / www.conad.it; *Chelsea Building Society* / Div. Financials /UK 1875 / www.thechelsea.co.uk; *Dairylea Cooperative Inc* / Food&Agri / U.S., 1907 / www.dairylea.com; *Groupe Euralis* / Food&Agri / France 1936 / www.euralis.fr; *Dairy Farmers of Britain* / Food&Agri / UK 2002 / www.dfob.co.uk; *Skipton Building Society* / Banks / UK 1853 / www.skipton.co.uk; *Affiliated Foods Midwest Coop. Inc* / Retailing / U.S., 1931 / www.afmidwest.com; *Groupe Coop Alsace* / Retailing / France 1902 / www.coop-alsace.ccop; **Markfed Andhra Andrah Pradesh** / Food&Agri / India 1958 / www.apmarkfed.org, *Conad Romagna* (Commercianti Indipendenti Associati) / Retailing / Italy 1959 / www.conadfo.it; *National Rural Utilities Cooperative Finance Corporation* (CFC) / Div. Financials / U.S., 1959 / www.nrucfc.org; *Farm Credit Bank of Texas* (FCBT) / Banks / U.S., 1916 / www.farmcreditbank.com; *American Crystal Sugar Co.* / Food&Agri / U.S., 1973 / www.crystalsugar.com, *Milk Link* / Food&Agri / UK 2000 / www.milklink.co.uk; *CCPL* (Consortium Coop. Produz Lav. S.c.r.i) / Commercial Services / Italy 1904 / www.ccpl.it; *ENAP, Inc.* / Utilities / U.S., 1967 / www.enap.com/; *Bayernland* / Food&Agri / Germany 1930 / www.bayernland.de; *Amul* (Gujarat Cooperative Milk Marketing Federation) / Food&Agri / India 1966 7 www.amul.com, *ReWe Group* (Zentral-AktiengesellschaftFU) / Retailing / Germany 1927 / www.rewe.de; *Saitama Coop* / retailing 7 Japan 1947 / www.coopnet.or.jp/saitama/; *Coop Consorzio Nord Ovest* / Retailing / Italy; *MFA Incorporated* / Food&Agri / U.S., 1914 / www.mfaincorporated.com; *MFA Oil Company* / Energy / U.S., 1929 / www.mfaoil.com/; *North Carolina Electric Membership Corp.* / Energy / U.S., 1949 / www.ncemcs.com/; *Avebe Group* / Food&Agri / Netherlands 1919 / www.avebe.name; *Milkobel* / Food&Agri / Belgium 1887 / www.milcobel.be; *Capsa* / Food&Agri / Spain 1959 / www.capsa.es; *Wesleyan Assurance Society* / Insurance / UK 1841 / www.wesleyan.co.uk/; *Maryland & Virginia Milk Production Cooperative Association* / Food&Agri / U.S., 1920 / www.mdvmilk.com; *Sunkist Growers* / Food&Agri / U.S., 1893 / www.sunkist.com; *Miyagi Coop* / Retailing / Japan 1932 / www.miyagi.coop; *Coopercentral Aurora* / Food&Agri / Brazil 1973 / www.auroraalimentos.com.br; *Cehave Landbouwbelang* / Food&Agri / Netherlands 1993 / www.cehave.com; *Riceland Foods* / Food&Agri / U.S., 1921 / www.riceland.com/; *Fairprice* / Retailing / Singapore 1973 / www.fairprice.com.sg; *Brazos Electric Power Cooperative* / Utilities / U.S., 1942 / www.brazoselectric.com; *Manutencorp* / Materials / Italy 1938 / www.manutencoop.it; *Coop Liguria* / Retailing / Italy 1945 / www.legaliguria.coop; *Lactogal* / Food&Agri / Portugal 1994 / www.lactogal.pt; *Associated Electric Cooperative inc.* / Energy / U.S., 1961 / www.aeci.org/; *Dairy Farmers Group* / Food&Agri / Australia 1900 / www.dairyfarmers.com.au; *Tri State*

Collective Farming: An organisation of agricultural production which several farmers run as a joint enterprise based on common ownership of resources on pooling of labor and income in accordance with the theoretical principles of cooperative organisations.

-

Worker Cooperative: A worker cooperative is owned and democratically controlled by its owners. No outside owners in a "pure" workers' cooperative, only the workers own shares of the business. Membership is not compulsory for employees, but generally, they can become members. In India, there is a form of workers' cooperative which insists on compulsory membership for all employees and compulsory employment for all members, as for the example, the Indian Coffee Houses.

-

Business & Employment Cooperatives (BEGs): These are a subset of the worker cooperatives that represent a new approach to providing support to the creation of new businesses: BEGs enable budding entrepreneurs to experiment with their business idea while benefiting from a secure income. The innovation BECs' introduction is that once the business is established, the entrepreneur is not forced to leave and set up independently, but can stay and become a full member of the cooperative. The micro-enterprises thus combine to form one multi-activity whose members provide mutually supportive environment for each other. BECs thus provide budding business people with an easy transition from inactivity to self-employment, but in a collective framework. They open up new horizons for people who have ambition but lack the skills or confidence needed to set off entirely on their own or who simply want to carry on an independent economic activity but within a supportive group contact.

-

Social Cooperative or *Social Structure Cooperative* or *Italian Social Structure Cooperative:* This is a particularly successful form of multi-stakeholder cooperative. The 7000 social cooperatives in existence are grouped into two types: Type "A" bring together providers and beneficiaries of social service members, with various categories of stakeholders as for example, paid employees, beneficiaries, volunteers

Generation and Transmission Association Inc / Utilities / U.S., 1952 / www.tristategt.org; *Camif Coollectives* / Retailing / France 1987 / www.camif.coop; *CAMST* / Food&Agri / Italy 1945 / www.camst.it; *First Milk* / Food&Agri / UK 2001 / www.first-milk.co.uk; *Old Dominion Electric Cooperative* / Utilities / U.S., 1948 / www.odec.com; *Alliance Agro-Alimentaire* (3A Groupe) / Food&Agri / France 2002 / www.group3a.fr; *Ecclesiatical Insurance* / UK1905 / www.ecclesiastical.co.uk; *OMG* (Cooperativa Muratori e Cementisti) / Capital Goods / Itali 1901 / www.cmc.coop; *Coop Centro Italia* / Retailing / Italy 1861; *Groupe Cooperatif Maisadour* / Food&Agri / France 1936 / www.maisadour.com; *Calgary Cooperative* / Retailing / Canada 1956 / www.calgarycoop.com; *Associated Growers inc* / Retailing / U.S., 1934 / www.agsea.com; *East of England* / retailing / UK 1844 7 www.eastengland.coop; *Central Electric Power cooperative* / Utilities / UK 1884 / www.cepc.net/, *Agrifirm* / Food&Agri / Netherlands 1909 / www.agrifirm.nl; *UNEAL Group* / Food&Agri / France 1992 / www.uneal.fr; *Midcounties Cooperative* / Retailing / UK 1860 / www.midcounties.coop; *U.S. AgBank, FCB* / Div. Financials / U.S., 1916 / www.usagbank.com/; *State Employees Credit Union* / Div. Financials / U.S., 1937 / www.secunmd.org/html/index/1...; *Groupe Intersport* / Retailing / France 1927 / www.intersport.fr; *Dairygold Cooperative Society Limited* / Food&Agri / Ireland 7 www.dairygold.ie; *Great River Energy* / Utilities / U.S., 1999 / www.greatriverenergy.com; *Associated Growers, Inc.* / Retailing / U.S., 1934 / www.agbr.com; *Conad Centro Nord* / Commercial Services 7 Italy 1997 / www.conad.it; *Eurial Poitourine* / Food&Agr / France 1966 / www.eurial-poitouraine.fr; *Piggy Wiggly Alabama Distributing Company* / Retailing / U.S., 1947 / www.pwadc.com/; *Alliance Group Limited* / Food&Agri / New Zealand 1948 / www.alliance.no.nz; *Berglandmilch* / Food&Agri / Austria 1996 / www.berglandmilch.at; *Agralys* / food&Agri / France 1992 / www.agralys.fr; *Associated Press* / Media / U.S., 1848 / www.ap.org; *Conad Adriatico* / Retailing / Italy 1972 / www.conad.it; *PAX Holding* / Health / Switzerland 1959 / www.pax.ch; *Scottish Midlands Cooperative Society Limited* (SCOTMID) / Retailing / UK 1859 / www.scotmid.com, *Groupe Alliance* / Food&Agri / France; *HBF* / Div. Financials / Australia 1941 / www.hbf.com.au; *East Kentucky Power Cooperative Inc.* (EKPC) / Utilities / U.S., 1941 / www.ekpc.com; *La Capitale* / Insurance / Canada 1940 / www.lacapitale.com; *Anglia Regional Cooperative Society Limited* / retailing / UK 1876 7 www.arcs.co.uk/; *Selectour voyages* / Consumer Services / France; *South Mississippi Electric Power Association* / Utilities / U.S., 1941 / www.smepa.coop; *Basin Electric Association* / Utilities / U.S., 1961 / www.basinelectric.com; *Grandi Salumifici Italiani* / Italy; *West Bromich* / Div. Financials / UK 1849 / www.westbrom.co.uk; *Group Orcab* / Capital Goods / France; *Consorzio Nationale Servizi-CNS* / Consumer Services / Italy / www.consoline.it; *GLAC* / Food&Agri / France 1966 / www.glac.fr; *National Grape Cooperative Association* / Food&Agri / U.S., 1869 / www.nationalgrape.com/. ICA was formed in 1895, with 224 active national federation members, and in 2006, published its first major index of the world's largest cooperatives and nutual enterprises, which demonstrates the scale of the cooperative movement globally. ICA has 224 active globally; ICA motto: cooperating among cooperatives, to "serve their members most effectively and strengthen the cooperative movement by working together through local, national, regional and international structures" (The Rochdale Principle); ICA, 15 Route des Morillons, 1218 Grand Saconnex, Geneva, Switzerland; www.ica.coop; global300.coop; icanews.ccop.

(up to 50%), financial investors and public institutions; and Type "B" bring together permanent workers and previously unemployed people who want to integrate into the labor market, provide health and social services, and at least 30% of its members must be from the disadvantaged target groups: people with physical and mental disability, drug and alcohol addiction, developmental disorders and problems with the law. Factors like race, sexual orientation or abuse are not included.

- *Consumers Cooperative:* This is a business owned by its customers, where employees can generally become members, and members vote on major discussions and elect the board of directors from amongst their own number. Consumer cooperatives offer a variety of retail and financial services.

- *Agricultural Cooperatives:* These are widespread in rural areas and offer both marketing and supply cooperatives, promote and actually distribute specific commodities.

- *Cooperative Banking* or *Credit Unions* or *Cooperative Savings Banks:* These provide cooperative banking, credit and savings services.

- *Federal* or *Secondary Cooperatives:* This is a form of cooperative wholesale societies, and cooperative unions, and are a means through which cooperative societies can fulfil the sixth Rochdale Principle: cooperating among cooperatives, with the International Cooperative Alliance (ICA).

- *Cooperative Wholesale Sociaty:* Aims to arrange bulk purchases, and if possible, organised production.

- *Cooperative Union:* This is a second form of cooperative federation whose objective is to develop the spirit of solodarity among societies, and exercise the functions of a government whose authority is merely moral.

Regenerative, organic and cooperative farming methods encompass the use of renewable energies and sustainability of agriculture and ecosystems. Agro-ecosystem biodiversity can be classsified in broadly six ways for agricultural quality sustainability, according to Larry Harrington, thus:[152]

- *Crop Genetic Diversity:* This embraces such factors as varietal concentration; pace of varietal change over time; genetic similarity among major cultivars; the conservation and pyramiding of favourable genes in breeders' varieties; the conservation and use of important genes present in folk varieties; land races, and wild relatives; and opportunities for expanding crop genetic diversity through wide crosses and biotechnology.

- *Crop Species Diversity Over Space:* Spatial species diversity maybe exceedingly narrow (e.g., monocropped rice field) or exceedingly broad (e.g., a home garden featuring simultaneous cultivation of fruit tree, banana plants, coffee, spices, and several food crops). Plots with low species diversity and high species diversity often are found within the same farming system.

---

[152] Larry Harrington, Manager, Natural Resources Group, Centro Internacionale de Mejoramiento de Maizy Trigo (CIMMYT): *"Diversity by Design."* Consultative Group on International Agricultural Research CGIAR) News, vol 4, Number 3, June 1997.

- *Crop Species Diversity Overtime:* Temporal species diversity maybe narrow (e.g., one maize monocrop crop per year); broad within a year (e.g., an annual sequence of multiple cropping involving cereals, legumes, and horticultural crops); or broad over several years (e.g., rice-potato-wheat patterns, broken every few years by a sugarcane crop). Crop species diversity over space and over time are not necessarily related.

- *Agro-ecosystem Diversity Through Crop-Livestock Interactions:* The presence of livestock in a system tends to greatly enhance the value of non-crop components crop residues, grazing lands, forest resources and typically features nutrient cycling between rangeland and cropland. This fosters improved productivity and sustainability of cropping systems and a higher potential for spatial and temporal crop species.

- *Natural Biodiversity Within Agro-ecosystems:* More diverse agro-ecosystems – particularly those with greater spatial diversity, and those with trees – may provide habitat for a wider array of wildlife.

- *Natural Biodiversity As Indirectly Affected By Agro-ecosystems:* Highly productive agro-ecosystems can indirectly foster natural biodiversity by making it unnecessary to farm marginal or fragile areas, or to clear new forest areas for agriculture. System diversity maybe broadened by increasing crop genetic diversity, expanding crop species diversity over space and time, fostering crop-livestock interactions, or improving productivity in favoured agricultural areas to protect biologically diverse fragile, marginal, or forested areas from agriculture.

Agro-ecosystem diversity – by design and demand-led - often can help achieve sustainable improvements in agricultural systems productivity. Researchers and farmers design more biologically diverse agro-ecosystems, which are widely adopted by farming communities. The process includes participatory research on indigenous technical knowledge about system diversity, with a view to exploiting such knowledge to comparable areas. Higher incomes and reduced poverty generated by more productive agricultural practices shift the structure of food demand towards a more diverse array of products, such as fruits, vegetables, and animal products. Farmers follow market signals and diversify their farming systems. The goals of sustainable food security, reduced poverty, improved public health, and conservation of natural bioligical diversity and resources - will be attached through widespread use of more productive, stable, resilient agro-ecosystems and conservation of crop genetic diversity. The Geripa Project in Brazil is a showcase for regenerative agriculture.

2.2.2 The GERIPA Concept of Food Production

THE GERIPA Project (GP)[153] is a new concept for renewable energies and food production with balanced environmental and social consequences. It is a proposal concerning a new social and ecological agro-industry structure to

---

[153] Dr. Romen Corsini: Idealizer of the integrated mini-sugar plant (GERIPA) Concept, sponsored by FAPESP. [www.unicamp.br/fea/ortega/energy/Omett.z.pdf] Editors: E. Ortega/S. Ulgiati: Proceedings of a IV Biennial International Workshop *"Advances in Energy Studies."*Unicamp,Campinas, SP, Brazil, 16-19 June, 2004, pp. 323-328; A.R. Ometto/P.Ramos/G. Lombardi represented Embrapa Satellite Monitoring, Brazil, Brazil/Instituto Politecnico de Havane, Cuba/USP Escola de Engenharia de Sao Carlos, Brazil.

produce energy and food in a sustainable way. It integrates the production of alcohol, food and electricity with a working period of 12 months – eight with sugarcane, and four with sorghum. The transport of materials runs on engines with vaporised alcohol. It considers thermodynamics optimization, with social and environmental quality, based on the Kyoto Protocol and the premises of Sustainable Development. The project occupies 4310 hectares, integrating the whole productive process like a *live organism.* It uses 40,000 litres of alcohol produced daily, as well as 5.2 MW of electricity, and 4760 tons per year of food. Once under economically stable operation, it can generate up to 5600 permanent jobs. With these characteristics, the project could provide the basic needs of a city of 17,000 inhabitants, a strategic setting for autonomous regional development.

The GERIPA experiment can replace the world's fossil-based food production methods. Brazil like the rest of the conventional world food industries, adopts the conventional sugar and alcohol production system (SAPS). This involves the use of sizeable agricultural areas, devastation of forests and biodiversity loss resulting from single-culture techniques, where fertilisers, pesticides, water and fire are intensely used. SAPS's environmental impacts include: burning before harvest, intense use of pesticides, excessive mechanical tillage of soil, unfair treatment of rural workers with low wages and strenuous work, land concentration, unfair land tenancy revenues, and inept use of distillery effluents. Proposed new methods in pursuit to overcome these problems are expressed in the creation of a project of social and ecological agro-industry entity in order to *generate renewable energy integrated with food production* or GERIPA.

Project GERIPA offers an *alternative* to the alcohol and sugar production in Brazil, and most probably, elsewhere in the world as well: the model is medium-sized and could be implanted as integral systems distributed throughout the country. GERIPA can be viewed as a genuine, innovative transformation, compared to conventional production models based on the large plantations. The actual challenge to implement GERIPA project rests in structuring a framework of institutional arrangements and regulatory order that would ensure the successful implementation of rural projects. The GERIPA procedure meets requirements for financing means such as the *Clean Development Mechanism* (UN's Kyoto Protocol). It is envisaged and designed to provide the basis to decentralise the rural development nodes in suitable areas for sugarcane production, provide acceptable quality jobs, and technical solutions in which to allow the feasibility for sustainable development. There is also the additional possibility for productive activities derived from their outposts, such as electricity, agricultural and animal farming products. And most importantly, GERIPA offers a feasible solution to face *exhorbitant* petroleum prices and concerns, and create jobs.

The area occupied by the GERIPA Project has the following distribution: *1589ha.,* for sugarcane; *1812ha.,* for sorghum rotated with fruits and other alimentary crops; *188ha.,* for semi-confined cattle; *930 ha.,* for forest reserve; *6 ha.* for industrial sectors, facilities, infrastructure, etc; and *35 ha.,* for other uses – altogether a total of *4,540 ha.* The process from sugarcane and sorghum to final products closes loops as a pedagogical tool, and GERIPA's characteristics and procedures are:

- *Farming:* The sugarcane is *organically* fertilised by *recycling* bio-fertiliser using dropping irrigation. Excellent conditions are created for cultivation without need for rotation crop. The time span of sugarcane doubles to 8 grinds, production increases up to 120 t/ha., (8 year average). This project assumes the productivity of 90 ha., t/ha., for sugarcane and 50 t/ha., for sorghum. The sugarcane and the sorghum are cut at its base. The production of sorghum lasts only 4 months and, in the remaining 8 months on the same area, food crops are produced in convenient soil conditions. The occupied agricultural area is smaller than the traditional one for the same amount produced. It does not promote *burning* during harvest, as in the traditional sugarcane production, maximising the use of biomass. It avoids using chemicals and biological control is adopted. It also considers the environmental and social legislation, as well as techniques and technologies of environmental control. All the transportation is powered by *vaporized* alcohol with no *petrol* consumption. GERIPA destines 20 of the total area for legal reserve, in addition to Permanent Preservation Area according to legislation.

- *Raw Material Preparation:* The whole thatch is sent to the qualification, preparation and cleaning section. Afterwards, the tips are cut and dispatched to the cattle feeding lot. The cleaning of the thatches is performed by pneumatic equipment, 60% of the straw is used for soil protection and 40% for boiler burning.

- *Juice Preparation:* The thatches are chopped and then shredded. Juice extraction is carried out in a high performance diffuser with efficiencies of 98%. The juice treatment is conventional, with cleaning, sterilisation, adjustment of concentration and temperature, and addition of yeast. It also produces wet *bagasse*[154] and low pressure steam.

- *Alcohol Production:* The section has four containers, a large one for containers operation where fermentation begins and three others that are batch-operated, until fermentation is complete. The wine formed is centrifuged for yeast extraction and, afterwards, it feeds into two columns distillery, where hydrated alcohol, vinasse and other sub-products are produced. GERIPA equipment is electrically-powered.

- *Bio-digestion:* The solid waste, part of the bottom of the containers, the dejects from cattle raising, dilated and treated, vinasse with pH ranging from 3.5 to 4 and the water from the dehydrator are mixed, adjusting the concentration, where methane-rich gas and pH-neutral is produced. The ashes from the boiler and other nutrients are added to the bio-fertiliser, producing an ideal fertiliser for the planting of sugarcane and sorghum. Anaerobic treatment of vinasse enables the transformation of a pollutant into a bio-fertiliser.

- *Steam and Electric Energy Generation:* This production uses all the bagasse and biogas to efficiently produce electric energy with 60-80kgf/cm² boilers and various stage turbines. The electric energy enables its own supply for industrial operation, and what is in excess is sold. The bagasse from the outlet of the diffuser grind, with 50% humidity goes to a suspended bed dryer in the boiler, O² free, which operates with a positive pressure. The bagasse comes out with 30% of moisture. The gas exit temperature is close to 80%, in the best *regenerative* process of the cycle. The excess of bagasse is packed to be used in the road-harvest season. The GP concept foresees the burning of 100% of the bagasse and 40% of the dry straw (40% of sugarcane) to produce superheated steam (p=62 atm and T=450°C), in a suspension-

---

[154] Bagasse: Fibrous residue remaining after sugarcane or sorghum stalks are crushed to extract their juice, currently used as a renewable resource in the manufacture of pulp and paper products and building materials; Algave/tequila weed bagasse: a similar material which consists of the tissue of the blue agave, or tequila weed of the *Algave tequiliana* species plant – a base ingredient of tequila, a popular Mexican alcoholic drink – after the extraction of the sap.

burning boiler, primordially, to be used for electricity production in a high power and efficiency turbo-generator. The boiler is capable of using natural or bio-digester gas.

- *Yeast Treatment:* The yeast from the bottom of the first cube and from the centrifuge is pumped to the termolizer to remove and reuse the residual wine. Afterwards, the yeast is concentrated in a vaporizer and dehydrated with a dryer, and only then stored.

- *Cattle Husbandry:* A total of 2800 animals are raised in semi-confinement, five months out of confinement with pasture turn. Partial confinement makes it possible to double the number of animals, avoids stress, reduces disease incidencies and increases product quality. The GP has a corral area of 4630m² for milking, and cattle food is prepared with sugarcane tops, sorghum seeds and part of the yeast resulting from the broth fermentation, hence seeking a protein-adjustment. All wastes pass through an anaerobic bio-digestion process conditioning them to achieve good standards, where the nutritional properties are kept while it is isolated from air. Also, there is a semi-confined cattle rising that uses sugarcane and sorghum tops, and also sorghum seeds, a non-aerobically-silaged vegetable feed; the animal husbandry supplies milk, cheese and other such dairy produce, as well as meat.

- *Futures Possibilities:* Presently, the spent yeast is fed to animals; it could be processed to render human food; Potential development of alcohol-chemical industry, aiming to reduce oil dependence; The superior productivity and quality of sugarcane-sorghum plantation make it an ideal *alternative* to fossil fuels, especially when considering the excellent physical-chemical and thermodynamic properties of alcohol. Mechanised and mannual picking cases were studied. The mannual picking case is hereby presented because it generates greater number of jobs. The GERIPA Project is also founded on methods and practices of leading-edge technology, and praised for its use. The sugarcane-sorghum association elevates the animal work period to 12 months against 8 months of the present sugar plants. The final products are: *2800 animals* (cattle), *2000* tons of *horticulture* per year; *2760* tons of *fruits* per year; *40,000* litres of *fuel alcohol* per day; *5.2* MW of *surplus electricity;* and the growth and maintenance of forested are *93* ha.

- *GERIPA and Conventional Ethanol Process:* (a) GERIPA PROCEDURE: Rural Products: sugarcane, sorghum, cattle, fruits and other food crops; Rural Procedure: No chemicals for cultivation and no burning for harvesting; By-products used for recycling: straw, bagasse, sugarcane tops, yeast and cattle feces; Industrial Products: Fuel ethanol, bagasse, electricity, yeast, fruit and other food products. (b) CONVENTIONAL ETHANOL PROCESS: Rural Products: sugarcane; Rural Procedure: chemicals for cultivation and burning for harvesting; By-products used for recycling: bagasse; Industrial Products: sugar and fuel ethanol.

GERIPA environmental and social aspects initially generate 241 permanent jobs and 5,400 workers will be engaged in the farming area devoted to sorghum (1812 ha) during the 8 months between sorghum planting and harvest. The area will produce several vegetables. It is important to emphasise that all GERIPA planting is *organic,* thus eliminating the use of *pesticides.* The procedure of *not* using chemicals *nor* performing any sugarcane burning prior to harvesting eliminates many health problems due to the *dioxins* and *furane emissions, high concentration of O³ and CO² in the air, loss of organic matter* and *micro-biota* from the soil, which leads to increased rates of *erosion,* and *loss of vegetal* and *animal* species. In addition, the dense fumes produced in cane-burning can cause road accidents and serious health problems to picking workers. As a matter of fact, a Brazilian jurist considers the manual harvest a criminal activity because it causes human and animal mortality. GERIPA's electric co-generation enables to claim benefits from the Kyoto Protocol Clean Development Mechanism as substitutes *fossil fuel* used in conventional electrical installations. GERIPA's carbon dioxide emission reduction corresponds to 30,000 tons of carbon yearly.

Considering negotiation of carbon credits (CC) as USD5.00/CC, GERIPA has an additional revenue of USD150.00/year, not accounted in economical calculations due to this market's uncertainty. The investment of this revenue in agro-forests enables to generate 300 jobs every six years of the agro-forest cycle. The agro-forest also generates CCs and, therefore more jobs. Such feasibility is another positive characteristic of GERIPA. The production cost of alcohol is USD0.00/litre, against USD0.8 of the conventional alcohol distilleries. The cost of GERIPA-alcohol in the market at present is 35% and 65% respectively less than diesel and gasoline for every km driven. Furthermore, GERIPA relies on exemplary civic and social facilities, whose purpose is to provide the entire workforce with social, medical and cultural assistance. The residential village consists of houses that come with a living room, a dining room, a kitchen, a bathroom, one, two or three rooms and projected by an architect specialised in using local resources potentially. The GERIPA unit has two infirmaries, each with a physician and two nurses, ambulance, driver, safety bridge and other medical centre services. It also contains a nursery, pre-school, elementary school and adult alphabetisation courses, applying an excellent-level methodology. Sports, cultural and social practices with convenient resources, such as covered courts and outside activities will be stimulated. Also foreseen in the GERIPA administrative and operational departments costs, is the demand of a work group made up of highly qualified technicians.

The current conventional technological operational level in the sugar and alcohol industry inhibits the implementation of GERIPA procedures. The GERIPA process, like the introduction of renewable energies to replace fossil energy systems – is *resisted by the established conventional methods and systems. Commonsense that is always common,* tells and demonstrates to us, that the *substitution* of fuel derived from *petrol* to *alcohol, will* improve the Brazilian and other world economical and environmental balance; it will present a concrete response to the Kyoto Protocol, and confer Brazil an all-embracing leadership in the treatment of the Earth. Industrialised animal farming, cultivating agriculture and sugarcane-sorghum integration are the main responsibilities for the success of GERIPA. This presents rewarding productivity, environmental quality and social growth, in addition to regional development. By means of cooperatives and companies with shared participation, the implementation of small rural producers or rural settlements, becomes a wealth distribution and enables to corporate Brazilian territories (and other countries' territories) that are currently economically marginalised. The creative concept of GERIPA project takes advantage of the technical, economical and ecological potentialities of *regenerative-recycling* procedures within the productive context of alcohol and the excellent eoc-social fuel as adequate substitutes to fossil fuels.

GERIPA's positive results are due to *integrative recycling* and adequate organisation of the diversified production of *alcohol, electric power, fruits, garden vegetables, milk,* and *meat.* This project can be seen as a regional rural development tool that embodies social and ecological concerns, which assures conceptual superiority to integrate such a combination in the form of agro-industries, in which all leading-edge technology is explored through optimised processes and efficient procedures. It thus makes the production goods within an excellent context of feasible sustainable development.

2.2.3 Alternative Currencies, Communities

MANY RURAL communities hit hard by the global economic crunch resort or take cover in using alternative means of purchasing and selling their commodities. It is a means of trade that often emerges during tough economic times, when money gets dried up, and there are needs still to be met in societies, and creative means for these unmet needs to be met. Like the GERIPA procedures now coursing through rural Brazil, alternative currencies and the communities that use them throughout the world, sustain rural economies; they provide the badly needed means of living a decent life, and minimise the use of energy. This dovetails with the urgent need to re-orientate agricultural, transport and energy consumption and production towards renewables. Alternative or "contemporary" currencies are generally used in conjunction with conventional money; they are most useful in remote geographical areas and social sectors where money doesn't flow sufficiently.

Alternative currencies grant economic agency to people and societies languishing on the margins of fiscal life, including the elderly, sick and disabled, underemployed and those steeped in seeming eternal poverty. Communities using alternative currencies established economies in recent years, stem in part from "localisation movements...periodically ditching the dollar (or the pound, or yen) in favor of homegrown currency that doesn't merely fortify the local economy, but also builds community: people have a stake in their neighbors' well-being because that neighbor represents both market and supply chain. Some argue that such transactions are more secure than others because knowing the person you're dealing with (and his family and friends) serves as a kind of social collateral," writes Schartz.[155]

The Community Exchange System (CES)[156] is a community-based, global trading network using money other than our familiar conventional ones – an alternative, parallel, local, community or complementary currency system, in short, a new money system. CES include the *Local Exchange and Trading Systems* (LETS), *Mutual Credit Trading Systems* (MCTS) or *Time Banks*, and *United Nations International Local Employment Trading System* (UNILETS),[157] and *Community Exchange Network* (CEN). Community currencies are as versatile as conventional ones. The main difference between these new money systems and conventional money systems is that, the scope of the formers' money is *usually restricted to a*

---

[155] Judith D. Schwartz: *"Alternative Currencies Grow in Popularity."* TIME/CNN, 14 December 2008.
[156] Community Exchange System (CES): http://www.community-exchange.org/; info@community-exchange.org; 'Officials': Les Squires and Tim Jenkin. The CES Logo created by Tim Jenkin depicts three hands, representing the continents around the world.
[157] United Nations International Local Employment Trading System (UNILETS): This is a newly developed local currency system aimed as a mechanism for linking Local Exchange and Trading System (LETS) in communities around the world. The connection between UNILETS and the United Nations (UN) is unclear. The *Ripple Money System* has been proposed to connect the diverse LETS systems. [UNILET ONLINE].

geographical area or organisation. CES, LETS, MCTS or Time Banks' money does not 'exist' like the conventional money does. There is no need for a *supply* of it and you don't need any of it to start trading.

Electronic money, Digital money and *Digital gold currency; Scrip money, Private currency* - are all forms of local currency or *scrip* in that they are not government-backed. These include *e-money, electronic cash, electronic currency, digital cash, digital currency, subway tokens, tickets, e-gold, GoldMoney, Liberty Dollar, Disney Dollar, Microsoft Points, "points"* – substitutes for currency which is not legal tender; it is often a form of credit created as company payment in times where regular money is unavailable, such as remote coal regions or countries occupied in war times. *Commodity money* is one whose value comes from a commodity. For example, *gold, silver, copper, salt, peppercorn, large stones, decorated belts,shells, alcohol, cigarettes, cannabis, candy, barley* – have been used as mediums of economic exchanges.

The 'money' in these alternative systems is a *'retrospective score-keeping'* that keeps a record of *who did what* for *whom* and *who sold what* to *whom*. In these alternative money systems, there can never be a *'shortage of money'* and money does not have to be created by a third party (i.e., banks or governments) outside the circuit of buyers and sellers. For this reason, money and credit are free, for the buyers and sellers *'create'* it at the moment of a trade. About 16 countries with 109 exchanges use the CES networks under the global network of complementary currency exchanges statistics: Community Exchange Network (CEN) – Argentina (1); Australia (30); Canada (3); China (1); India (4); International (2); Italy (1); New Zealand (20); Norway (1); South Africa (22); Spain (2); Sweden (1); UK (1); U.S. (15); Vanuatu (4); and Zimbabwe (1).

More than 2,500 different existing alternative local, community, private currencies and scrips are currently legally used alongside the conventional currencies in the following countries:

- U.S.: *Ithaca Hours; Time Dollars; Greens; Disney Dollars; Liberty Dollar; PEN Exchange; Fourth Corner Exchange Inc; SVM; Truck System; BerkShares; PLENTY's.*

- Canada: *Canadian Tire Money (CTM); Brampton Dollars; Calgary Dollars; OUR Community Dollar; Saltspring Dollar; Toronto Dollar; Steinbucks; Tamworth Hours; Pioneer Bonus Bucks; Bow Chinook Hour; Holey Dollar; Prosperity Certificate; Saskbucks.*

- European Union and Europe: UK*: Totnes Pound; Lewes Pound; Locks; Readies; Scotia; Rheidol; Calderdale Green Currency; Palm; Credit Note*; France: *Sol; SEL; Troc de Services*; Germany: *Chiemgauer; Talent; Berliner; Joytopia; Tauschring; RGT, STERNTALER*; Hungary: *Kör*; Latvia: *Pilsetas Nauda*; Spain: *Kroonos.com; Banco de Tiempo; Axarco*; Italy: *Banca del Tempo; EcoAspromonte; Ecoroma; REL; SCEC; Simec; Sistema di Reciprocita Indirett.*

- Africa: South Africa: *Talents; Ora*; Senegal: *Bon de Travail*; Nigeria: *Green Naira*; Cameroon: *Flash Cash; Tontines*.[158]

- Latin America: Argentina: *Argentino; Credito; Red Global de Trueque*; Brazil: *Curitiba Bonus; Sabre/Saber.*

- Asia and Pacific: Japan: *Hureai Kippu; Eco-Money; WAT System; COME; Notebooks; Peas; Peanuts; Poran; Ohmi; Garu; Wat* - Watt System; Indonesia: *Community Coupons; PLAN International; Gotong Royong,*; Thailand: *Bia Note; Bia Kud Chum; Bia Kutchum notes; Bia Kutchum Pence; Jai*; Korea: *Duru* – and many local currencies only in Korean language; Hong Kong: *Shifenquam*; Papua New Guinea: *Traditional Shell-money; Tabushell; Mis; Kakal; Dog Teeth*; Philippines: Many local currencies only in Filipino languages.

- Australia and New Zealand: New Zealand: *Golden Bay.*

- Russia and Former Republics: [Many yet to be documented.]

- Local Currencies with Countries Unidentified As Yet*: Tianguis Tlaloc; Thay Gerh; Ruhmihuaico; Toctiuco; Salta Creditos; Ndajem-Wecco; Interser.*

- Middle East: *Hawala/Lundi.*

- *"Terra" has been proposed as a transnational alternative currency by Belgian economist Bernard Lietaer.*

- *BBX:*[159] Is a marketing-based financial institution with interest-free lines of credit and guaranteed sales for the business owner of today. In essence, BBX is a credit/debit card system, similar to other card systems, but *without* charging the interest rates charged by other card. BBX can help grow your business with new sales through the network of business owners involved in BBX, allowing them to use their interest-free credit card or the funds earned with new sales generated by BBX. BBX can also help grow your business witt new sales and customers as well as help move excess and idle stock at retail or near retail prices. BBX currently serves over 5,000 member businesses in Australia and New Zealand representing over 10,000 individual card holders, since its inception in 1993.

- *Barter Card* or *Bartercard:*[160] Is the premier business transaction card for businesses and

---

[158] Stephen DeMeulenaere:*"2007 MicroGrant Fund for Complementary Currency Sysstems in Asia, Africa and Latin America."* The MicroGrant Fund for Complementary Currency Systems began 2005 as an initiative by Stephen DeMeulenaere of the Strohalm Foundation, to encourage the implementation of systems in South-East Asia, Africa and Central-South America. In 2005, the Fund began to support the promotion of Complementary Currency Systems in the Philippines in the Filipino Language. In 2006, the Fund did the same with NGO Sarvodaya in Sri Lanka during the post-Tsunami recovery efforts. In 2007 up to US$500 were available in funding. [Stephen DeMeulenaere, Asian Programme Coordinatore, Strohalm Foundation, Bali/Indonesia; stephen@complementarycurrency.org; stephen_dem@yahoo.com, www.appropriate-economics.org/materials.html; www.complementarycurrency.org; www.strohalm.org.]

[159] *BBX:* BBX, phone 02 9622 9000, Fax 02 9622 9011 Australia; chris@bbxaustralia.com.au; BBX sponsors Bulldog, Rabbitohs, Sharks, North Hobart Demons, NITR Boxing and other community organisations.

[160] *Barter Card* or *Bartercard:* Was founded in 1979, a non-profit International Reciprocal Trade Association (IRTA) member to promote just and equitable standards of practice and operation within the Modern Trade and barter and other Alternative Capital Systems Industry. Batercard sponsors the following organisations and personalities: Australia's Master Summit; Restaurant & Catering Australia; Australian Rainforest Foundation; Queensland Reds Rugby Union; Bond University; Waratahs Rugby Union; Gold Coast Titans; Canberra raiders; Ipswich Jets; East Perth Football Club; Bartercard Cairns Marlins; Surf Life Saving

decision-makers that boosts their businesses and lifestyle, and provides a platform enabling SMEs and corporate Australia to convert their idle capacity into usable goods and services. This idle capacity can include downtime within labour-based and professional services industries, additional machinery capacity for manufacturers and excess, slow moving or end-of-run stock for the wholesale or rental industries. Since its introduction in 1991 in the Australian economy, Barter Card has helped revolutionise the way businesses across Australia conduct their day-to-day transactions through its innovating Trade Exchange System. It has enabled 55,000 trading members in six countries to benefit from the cashless economy of barter. Bartercard actively supports local communities, events and personalities.

Some of the advantages and benefits of using an alternative local, community currency, side-by-side the conventional currency include:

- *Mobilising the Real Wealth of a Community:* The knowledge and skills of its people is the real wealth of a community. Conventional money *drains away* while a local currency keeps this wealth working for all involved. People who have accumulated a wide range of skills and abilities suddenly become once again highly valued members of the community.

- *Fostering Self-Reliance & Self-Esteem:* In our communities, unemployment is growing and increasing numbers of people are unable to get their needs met. Single-parents may need respite care or other services for their children. Elderly pensioners also need a range of specialised services or may simply require company to combat loneliness. At present, a person's ability to access these and other services is proportional to their purchasing power. The community currency system *breaks this bottleneck,* by making it more possible to match someone's need with another's available labour. People are no longer dependent upon welfare or charity, and everyone's self-esteem benefits.

- *Increased Personal Savings & Disposable Income:* Because members can get local goods and services through a community currency, they can substitute it for the national (conventional) currency. Disposable income in conventional money, available after basic needs are met, actually increases. Those who regularly trade with community currencies will find they have more money left in their pockets at the end of each week. The rate of community savings, and therefore, of community investment and capital generation, will improve. This will result in an improvement in the quality of life for everyone.

- *Creating Local Economic Control:* Local currencies help *plug the leaky bucket* of the local community. By creating a local currency that *cannot leave* the community, *uncontrolled* and *activity-limiting capital outflows* are reduced. As a community currency only has value in the community in which it is generated, it continues circulating to create more wealth for everyone. They give community members a *powerful new tool* to "steer" the local economy in directions which benefit everyone.

- *Building Community Support Networks:* Because community currency systems *plug* members into a local information network, they provide new or isolated residents with an instantaneous coummunity support system. This avoids the embarrassment of introductions to strangers. Through a CES/LETS network all members have a ready reason for calling for support or help. Elderly pensioners, unemployed youth, supporting parents, new arrivals, and single-income families with partners trapped in a dormitory suburb can all build firm friendships on relationships established through a functioning network.

- *Fostering Social Justice & Equality:* Because the value attached to one's time and

Australia; Western bulldogs Football Club (AFL). Barter card 1300 barter, 1300 227837, Australia.

commitment is set individually by participants, a community currency *equalises* the wage differentials that exist in the conventional economy between the work of a woman and the work of a man. This greater equality prevents the *polarisation* of the community *"haves"* and *"have-nots"*. There is no point in accumulating community currencies as they *do not earn interest.* It is only by *putting them to productive work* that the individual or community benefits. Community currencies foster participation at all levels in the local community.

- *Building a Sense of Community:* The increasingly transient, temporary and mobile lifestyle in the world today has *seriously damaged* our sense of belonging to a meaningful community. Because a community currency builds local relationships, it is a powerful means of *regenerating* a sense of trust among members, a necessary component to the health of any community.

- *Keeping Wealth Where It is Created:* National or conventional Currencies *always leak away* to the *'money centres'*, creating *money deserts* and the cessation of local economic activity. Complementary currencies, on the other hand, are *community-based* and so *keep wealth where it is created; where previously economic activity was stagnant,* the local currency *stimulates* trade and permits things to happen where formerly there was no economic activity. By circulating in a community, the entire community becomes *self-sufficient* and does not have to rely on external businesses to provide what is required.

- *Bringing the 'Money Power' Back to the Commons:* The money we use in our daily lives is provided by the corporate financial system as a *profit-making* enterprise, not by the government as a public service to the community. As such, the money we use does *not belong* to the commons and so we have *little control* over how it is spent and who it benefits. A community currency brings the *'money power'* back to the people because its users can *decide* how that power is exerted.

- *Stream of Income:* Provides another stream of income, bridging the 'money gap' between the skills, offers, talents, and gifts of sellers on the one hand, and the wants, needs, and requirements of buyers, on the other hand. Conventional money usually can't bridge this gap because its supply is limited or non-existent.

The only problems that can be ascribed to these local, community currencies, is that they are localised and its workers, volunteers and organisers are just that.

Community currency trade and services are not a form of *'trade by barter',* nor are they a *'tax dodge'.* Trade by barter always involves bargaining between two individuals to establish the relative worth of the goods or services they wish to exchange. There is no bargaining with a community currency, and when you purchase something you are in no way obliged to the seller. You 'pay' for what you have received by delivering or selling something to another trader in the community. The use of community currencies is also definitely not a tax dodge. Their motives are noble: to create a *more equal society* where wealth is distributed according to *contribution,* not according to one's activity to *'make money'.* In other countries where these alternative money systems have become big, the state has either ignored the tax angle because it saves state expenditure on welfare payments, or there is an agreement to provide services to the state. The CES approach is that, when it becomes big, the state should become a CES-user and participate in the normal way. In this way, the state could *credit* itself through the services it provides to all members and *debit* itself by purchasing the services of CES-users.

Why in the first place, does anyone need a New Money System, when our official conventional currencies – *dollars, euros, yens, ringgits, yuans*, etc., seem to do the job for us? The reason is chiefly because our conventional money systems are at the *root of most of our misery, suffering* and problems faced by humanity. Our present money systems are the *prime factor* behind the environmental crises we face, because they are not *neutral, non-partisan* services provided by our governments, but by private financial institutions (banks) *specifically for their own benefits* - rather than to the benefits of those who use them. Our conventional money systems only work for those who *already have money,* and *it marginalises the rest.* Our money systems is also the fuel that *powers* the growth imperative of our economies, *forcing* us all to *compete* and having disastrous consequences for the health of our planet. Conventional money "exists" (or at least so we are encouraged by commercial banks to believe) so they can "lend" it to us at a price! This money has to be created, distributed and its amount restricted and controlled; and as money comes into existence when commercial banks grant loans, every unit in existence is based on a *unit of debt.* This unit of debt determines the quantity of money, which has nothing to do with the amount of money people require to live decent lives. Such money is also based on *speculation*, because it is *localised* into existence on the premise that it *"will be paid back in the future with interest."*

The debt-based money system was developed for the industrial revolution to provide a rapidly expanding money supply that could not be provided by a money system based on the *quality* of precious metals. This introduced *intangible* money that did not exist in the same way as earlier *'hard monies',* but people continued to treat it as a tradable commodity. Money that exists can thus be accumulated like any other commodity. And it can be stolen, traded, collected, destroyed and lost. Its distribution is not based on the delivery of *value* to others but on the *ability* of people to 'make money'. Therefore, conventional money has no restraints and *always flows away* from where it is created and needed towards the *'money centres'*, that is, the private commercial banks and industries.

The Community Exchange System (CES) *breaks out* of this paradigm. It recognises that the electronics revolution has eliminated the need for an exchange medium. Never before in the history of humankind, has it been possible to record accurately who *delivers value to whom.* Now that this is possible, there is no longer a need for an *'existential'* money; money can at last *truly measure the delivery of value and be based on nothing other than the expenditure of effort by people for others.* Money is information – a *unit of measure* – not a thing. If money does not need to exist as a thing, it does not have to be created and distributed. People will earn money solely on the basis of their *delivery of value to others,* not through *charging interest, trading it in money markets, and a multiple of other ways without delivering anything of real value.* Money that *does not exist* can never be in *short supply,* but no one can have more of it than the *value* they can deliver. No one will be able take more of the social product than they can contribute to it, as they do under our current money system. *Wealth will remain where it is created and needed, and not leak away to the 'money centres'.*

To summarise the problems with Conventional Money and Systems:

- *Money is Partisan and Debt-Based:* Money as we know it is not a *neutral service* provided by the government. Our money supply is created by private financial institutions on a for-profit basis. The money system is designed to benefit those who provide it, and *not to benefit those who use it.* Money is based on debt: it is created when banks grant loans. Thus for every unit created, there is *one unit of debt.* Money is expensive: every unit of conventional money is based on a *unit of debt.* This debt has to be paid back with *interest,* and the interest is *compounding.* Interest is built into the prices of everything we buy, resulting to *higher* consumer prices.

- *Money is Information, Not a Thing:* We are encouraged to think of money as a thing, but money is essentially *information,* and has no physical existence. Yet, banks encourage us to think of money as a 'thing' so that they can 'lend' it to us and thereby make a profit by charging interest. 'Thing' money also has to be created, distributed and controlled, so that there is not much of it. It can also be stolen, lost, bought, sold and counterfeited, with serious consequences for everyone.

- *Money's Cancerous Growth and Permanent Shortage:* Banks *continuously* need to *create* money than is required to pay back their loans so that borrowers can pay back the interest on those loans. This is the *source* of the growth of imperative of our economies. There must be a *continuable expansion* of bank credit or else the economy goes into recession. Systemic growth leads to the environmental problems we now face. Money's value is based on its shortage. The shortfall of money keeps it *valuable.* There only needs to be enough of it to buy the goods and services available. This has *nothing* to do with the monetary requirements of people. Those who have *none* are not seen by the market and are *marginalised.* The money to pay the interest on debt-money is never created. There is, therefore, a permanent shortfall of money to pay back both the principal and the interest.

- *Money Distributes Wealth from Poor to Wealthy, Fosters Competitiveness:* Money distributes wealth from the poor to the wealthy. *Usury* is the tool used by the wealthy to suck wealth from the poor and middle classes to the moneyed classes. *Parasitism* and *antagonistic* social classes are the product of this. Money *always* leaks away from where it is created to the 'money centres'. Conventional money knows no *bounds* and *loyalty.* The shortage of money means we all have to fight for a share of an amount that is too small to go around. The need to repay interest means that we have to *ea t others* to prevent ourselves from going under.

- *Money Promotes Dishonesty and Corruption:* You can get money *without* delivering anything of value (e.g., speculation, interest, gambling, etc.). So people concentrate on 'making money' rather than producing or delivering anything of real value. It is virtually far easier to get money through *dishonest* means than by *honest* work. When you have no money, you have no choice but to try and get it dishonestly.

- *Money Destroys Local Economies:* Goods produced cheaper elsewhere replace locally produced goods. This creates a local shortage of money and reduces the market for local sellers. This also results in the *irrational* transportation of goods all over the world, consuming precious fossil fuels and creating pollution.

- *Money Creates Poverty and Causes Social and Cultural Degradation:* While it makes some super rich, money makes most people poor. Poverty is caused by a lack of money – not by a lack of jobs. *Usury* and the need to keep money *scarce* ensures that money constantly moves to those who already have money. The elimination of local opportunities to exchange and relate to one another focuses attention on ways and means of getting money outside the community. Communities fall apart as they become *indebted* to entities outside their communities.

Some notable characteristics of CES and Community Money include:

-
    It's In Abundance and Discriminates Against No One: CES money is in abundance, can never be in short-supply or in over-supply, and causes no inflation; it is democratic – it returns 'money power' to the people and allocates credit democratically; it promotes honesty because one can never 'run-out of money'; it levels the playing field: everyone starts from zero and those who deliver *real value* to others are the ones who get 'rich', not those who deliver *nothing* but acquire their wealth by manipulating the currency.

-
    Eliminates and Destroys Selfishness: CES money eliminates and destroys usury (i.e., it is 'free' of interest); cheating, theft, fraud (of money), and the problem of sellers not being paid for what they supply; the wasted effort of transporting goods all over because its focus is local; the notion that the source of money is a job: the source of money is the *delivery of value* from one entity (person, company, etc.) to another; streamlines transactions, as for example, sending accounts, and debt collection; it reconciles the accounts of buyers and sellers *immediately.*

-
    Leaves Money at Its Source: CES money keeps wealth where it is created; mobilises the real wealth of a community: the knowledge and skills of its people is the *real wealth* of a community; it gives local suppliers the preference.

-
    Fosters Human Dignity: CES fosters self-esteem, self-reliance, and social justice and equality.

The object of the CES is to facilitate trade *without* using our conventional national currencies, and build a sense of community at the same time. By 'trade' is meant the normal economic activity of selling goods and providing services by 'producers', 'sellers' or 'providers' and the purchase of these by 'buyers', 'customers', 'clients', 'patients', 'consumers', etc., generally using an *Online Money* and *Banking System,* over a 'marketplace' where people sell goods and services. Those who don't have computers can also participate - by interfacing with the system through 'branches' where everything is done manually and information is available on paper. Each user gets an *Account Number* and a *Password,* and this gives them *access* to their account on the CES website, which works like a true *online banking service.* Participants can view their *Current Balances* and obtain *Statements of Account.* They can also keep track of the trading positions of others.

Goods and services are advertised on the website through an *'Offerings List'.* There is also a *'Wants List'* where participants can advertise for goods and services they require. Participants look through these lists, or do a search, and if they find anything they want or want to offer, they contact the seller or buyer who then provides the goods or service. Payment is effected through a *Trading Slip* which serves both as a *means of Payment* and *Receipt of the goods or service.* The information on the Trading Slip is entered by the seller into a *Transaction Form* on the website. This *credits* the account of the seller and *debits* that of the buyer. Accounts record these debits and credits, giving a balance after each transaction. To ensure that unscrupulous buyers do not exploit the system, details of each user's overall trading position are available to all. General trading standards are also available to show how much trading is taking place. The website also provides all the information needed to contact other participants. Trading in this system requires *no supply of money,* either by the community as a whole, or by each participant. Instead of using a 'hard' currency which then has to be allocated according to some formula, the *currency* of this system is the pure recording of the *values* exchanged in trade. Money in this system is thus *created* at the *trading interface* and recorded as *credits* for sellers and as *debits* for buyers.

To operate a UNILET to connect internationally with LETS and other community currencies and exchanges, each member has a *'Record Book'* that will replace currently used *bi-directional emails.* There are no physical tokens – it's more like earning gold stars or brownie points. You can offer anything or everything that is legal, and members can advertise items for sale in the *Newsletter* at no cost. All kinds of services, skills or even goods are offered; everyone can do something useful, whether it's a trade or profession like plumbing or legal advice, or just help with ironing, gardening or baby-sitting. Maybe you have a van to hire, a ladder to lend or just some muscle to get a wardrobe upstairs? You can also sell unwanted items.

There are CES exchanges in a number of countries around the world. Each exchange has its own currency but members of one exchange can trade with members of other exchanges making CES currencies as versatile as our conventional currencies. Credits earned in one exchange can be spent in another if you are visiting another area, and you can trade with local CES traders. New exchanges are starting in new areas all the time, and existing ones are growing steadily. The local currencies are *units of measure ration,* rather than tradable commodities like conventional currencies. However, to make these currencies meaningful to users, their value is based on the national currencies. This is purely to give them reference, and in no way tied to the national currencies, and will deviate from them over time. There are no rules for pricing in the CES. The 'law' of supply and demand prevails. However, within the context of the CES, certain services that otherwise would not be highly valued, might increase in value of their relative shortage. Other services outside the CES that are *expensive* might be *cheaper* in the CES because the provider wishes to attract customers.

An example of a trade in CES Requirement: *Your Car Needs an Oil Change:*

-
    Step 1: You either look through the *Offerings List* or do a search to see if anyone is offering oil changes or car maintenance. Someone is offering oil change for T80, but you must bring your own oil and oil filter.

-
    Step 2: In the *Offerings List* you click on the person's name to obtain contact details. You phone the person (the 'seller') and agree on a time and place for the oil exchange.

- *Step 3:* The oil change takes place and then you (the 'buyer') fill in a *Trading Slip* (obtained from the website), giving the date, your name, your account number, the amount (T80) and your signature. You will fill in the same details on the counterfoil and get the seller to sign it. The *counterfoil* is then separated from the slip and you hand the main part to the seller, keeping the counterfoil for yourself. For the seller, your *Trading Slip* represents your *payment* and your acknowledgment of the service or goods delivered; for you, the *counterfoil* is your record of *payment.*

- *Step 4:* You leave, satisfied that your car has fresh oil. The seller then goes to a computer and enters the details of the trade into the the the *Transaction Form* of his or her *CES Bank Account.* This becomes a *credit* for the seller and a *debit* for you. You are now obliged to provide goods and services to the community worth *T80.*

To become a CES participant - whether as a buyer or seller - you have to register with an existing CES group in your community or country, or create one yourself. Presently there are about 19 CES groups worldwide: *Cape Town Talent Exchange:* 9 members; *Prospective Users Maillist:* 7 members; *Christchurch New Zealand – Mark II:* 7 members; *CES Users Maillist:* 5 members; *New Zealand:* 5 members; *Christchurch New Zealand – CCD:* 5 members; *CES Admin Maillist:* 5 members; *CESwatauga:* 5 members; *Johannesburg Talent Exchange:* 5 members; *Saint Louis Community Exchange:* 4 members; *Joiner's Group:* 3 members; *Shoalhaven:* 3 members; *Sydney and Illawara LETS, Australia; Brisbane LETS-BrisLETS:* 2 members; *Transition Towns ZA:* 2 members; *Central Coast LETS:* 1 member; *Wairarapa Green Dollar Exchange Inc:* 1 member; *Goshen, India Exchange:* 1 member; *Byron, Mullumbimby Tweed LETS:* 1 member.

Websites that deal with alternative money systems and related issues are:

Access Foundation (Alliance of Contemporary Currencies Enabling Sustainable Societies) – Alternate Money Systems (AMS)[161] that proposed the establishment of a Global Resource Bank (GRB)[162] to the UN Conference on Social Welfare;

---

[161] Access Foundation: *"First International Complementary Currency Summit: Expanding the Impact of new monetary innovations worldwide.."* 31 July-5 Aug., 2005, co-sponsored by ACCESS Foundation, Boulder Colorado, and the MARPA CENTRE for Business & Economics, Naropa University, U.S.; Jacquidb@earthlink.net; www.terra-trc.org.

[162] Betina Corke, co-chair of the International Council on Social Welfare/Lennart Bylund: *"Global Resource Bank* (GRB)": Article of Association: Art. 1: *Shareholders:* The Bank is organised by all people. Each person owns one share of the whole. A share is a non-transferable birthright. The Bank is a free public enterprise directed by the shareholders; Art. 2: *Bank Access:* Free access to the Bank is by telecommunication; Art. 3: *Bank Assets:* The assets claimed by the Bank are the resources of nature outside national jurisdiction. The shareholders value the national production capacity of ecosystems in this region at 6,000 trillion Global Resource Bank dollars on a par with US dollars; Art. 4: *Shareholders Capital:* Shareholder capital is transferred from bank assets to shreholder accounts in the amount of 10,000 dollars per year for ten consecutive years. Electronic cash transfers between accounts are cleared by the transparent cash transfer programme described in articles five through eight; Art.5: *Bank Income:* The initial rate of bank income is 2% of cash transfers.This transfer fee is divided equally between the accounts taking part in cash transfer. The Bank's initial percentage rates are adjusted to the average percentage rates that result from shareholder/director choices; Art. 6: *Shareholder Dividend:* The initial shareholder dividend rate is 80% of bank income divided equally among shareholder accounts. Upon death, shareholder dollars revert to bank assets; Art. 7: *Environmental Investment:* The initial environmental investment rate is 10% of bank income divided equally among shareholder environment accounts. The shareholders are entrusted to allocate this money to investments that protect and/or renew ecosystems; Art. 8: *Communicattion Account:* The initial rate the communication account receives is 10% bank income. This money is to pay for a full service telecommunication network. The communication account manager is hired by the majority of voting shareholders to buy telecommunications at the market; Art. 9: *Start-Up-Capital:* The Bank transfers from bank assets 5,000 dollars to each shareholder's environment account and 30 trillion dollars to the communication account; Art 10: *Creditor Capital:* The Bank transfers from bank assets to creditors' bank accounts sufficient Global Resource Bank dollars to pay off the world's public debt at the US dollar exchange rate; Art. 11: *Steady State:* The dollars required to maintain the ratio of total capital per shareholder at a steady state are automatically transferred between bank income and the Bank's dollare reserve; Art. 12:*Amendments:* These Articles of Association may be amended by the majority of shreholders. Guardians of shareholders have proxy rights; General Prospects: The Global Resource Bank (GRB) is a free public enterprise that earns all people financial freedom, invests in a natural environment, buys telecommunications and pays off the

Altruists International; Pedia – an open collective knowledge centre on complementary currency systems; Crom Alternative Payment System; Local Exchange Systems in Asia, Africa and Latin America; Community Information Resource Centre (CIRCA); Complementary Currency Resource Centre; Creating Community Currency; EF Schumacher Society; Forum for Stable Currencies; Ithaca HOURs Online; International Journal of Community Currency Research; Letslink UK; LETS – Linkup; Living Economics; LETS.net; Local and Interest-Free Currencies, Social Credit and Micro-credit; The Money Fix; Online Database of Complementary Currencies Worldwide; The Open Money Project; P²P Foundation; Reinventing Money.com; Community Exchange System; Strohalm for a change; The Terra – Trade Reference Currency (TRC); Time Dollar Institute; Transaction Net-Enabling Markets Online; The Transitioner – The Transitioner.org; UniteDiversity Wiki; Workgroup on Solidarity Socio-Economy – Social Money.

Social Money or alternative, local currencies is "Non-Zero-sum" structured, while Modern Money or conventional currencies is "Zero-sum" structured. "Zero-sum" money is *relative* and not *absolute value,* and its zero-sum nature – where an *extra* zero can be added on the end of all the prices - may be the single hazardous problem of the centralised money system. The wealthy and rich are wealthy and rich only as the poor are poor, and until a *non-zero sum* money system is established, 'money-making' will inevitably mean *depriving* someone else of it, since a zero-sum money system only has so much to go round. Zero-sum centralised money systems such as hold sway presently in our world, produce fear, scarcity and conflict, according to Upton and Beresford:[163]

---

world's public debt. The GRB issues global dollars that measure the value of ecosystem production outside national jurisdiction. The economies of ecosystems within this region produce the majority of commodities that sustain life of Earth. Together all people value the production capacity of these ecosystems at six thousand trillion GRB dollars. The Bank is the clearinghouse for GRB dollar transfers. GRB dollars store the primary value of ecosystems which is the substance of life. The Bank's money coexists with all other mediums of exchange; Prospects for each Article: 1. The Bank is a direct democratic business that includes everyone. As shreholders and directors, the people adjust the Bank's percentage rates, earn shreholder dividends, invest in the environment, hire the communication account manager and amend the charter. How these responsibilities are accomplished is described in articles 5, 6, 7, 8, 9, 10, 11, and 12; 2. Free access to the Bank is through a 24 hour toll free number paid for out of the communication account by the communication account manager; 3. The assets the people claim and value are the economies of ecosystems outside the jurisdiction of nations. The Bank quantifies the capacity of the people's commonwealth to produce the commodities that sustain life on Earth at one million GRB dollars for each one of six billion people. This comes to a total of 6,000 trillion GRB dollars; 4. Over 10 years the Bank liquefies $100,000 per person for a total of 600 trillion GRB dollars in personal working capital. Transparent, telecommunicated cash (telecash) transfers ensure open accounting and instant settlement. The free flow of GRB dollars is unregulated; 5. The amount of bank income varies with the transfer fee and the amount of cash transferred. The amount of cash transferred is a function of the market. The people manage the system by adjusting the transfer fee and choosing what percentage of the income goes to the shareholders personal accounts, shareholdes' environment accounts and the communication accounts; 6. The shreholder dividend earns every man, woman and child lifelong economic security; 7. The shareholder's environment account gives each person direct responsibility for Earth's natural environment. This investment in nature grows the global economy by increasing ecosystemproduction. The large magnitude of the investment reverses negative trends produced by lesser monetary systems that value the consumption of natural capital; 8. The cash flow to the communication account pays for an interactive communication network that fully employs communication technology. Free access to the Bank is a universal condition. The Bank's home office is with the communication account manager; 9. The start-up capital is for initial investments in the environment and telecommunications; 10. The creditors of the public debt, including individuals, U.S. Federal Reserve Banks, the Bank of International Settlement, the World Bank, other banks, the International Monetary Fund, municipal bond holders, etc.m, are paid off with GRB dollars. When the people establish the Bank an amount of GRB dollars equal to the world's public debt in US dollars is put in escrow. The acceptance of the pay-off is at the discretion of the creditors. The lending GRB dollars is the domain of creditors in free capital markets; 11. To nullify inflation or deflation of GRB dollars, an automatic ready state mechanism specifically maintains the approval of a majority of shareholders. Parents vote for children and manage their accounts until they reach adulthood.

[163] Robin Upton, internet consultant/Clive Beresford, community activist: *"Modern Money is Zero-Sum"; "Re-establishing altruism as a viable social norm: Altruistic Community Working for Love, Not Money."* Altruists International (AI), a grassroots movement began in 2001 by Robin Upton, for social change by helping volunteers to work together and by encouraging people to think about the welfare of others, not about what they can get from other people. AI is an optimistic community with members from both under- and over-developed world, united by their commitment that a money-centered struggle for personal gain is no way to make the world a better place. AI tries to ignore money but put people at the heart of what they do. [http://www.altruists.org/.]

- *Discourage Altruism:* Zero-sum money models of every transaction are as a *win-lose*, rewarding *competition* rather than *cooperation*. It discourages altruism[164] by perpetuating destructive and wasteful social cycles which lead to fear and selfishness that creates social conflicts that produce scarcity and reinforce fear - the kind of thinking that people need to get away from if they are to make the transition from scarcity to abundance. Positive social cycles create love which in turn inspires sharing and cooperation that generates abundant provision and refurbishes love.

- *Amount of Money in Circulation:* Strictly speaking, the amount of money in circulation does increase, by expansion of debt owed to central banks. However, individuals cannot increase the money supply, although it is their labor, not that of banks, that increases the wealth of society. Central banks have a monopoly over money creation. Although no effort or resources are needed to create new money, central bankers are entitled to collect rent on it in the form of *interest*.

- *Altruistic Economics:* The developing altruistic economics, is a system for scoring human interactions, which needs no central authority that issues currency. The amount of goodwill in circulation is not fixed, but changes automatically by statements of accreditation that individuals issue to one another. By reconnecting people with the consequences of their actions, we believe it can rekindle the spirit of altruism within them, so long repressed by a zero-sum money system.

- *Global Consciousness:* Until a fundamental shift of consciousness occurs, disasters, such as ecological degradation, can get worse. Communications technology – and *WWW* in particular – is boosting altruism and establishing a global consciousness. Individual acts of altruism can have a great impact, as for example, *Wikipedia, free software*[165] or *giveaway websites.*

Global Consciousness and Environmentalism are key structures of globalisation. They link the world's growth in which the production and use of renewable energies, and the production of food, preservation and sufficiency, are and must be the sustainable sectors of lifestyles in the 21st century.

2.3 Renewable Energies' Production, Food Preservation and Sufficiency

---

[164] Altruism: Altruism means: 1. Loving others as yourself; 2. Behavior that promotes the survival chances of others at a cost to one's own; 3. Self-sacrifice for the benefit of others; altruists choose to align their well-being with others – so they are happy when others thrive, sad when others are suffering. Essential is establishing strong relationships, most societies acknowledge the importance of altruism within the family. By motivating *cooperation* rather than *conflict*, it promotes harmony within communities of any size. Of course, peace within communities does not necessarily herald peace between communities, and the two may even be inversely related – witness for example, the way in which social strife tends to decrease within countries at war. Altruists broaden their perspectives in an effort to overcome the artificial categories that break up the complex web of life. Altruism is the abdication of claims of power over others. To state that "none of us are worth more and none are worth less" is almost a truism, but modern technology has given a new urgency to all such appeals of altruism. Life on Earth is being destroyed at an alarming rate, and evidence is mounting of impending disasters such as *ecological collapse* and *climate change* that threaten us all.

[165] GNU/Linus: Is a project launched in 1984 to develop a complete Unix-like operating system which is FREE SOFTWARE: the GNU System., whose kernel is used with the Kernel Linux – a combination of GNU and Linux (GNU/Linux) operating System, now used by millions. There are many variants or "distributions" of GNU/Linus, but those that are 100% free software, entirely freedom, an alternative to Microsoft Windows, among many, include: *PHP* – freeware script language; *MySQL* - Free Database; *SysInternals* – A collection of top-notch diagnostic utilities, including Filemon, Regmon, Process Explorer, and Autoruns; *Open Office* – Replacements for MS Office. Word Processing, Spreadsheet, Database, compatible with DOC, XLS formats; *XOL* – allows you to run multiple Operating systems on the same computer; *Firefox* – Safer alternative to internet Explorer; *CDEX* – CD Ripper/Audio file manipulation tool; *The Tower of the Moon* – Beautiful PC-conversion of Mike Singleton's classic ZX Spectrum strategy game; *HTTrack Website Copier* – Wibsite downloader, automatically downloads entire sites onto your computer; *E-Sword* – Very flexible Bible study tool; *XNView* – Very flexible image viewer that loads over 100 filetypes; *Skype* - Free Voice-Voice and SMS communication over the Net.

POVERTY REDUCTION includes the increased use of CES money along side conventional currencies; enhancing energy security through the use of renewable energy; agro-ecosystem diversity; and active development of renewable energy sources. These are some of the crucial imperatives that face humanity. Other imperatives that glare into our faces are: *facing-off* as rapidly as possible, the production and uses of fossil fuels. This is the core message that emerged from the Global Renewable Energy Forum (GREF)[166] in Brazil in 2008. Summaries of the three plenary sesions read:

- *Energy Security in Latin America – Renewable Energy as a Viable Alternative:* Discussed in September 2006 on energy security by and regional cooperation to *harness* the potential of renewable energy technologies. The main outcome was a Ministerial Declaration, which encouraged governments to assess the feasibility of establishing an Observatory for Renewable Energy in Latin America and the Caribbean, and a proposal to organise the GREF jointly by the Brazilian Ministry of Mines & Energy and UNIDO in 2008.

- *Global Developments on Biofuels:* Discussed in Kuala Lumpur, Malaysia in July 2007, on global developments in biofuels, supply, demand, economics and sustainability of biofuels; production of feedstock; technology and applications on biofuels from oils, fats and biomass; and the sustainability and viability of biofuels.

- *Sustainable Biofuels Development in Africa – Opportunities and Challenges:* Discussed in Addis Ababa, Ethiopia, in July 2007, to explore biofuel development possibilities in Africa, while simultaneously ensuring its potential risks and trade-offs. The main outcome was the Action Plan for Biofuels Development and the Addis Ababa Declaration on Sustainable Biofuels Development in Africa: developing enabling policy and regulatory frameworks; participating in global sustainability discussions; formulating guiding principles on biofuels to enhance Africa's competitiveness; minimising the risks of biofuels development for small-scale producers; North-South cooperation; and the establishment of a forum to promote access to biofuels information and knowledge.

- *Sustainable Biofuels Production and Use in Central & Eastern Europe:* Discussed in Dubrovnik, Croatia, in November 2007, with the objective of promoting the sustainabl production and use of biofuels; capturing the environmental and economic benefits of biofuels; increasing the security of energy supply.; drivers and concerns about the production and use of biofuels; country-specific biofuels-related issues in Central/Eastern European countries; sustainable biofuels; regional and international trends; and industrial and technological development.

- *Making Renewable Energy Markets Work for Africa – Industries and Finance for Scaling Up:* Discussed in Dakar, Senegal, in April 2008, emerged with a Plan of Action on Scaling-Up Renewables in Africa, and the Dakar Declaration on Scaling-Up Renewables in Africa. Partners agreed to an African continental target for governments, with support from development partners to scale-up annual renewable energy investments to $10 billion or more between 2009-2014; adopt the Plan for African governments, their international development partners, NGOs and the private sector to support implementation with adequate

---

[166] Global Renewable Energy Forum Bulletin: Brief History of Renewable Energy Meetings under UNIDO and Other Processes: *"A Summary Report of the Global Renewable Energy Forum* (GREF):" The GREF met at Bourbon Cataratas Hotel, in Foz do Iguacu, Brazil 18-21 May 2008, jointly organised by the UN Industrial Development Organisation (UNIDO), the Brazilian Ministry of Mines & Energy, Eletrobas and Itaipu Binacional. The GREF brought together over 1,400 participants and 80 accredited high-level representatives and experts in energy and industry from Latin American, African and other countries, and representatives from UN agencies, bilateral organisations, the private sector, academia and regional and international non-governmental organisations. The GREF Bulletin is published by the International Institute for Sustainable Development (IISD), vol. 128 No.2, Saturday 24 May 2008; http://www.iisd.ca/ymb/greb/2008.

reserves; the African Union, UNIDO and other relevant development partners to establish a ministerial-level policy advocacy group to be supported by a coordinated unit.

- *Renewable Energy in the Carpathians:* Discussed in Liv, Ukraine, in May 2008, to prepare a baseline report on renewable energy policies and financial instruments in countries of the Carpathian Convention; identifying a regional cooperation programme to promote renewable energy development; and promoting a network of institutions and agencies involved in renewable energy.

- *Private Sector & Renewable Energy Production and Market Adoptionn:* The UN Commission on Sustainable Development (CSD) on April 2001 encouraged the private sector's role in energy provision, strengthen research and development and institutional capacities, develop and use *indigenous* sources of renewable energy, and to strengthen financial support to developing countries; noted: that rural energy accessibility and services is a *prerequisite* to *halving* the proportion of people living on less than a dollar per day by 2015; the *World International Renewable Energy Conference* (WIREC) 2008, held in Washington, U.S.; the *Beijing International Renewable Energy Conference,* held in Beijing, China in November 2005; the *Bonn Renewables 2004* (the first International Conference on Renewable Energy), held in Bonn, Germany in 2004; and the *G8 Renewable Energy Tastforce 2000,* in July, respectively, stressed: market adoption and finance; agricultural, forestry and rural development; state and local authorities; research and development; practical measures, success stories and effective legislation and policy measures to create institutional and technical infrastructure; significant public and private financial resources and investment in renewable energies and energy efficiency; identify actions to provide change in the supply distribution and use of renewable energies in developing countries.

- *IPCC on Renewable Energies:* A Special Report issued by the Intergovernmental Panel on Climate Change (IPCC) at its 25th meeting in Port Louis, Mauritius, in April 2006, reported that energy efficiency, renewable energy sources and technologies represent a *major mitigating option* that remains to be addressed in-depth. Moomaw[167] in his "Conclusion" writes: "The transformation in the energy system that is needed to address climate change is indeed massive. But it is much more than an environmental issue. It is in fact, an economic issue where the energy-driven economy of the industrial revolution need to be *replaced* with a system that is *sustainable* for both developed and developing countries. The strategies of the past with all of its energy inefficiencies developed because *fossil fuels* were abundant and cheap. This is no longer the case, and our future economic well-being as well as the climate system will need to use traditional fossil fuels much more efficiently, to significantly reduce their release of *carbon dioxide* (and other industrial and agricultural gases) to the atmosphere, to provide energy services with *much less primary* energy and to develop technologies that produce *little or no heat-trapping* greenhouse gas emissions. *Renewable energy* technologies along with improved demand side efficiencies can make a substantial contribution to lowering emissions along with nuclear power, carbon dioxide capture and storage from existing fossil fuels, and other technologies."

Usually, follow-ups that emerge after these renewable energy forums, are *no-follow-ups* or *watered-down follow-ups* or *sanitised follow-ups.*

2.3.1 Food Sufficiency, Preservation, Speculation

---

[167] William Moomaw: *"Renewable Energy and Climate Change: An Overview."* The Fletcher School, Tufts University, U.S.

THE UN Alliance Against Hunger (IAAH) establsihed by the World Food Day (WFD) programme is celebrated each year on 16 October with different themes:[168] 2007 - *The Right to Food*; 2006 – *Investing in Agriculture and Food Security*; 2005 – *Agriculture and International Dialogue*; 2004 – *Biodiversity for Food Security*; 2003 – *Working together for an International Alliance Against Hunger*; 2002 – *Source of Food Security*; 2001 – *Fight Hunger to Reduce Poverty*; 2000 – *A Millennium Free from Hunger*; 1999 – *Youth Against Hunger*; 1998 – *Women Feed the World*; 1997 – *Investing in Food Security*; 1996 – *Fighting Hunger and Malnutrition*; 1995 – *Food for All*; 1994 – *Water for Life*; 1993 – *Harvesting Nature's Diversity*; 1991 – *Trees for Life*; 1990 – *Food for the Future*; 1989 – *Food and the Environment*; 1988 – *Rural Youth*; 1987 – *Small Farmers*; 1986 – *Fishermen and Fishing Communities*; 1985 – *Rural Poverty*; 1984 – *Women in Agriculture*; 1983 – *Food Security*; 1982 – *Food Comes First*; 1981 – *Food Comes First.*

In spite of, or despite these gigantic efforts, to significantly *reduce* hunger, poverty, malnutrition, resources waste, obesity and over-consumption, arbitrary rising prices, and commodity hoarding and speculation – *gross inequalities continue to be monstrous realities in our world.* Food Sufficiency and security calls for its responsible preservation as against unatruistic speculation and hoarding for profit-making's sake. There is a correlation between dependence on primary commodities and poverty, as measured by the human development index. This is due to the three prominent features of commodity markets, (i.e., *price volatily; the secular decline of long-term prices;* and *market concentration),* according to an ActionAid report edited by Asfaho.[169] Commodity price fluctuation is anathema to economic development for commodity-exporting developing countries as this translates into *export-earning fluctuations*, the report points out. These export-earning fluctuations in turn, lead to fluctuations - hence uncertainty, in domestic income, savings, and in government revenues – often largely dependent on taxes on the export. The result of this is an adverse effect on domestic investment in productive assets. Therefore, *commodity price volatilities* lead to macroeconomic *instability,* which is detrimental to economic development. The challenge is typically more onerus for countries that have difficulty in borrowing abroad to smooth short-term volatilities. Volatility in commodity markets has been exacerbated by the abolition of trading activities of national commodity boards and international commodity organisations.

[168] International Alliance Against Hunger (IAAH), established by the UN World Food Day (WFD) programme was launched on 16 October 2003, [first established by the UN Food & Agricultural Organisation – FAO – on 1 Nov., 1979 at FAO's 20th General Conference] to join forces in efforts to eradicate hunger; IAAH provides 5 ways you can stop people from dying of hunger: 1. See if any families living near you whose health and life are in danger because of rising food prices. Meet with them, discuss the issue and try to workout solutions together; 2. If you cannot solve the problem above, talk about it to friends, colleagues at work, others who attend the same church, mosque or temple. Decide what needs to be done – and do it; 3. If you or your parents have moved from your birthplaces, get in touch with relatives or friends from where you came. Find out about the local situation and how they think you could help. It may mean sending more money home – if you can – to ease things; 4. Find out about charities that assist the hungry in your own or other countries. If you see that what they are doing is good,try to help them. Maybe you can save cash by cutting on what you spend on your own food and drink, and by reducing food waste, and donate part of the savings. 5. Find out what your government is doing to cushion the impact of rising prices. If you think that they are not doing enough, or not doing the right things, raise your voice. Talk to your friends, your boss, your local newspaper, your mayor, your member of parliament.persist and eventually someone will wake up, listen and do something…

[169] Samuel Asfaho, South Centre Geneva: *"Suggestions to Tackle the Commodities Problems."* (Executive Summary), 33 pp ActionAid International, PostNet Suite #248, PrivateBag X31, Saxonwold, Johannesburg, South Africa / South Centre Geneva, CP 1211 Geneva 19, Switzerland; www.actionaid.org.

The persistent decline in long-term prices of commodities has dominated agricultural commodity markets for over five decades. This has proved devastating to food sufficiency, security and preservation. Nominal recovery in prices that has occurred in recent years because of structural commodities shift demand from China and India, was mostly in *concentrated demand* on oil and minerals; it has no significant positive impact on the poor who mainly depend on agricultural commodities, even though there may have been some benefits for producers of fruit and cotton. There has been little improvement in real terms. Instead, the purchasing power of most commodity-dependent developing countries has either *deteriorated* or has *not improved significantly.* This is particularly so in the case of those for which oil and foodstuff commodities account for a significant share of their import baskets.

In recent years, increased downstream market concentration in commodity-value chains or *monopsony power* has appeared to be an important contributor to *low prices* for traditional tropical commodities. Often, market power is concentrated in the hands of a few processors, traders and retailers; these companies have fantastic buying power or *oligopsony,* and have increasingly controlled, governed and dominated commodity markets making *immense profits* at the expense of suppliers and consumers. Theoretically, food sufficiency, security and preservation are supposedly under the direction and control of the following international policy instruments:

- *Adapted United Nation (UN) Codes:* UN Conference on Trade and Development's (UNCTAD) Set of Multilaterally Agreed Equitable Principles and Rules for the Control of Restrictive Business Practices, adopted by the UN General Assembly in 1980 (TD/RBP/CONF/10/Rev.1); Tripartite Declaration of Principles Concerning Multinational Enterprises and Social Policy, passed in November 1977 by the International Labor Organisation (ILO) Governing Body; the joint-World Health Organisation (WHO) and UN International Children's Emergency Fund (UNICEF or Children's Fund) Code for the Marketing of Breastmilk Substitutes; and the joint-Food and Agricultural Organisation (FAO)/World Health Organisation (WHO) Code of Ethics for International Trade in Food, adopted by the Codex Alimentarius Commission in 1979.

- *The Havana Charter 1948:* The UN Conference on Trade and Employment held in Havana, Cuba, in 1947, adopted *but did not implement,* the Havana Charter, but ended in creating a UN General Agreement on Tariffs and Trade (GATT). The Havana Charter was meant to establish a multilateral trade organisation. Purpose of the Havana Charter was to: assure a large and steadily growing volume of real income and effective demand to increase the production, consumption and exchange of goods, and thus to contribute to a *balanced* and expanding world economy; foster and assist industrial and economic development, particularly of those countries which are still in the early stages of industrial development, and to encourage the international flow of capital for productive investment; further the enjoyment by all countries, *on equal terms,* of access to markets, products and productive facilities which are needed for their economic prosperity and development; *promotion* on a reciprocal and mutually advantageous basis the *reduction* of tariffs and other barriers to trade and the *elimination* of discriminatory treatment in international commerce; enable countries by increasing the opportunities for their trade and economic development, to *abstain* from measures which would *disrupt* world commerce, reduce productive employment or *retard* economic progress; facilitate through promotion of mutual understanding, consultation and cooperation of the solution of problems relating to international trade in the fields of employment, economic development, commercial policy, business practices and commodity policy. The unimplemented Havana Charter was followed by various conferences in Anncy, France; Torquay, UK; Tokyo, Japan (The Tokyo Round – 1973-1979 – reduced tariffs and established new regulations controlling the proliferation of non-tariff barriers and voluntary export restrictions); Punta del Sol, Uruguay; Montreal, Canada; Brussels, Belgium; and, finally to Marrakesh, Morocco in 1994. On 1 January 1995, the World Trade Organisation (WTO) was created to replace the GATT established in 1947.

- *UN World Trade Organisation* (WTO): The World Trade Organisation (WTO) replaced the GATT in 1995; it deals with the rules of trade between its 153 member-nations at a near-global level, is responsible for negotiating and implementing new trade agreements, and is in charge of policing member countries' adherence to all WTO agreements signed by the majority of the world's trading nations and ratified by their parliaments. The *Uruguay Round 1986*, the *1996 Singapore Agenda,* and the *2001 Doha Round* - are still on the WTO discussion and negotiation table. The WTO trading system should be: *without discrimination* – a country should not discriminate between its trading partners [(giving them equally "most-favoured-nation" (MFN)] status; and it should not discriminate between its own foreign products, services or nationals (giving them "national treatment"); *freer* –barriers coming down through negotiations; *practicable* – foreign companies, investors and governments should be confiident that trade barriers (including tariffs and non-tariffs) should not be raised arbitrarily; tariff rates and market-opening commitments are "bound" in the WTO; *more competitive* – discouraging "unfair" practices such as export subsidies and dumping products at below cost to gain market share; *more beneficial for less developed countries* – giving them more time to adjust, greater flexibility, and special privileges.

- *Singapore Agenda 1996/97:* The Singapore Agenda of the WTO was implemented in 1997 during the WTO Director-General Renato Ruggeiro's address to the Asian-Pacific Economic Cooperation (APEC) Trade Ministers *("Implementing the WTO Singapore Declaration in 1997 and beyond")* in Montreal, Canada on 10 May 1997. Summary: the most thorny Labour standards and textiles issues; critical negotiations toward multilateral liberalisation in financial sevices and telecommunications; information technology agreement (ITA) and liberalisation of basic telecommunication services; accession negotiations; high-level conference on the Least-Developed Countries – a global trading system to embrace the poorest countries; 50th anniversary of the multilateral system; and investment for developiong countries.

- *Uruguay Round 1986/1993:* The Round began September 1986 and ended April 1994 in Punta del Este, Uruguay, its purpose being to: reduce agricultural subsidies; put restrictions on foreign investments; begin the process of opening trade services like banking and insurance; deal with copyright violation and other forms of intellectual property rights; its non-achievements included: paying insufficient attention to the special needs of developing countries; setting up too many constraints on policy-making and human needs.

- *Doha Round, Agenda* (DRA) *2001:* The 2001 4th Ministerial Council of this Round in Doha, Qatar, listed 19-21 subjects, missed its original deadline of 1 January 2005, so targeted 2006. The Cancun, Mexico 2003 5th stock-taking Ministerial Council ended in deadlock over agricultural issues including cotton. The 19-21 subjects include:

- *(1) Implementation-related Issues and Concerns* [para. 12]: More than 40 items under 12 headings were settled on or before the Doha Conference for immediate delivery, but the vast majority of the remaining items were under negotiation, as folloows:

- General Agreement on Tariffs & Trade (GATT)-balancing-of-payments; market-access commitments; *Agricultural* (rural development and food security for developing countries, leaset-developed and net-food-importing developing countries, export credits, export credit guarantees or insurance payments, tariff rate quotas); *Sanitary & Phytosanitary* (SPS) measures – more time for developing countries to comply with other countries' new SPS measures; "reasonable interval" between publication of a country's new SPS measure and its entry into force; equivalence: putting into practice the principle that governments should accept that different measures used by other governments can be equivalent to their own measures for providing the same level of health protection for food, animals and plants;

review of the size of the SPS agreement; developing countries' participation in setting international SPS standards; financial and technical assistance; *Textiles/Clothing:* "effective" use of the agreement's provisions on early integration of products into GATT rules, and elimination of quotas; restraint in anti-dumping actions; the possibility of examining governments' new rules of origin; members to consider quota treatment for small suppliers and least-developed countries, and larger quotas in general; *Technical barriers to trade:* technical assistance for least-developed countries, and revivals of technical assistance in general; when possible, a six-month "reasonable interval" for developing countries to adapt to new measures; the WTO Director-General encouraged to continue efforts to help developing countries participate in setting international standards; *Trade-related measures* (TRIMs): the Goods Council is to "consider positively" requests from least-developed countries to extend the seven-year transition period for eliminating measures that are inconsistent with the agreement; *Anti-dumping* (GATT Article 6): no second anti-dumping investigation within a year unless circumstances have changed; how to put into operation a special provision for developing countries (Artt. 15 of the Anti-Dumping Agreement), which recognises that developed countries must give a "special regard" to the situation of developing countries when considering applying anti-dumping measures; clarification sought on the time period for determining whether the volume of dumped imported products is negligible, and therefore no anti-dumping action should be taken; annual reviews of the agreement's implication to be improved; *Customs validation* (GATT Art. 7): extending the deadline for developing countries to implement the agreement; dealing with fraud: how to cooperate in exchanging information, including on export values; *Rules of origin:* completing the harmonisation of rules of origin among member governments; dealing with interim arrangements in the transition to the new, harmonised rules of origin; *Subsidies and countervailing measures:* sorting out how to determine whether some countries meet the test of being below US$1,000 per capita GNP allowing them to pay subsidies that require the recipient to export; noting proposed new rules allowing developing countries to subsidise underprogrammes that have "legitimate developmental goals" without having to face countervailing or other action; review of provisions on countervailing duty investigations; reaffirming that least-developed countries are exempt from the ban on export subsidies; directing the Subsidies Committee to extend the transition period for certain developing countries; *Trade-related aspects of intellectual property rights* (TRIPS): "non-violation" complaints: the unresolved question of how to deal with possible TRIPS disputes involving loss of an expected benefit even if the TRIPS Agreement has not actually been violated; technology transfer to lease-developed countries; *Cross-cutting issues:* which special and differential treatment provisions are mandatory? What are the implications of making mandatory those that are currently non-binding? How can special and differential treatment provisions be made more effective? How can special and differential treatment be incorporated in the new negotiations? Developed countries are urged to grant preferences in a generalised and non-discriminatory manner, i.e., to all developing countries rather than to a selected group; *Outstanding implementation issues:* to be handled under para. 12 of the main Doha Declaration; *Final provisions:* The WTO Dir-Gen is to ensure that WTO technical assistance gives priority to helping developing countries implement existing WTO obligations, and to increase their capacity to participate more effectively in future negotiations; the WTO Secretariat is to cooperate more closely with other international organisations so that technical assistance is more efficient and effective;

- *(2) Agriculture* [para. 13, 14]; *(3) Series* (para. 15); *(4) Market access for non-agricultural products* [para. 16]; *(5) Trade-related aspects of intellectual property rights* (TRIPS) [para.17-199]; *(6) Relationship between trade and investment* [para. 20-22]; *(7) Interaction between trade and competition policy* [para. 23-25]; *(8) Transparency in government procurement* [para. 26]; *(9) Trade facilitation* [para. 27]; *(10) WTO anti-dumping and subsidies* [para. 28]; *(11) WTO rules: regional trade agreements* [para. 29]; *(12) Dispute Settlement Understanding* [para. 30]; *(13) Trade and environment* [para. 31-33]; *(14) Electronic commerce* [para. 34]; *(15) Small economies* [para. 35]; *(16) Trade, debt and finance* [para.36]; *(17) Trade and technology transfer;* *(18) Technical cooperation and capacity building* [para. 38-41]; *(19) Least-developed countries* [para. 42-43]: *(20) Special and differential treatment* [para. 44]; *(21) Cancun 2003, Hong Kong 2006.*

- *Common Market for Eastern and Southern Africa* (COMESA) *Project:* Attained a Free Trade Zone Area (FTZA) on 31 October 2000, with the objectives to: strengthen the capacity of member states in public procurement; enhance competition in the procurement of goods and services and intra-trade within the FTZA area – Angola, Burundi, Comoros, DR Congo, Djibouti, Egypt, Eritrea, Ethiopia, Kenya, Madagascar, Malawi, Mauritius, Namibia, Rwanda, Seychelles, Sudan, Swaziland, Uganda, Zambia, Zimbabwe; harmonisation of public procurement rules and regulations; encourage more awareness of procurement opportunities; improve national procurement systems through transparent rules, regulations and procedures; promote efficient and optimal utilisation of government resources; and ensure sustainability of goods and services in public procurement.

- *Poor Farmers:* These are often sidelined by commercial landlords and big traders during the development of national public policies. It is essential that poor farmers' interests are represented through their strong and participatory process. This will ensure that any response to commodities' crises to reduce poverty will be effective. Poor farmers' active and effective participation in all agricultural improvement mechanisms by the World Trade Organisation (WTO) is crucial. Financial services such as insurance and credit should be made available to poor farmers, contrary to the present practices where banks refuse to lend money to them to manage between harvests.

Food sufficiency, security and preservation have proved utopic as the measures meant to enforce them have fallen prey to speculation, hoarding and protectionism.

2.3.2 Energy Security, Speculation

SIDELINED POOR farmers shut off from energy are among several deficits of our modern industrial system. This is due largely to the constantly rising prices, hoarding, and speculation of this commodity. Energy security or independence is far beyond those marginalised and are at the periphery of our many economies, communities and countries. These include especially sidelined poor farmers, the rural poor and economically disadvantaged. Without access to cheap and available energy, these economies and their dependants are doomed to eternal poverty, disease and exposure to the terrors of climate change. The advantage of producing and using renewable energies is that, these energy sources are available at the sources of its users – the rural poor.

The world fossil-petroleum energy system is at the heart of *scarcities, rocketing prices, hoarding and speculation, energy insecurity, plus a long list of vulnerabilities*. Fossil-petroleum energy supplies are unevenly distributed among countries. This breeds political instability and threats to producing countries; manipulation of, and limited energy supplies like gas and oil; competition over energy sources; terrorist and non-terrorist attacks on supply infrastructure; accidents and natural disasters; and pollution and climate change. Is there a way out of the near-total grip by the fossil-petroleum regime? *Yes, there is, and has always been!* Building electricity-generating capacity close to the source of demand, can improve combustion efficiency – where the captured wasteheat can reduce imports of natural gas and other fossil fuels; using on-site renewable powered energy can further reduce fuel imports and emissions responsible for climate change and air pollution; and legislating against energy price hiking, excessive energy and commodities speculation, and commodity futures markets.

Masters of *Masters Capital Management, LLC,*[170] testified on "Commodities Speculation" to the U.S., Homeland Security & Government Affairs Committee regarding energy banditry as follows:

---

[170] Michael W. Masters: *"Energy Speculation Causes Fuel Price Inflation."* 22 June 2008, 120 pp., published by Progressive Democrats of America (PDA)-mobilising the Progressive Vote – detailing Michael Master's testimony before the Committee on Homeland Security & Government Affairs in May 2008 on the subject of *Commodities Speculation.* PDA: Kimberly Buchan,

- *Are Institutional Investors Contributing to Food & Energy Price Inflation?:* "Unequivocally, YES. In this testimony, I will explain that Institutional Investors are one of, if not, the *primary,* factors affecting commodities prices today. Clearly, there are many factors that contribute to price determination in the commodities markets. I am here to expose a fast-growing, yet virtually unnoticed factor, one that presents a problem that can be *expediently* corrected through *legislative* policy action. What we are experiencing, is a *demand-shock* coming from a new category of participants in the commodities futures markets. Specifically, Institutional Investors, are *Corporate* and *Government Pension Funds, Sovereign Wealth Funds, University Endowments* and *Other Institutional Investors.* Collectively, these investors now account on average for a larger share of outstanding commodities futures contracts than any other market participant. These parties, who I call *Index Speculators,* allocate a portion of their portfolios to "investments" in the commodities market, and behave very differently from the traditional speculators that have always existed in this marketplace. I term them as "Index" Speculators because of their investing strategy: *they distribute their allocation of dollars across the 25 key commodities futures* according to the popular indices – the Standard & Poors – Goldman Sachs Commodity Index and the Dow Jones-AIG Community Index.

- *Index Speculators Demand Is Driving Prices High:* "Today, Index Speculators are pouring billions of dollars into the commodities markets, speculating that commodity prices will increase. Assets allocated to commodity index trading strategies have risen from $13 billion at the end of 2003 to $260 billion as of March 2008, and the prices of the 25 commodities that compose these indices have risen by an average of 183% in those five years!

- *Commodity Index Investment vs Spot Prices:* "In the popular press the explanation given most often for rising oil prices is the increased demand for oil from China. According to the DOE, annual Chinese demand for petroleum has increased to 2.8 billion barrels, an increase of 920 million barrels. Over the same five-year period, Index Speculators demand for petroleum futures has increased by 848 million barrels. The increase in demand from Index Speculators is almost equal to the increase in demand from China!. *Commodity Index Purchases Last Five Years:* "Index Speculators have now stockpiled, via the futures market, the equivalent of 1.1 billion barrels of petroleum, effectively adding eight times as much oil to their own stockpile as the U.S., has added to the Strategic Petroleum Reserve over the last five years.

- *Index Speculators Demand Characteristics:* "Demand for futures contracts can only come from two sources: *Physical Commodity Consumers* and *Speculators.* Speculators include the *Traditional Speculators* who have always existed in the market, as well as *Index Speculators.* Five years ago, Index Speculators were a tiny fraction of the commodities futures markets. Today, in many commodities futures markets, they are the *single* largest force. The huge growth in the demand has gone virtually *undetected* by classically-trained economists who almost never analyse demand futures markets. Index Speculator demand is dinstictly *different* from Traditional Speculator demand; it arises *purely* from portfolio allocation decisions. When Institutional Investors decide to allocate 2% to commodities futures, for example, they come to the market with a *set* amount of money. They are not concerned with the price per unit; they will buy as many futures contracts as they need, at *whatever* price is necessary, until all of their money has been "put to work." Their insensitivity to price *multiplies* their impact on commodity markets.

- *Commodity Futures Market Size:* "As money pours into the markets, two things happen *concurrently:* the markets expand and prices rise. One particularly troubling aspect of Index Speculator demand is that, it *actually* increases the more prices increase. This explains the accelerating rate at which commodity futures prices (and actual commodities prices) are

Admin. Dir.; kimberly@pdamerica.org; POBox 150064, Grand Rapids, MI 49515-0064; info@pdamerica.org; www.pdamerica.org.

increasing. Rising prices *attract* more Index Speculators, whose tendency is to *increase their allocation as prices rise.* So their *profit-motivated* demand for futures is the *inverse* of what you would expect from price-sensitive consumer behavior. Prices have increased the most dramatically in the first quarter of 2008. We calculate that Index Speculators *flooded* the markets with $55 billion in just the first 52 trading days of this year. That's an increase in the dollar value of outstanding futures contracts of more than $1 billion per trading day. Doesn't it seem likely that an *increase in demand of this magnitude in the commodities futures markets could go a long way in explaining the extraordinary commodities price increases in the beginning of 2008?* There is a crucial distiction between Traditional Speculators and Index Speculators: Traditional Speculators provide *liquidity* by buying and selling futures. Index Speculators buy futures and then *roll* their positions by buying *calendar spreads.* They *never* sell. Therefore, they consume *liquidity* and provide *zero-benefit* to the futures markets. Is this what Congress expected when it created the *Commodities* [Futures] *Trading Commission* (CFTC)?

- *The CFTC Invited Increased Speculation:* When Congress passed the Commodity Exchange Act in 1936, they did so with the understanding that speculators should not be allowed to *dominate* the commodities futures markets. Unfortunately, the CFTC has taken *deliberate* steps to allow certain speculators *virtually unlimited* access to the commodities futures markets. The CFTC has granted Wall Street banks an *exception* from speculative *position limits* when these banks *hedgeover-the counter* swaps transactions. This has *effectively* opened a loophole for *unlimited speculation.* When Index Speculators enter into commodity index swaps, which 85-90% of them do, they face no speculative position limits. The really shocking thing about the *Swaps Loophole* is that Speculators of all stripes can use it to access the futures markets. So if a hedge fund wants $500 million position in *Wheat,* which is a way beyond position limits, they can enter into swap with a Wall Street bank and then the bank buys $500 million worth of Wheat futures. In the CFTC's *clarification scheme,* all Speculators accessing the futures markets through the Swaps Loophole are categorised as "Commercial" rather than "Non-Commercial." The result is a *gross distortion* in data that effectively hides the full impact of Index Speculation. Additionally, the CFTC has recently proposed that Index Speculators be *exempt* from all position limits, thereby throwing the door open for *unlimited* Index Speculators "investment." The CFTC has even gone so far as to issue press releases on their website touting "studies" they commissioned showing that commodities futures make good additions to Institutional Investors' portfolios.

- *Congress Should Eliminate The Practice Of Index Speculation:* "I would like to conclude my testimony by outlining three steps to immediately reduce Index Speculation. Number One: Congress has closely regulated pension funds, recognising that they serve a public purpose. Congress should modify ERISA regulations to prohibit community index replication strategies as unsuitable pension investments because of the damage that they do to the commodities futures markets and to America as a whole; Number Two: Congress should act *immediately* to close the Swaps Loophole. Speculative position limits must 'look-through' the swaps transaction to the ultimate counterparty and hold that counterparty to the speculative position limits. This would curtail Index Speculation and it would force ALL Speculators to face position limits; Number Three: Congress should further *compel* the CFTC to reclassify all the positions in the Commercial category of the Commitments of Traders Reports to *distinguish* those positions that are controlled by 'Bona Fide' Physical Hedgers *from* those controlled by Wall Street banks. The positions of Wall Street banks should be further *broken down* based on their OTC swaps counterparty into 'Bona Fide' Physical Hedgers and Speculators; There are hundreds of billions of investment dollars poised to enter the commodities futures markets at this very moment. If *immediate* action is *not* taken, *food and energy prices will rise higher still.* This could have catastrophic economic effects on millions of already stressed U.S., consumers. It literally could mean *starvation for millions of the world's poor.* If Congress takes these steps, the structural integrity of the futures markets will be restored. Index Speculator demand will be *virtually eliminated* and it is likely that food and energy prices will come down sharply."

On 15 July 2008, the U.S. Congress voted on a *"Stop Excessive Energy Speculation Act"* (SEESA) which was neither passed nor signed into law by the Bush Administration. The Act seeks to reduce the amount of excessive speculation in the oil markets, increase the resources and authority needed by the Commodities and Futures Trading Commission (CFTC) to detect, present, and punish *price manipulation* and *excessive speculation,* and give the CFTC emergency authority needed to rapidly implement the legislation; the Act will also strengthen the amount and quality of information available to the CFTC so that the Commission can better regulate all aspects of the energy futures markets. This would also provide better transparency in the trading of energy derivatives by closing the "London Loophole" so that traders using a foreign exchange cannot manipulate the price of oil in the U.S. It will require the CFTC to implement position limits to *restrict* excessive speculation that would still allow for reasonable trading for price discovery, liquidity, and legitimate hedging purposes.

Major provisions of the SEESA include:

- *Speculative Limits And Transparency Of Offshore Trading:* This section would close the "London Loophole" by treating oil traders using a foreign exchange as if they were trading in the U.S., for regulatory purposes, in order to stop traders from manipulating prices and speculating excessively by routing oil trades away from U.S.-based exchanges.

- *Working Groups On International Regulators:* This section would require the CFTC to convene a working group made up of international regulators to develop uniform reporting and regulatory standards so that excessive speculation will not harm consumers, national economies, and energy futures markets.

- *Elimination of Manipulation and Excessive Speculation As A Cause Of High Oil, Gas, And Energy Prices:* This section would require the CFTC to eliminate excessive speculation while protecting and promoting legitimate hedge trading by adding a new definition of "legitimate hedge trading" to "trading by commercial producers and purchasers of actual physical petroleum and energy commodities for future delivery and direct counterparties to such trades (regardless of whether the counterparties are commercial producers or purchasers)." Additionally, the section would require the CFTC to impose by rule reasonable speculative position limits on trading that is not legitimate hedge trading.

- *Large Over-the-Counter Transactions:* This section would give the CFTC the authority to begin collecting data on large over-the-counter traders so that it has the data necessary to determine whether the price manipulation or excessive speculation is taking place. It would also require the CFTC to prepare a report on each person who enters into a "covered over-the-counter transaction."

- *Index Traders And Swap Dealers*: This section would require that the CFTC routinely collect detailed information from index traders and swap dealers, and be able to distinguish between the two. The bill will require a review of the CFTC's trading practices to ensure that index traders are not adversely impacting the price discovery process.

- *Disaggregation Of Index Funds And Other Data In Energy:* This section would require the CFTC to release data each month on the number and value of index funds in the energy markets as well as data on the speculative positions of those index funds in relation to normal market hedgers.

- *Additional CFTC Employees:* This section would authorise the CFTC to hire at least 100-full-time employees so the Commission can improve and increase its regulation and enforcement of the energy derivatives markets.

- *Working Group On Energy Markets:* This section would create a new interagency Working Group on Energy Markets, to be chaired by the Secretary of Energy. Other members include the Secretary of Treasury, Chairman of FERC, the CFTC, the SEC and the EIA Administrator. The purposes of the Working Group would include: investigating the effects of speculation on energy commodities and prices critical to energy security of the U.S.; recommending to the President and Congress laws and regulations necessary to prevent excessive speculation in energy commodities; coordinating the federal responses to energy emergencies; and reviewing energy security considerations to developments in international energy markets.

- *Study Of Regulatory Framework For Energy Markets:* This section would direct the Working Group On Energy Markets to conduct a study and report on the role of speculation on petroleum prices, and on any regulatory gaps that exist among relevant federal agencies that might hinder the effective oversight of markets critical to the energy security of the U.S., and report back to Congress within one year.

- *Collection And Analysis Of Information On Energy Commodities:* This section would amend the DOE Organisation Act, to require federal agencies to provide the EIA Administrator with information necessary to identify energy-producing companies. In addition, the legislation would direct EIA to collect on a weekly basis more complete information on company-specific ownership of oil and natural gas volumes, and storage and transportation capacity owned or leased. For every commodities physically delivered in the U.S., it would also require any entity holding or controlling energy futures or swap contracts above an amount specified by the Secretary to file monthly reports to the EIA detailing the quantity of physical stocks owned, physical purchases and sales agreements, and storage capacity owned or leased. Finally, the legislation would create within EIA a new Financial Market Analysis Office.

- *National Natural Gas Market Investigation:* This section would direct Federal Energy Regulatory Commission (FERC) to undertake a study of the role of financial institutions on natural gas markets, including trends in investment in storage and pipeline capacity, factors contributing to potential effects or wholesale natural gas prices; the character and number of related financial positions; and any international considerations the Commission deems relevant. The legislation also would reaffirm that federal agencies must cooperate with the FERC sharing data the FERC Commission deemed relevant to the investigation.

2.3.3 Virtues of Permaculture

PERMACULTURE IS not organic farming or energy and food security; it is not about regeneration or polyculture or sustainable agriculture; it is not about biodiversity or ecosystem preservation; it is not about protection of the environment from greenhouse gases' pollution and contamination; it is not about the use of renewable energies. *Permaculture is a combination of all of these attributes and more.* Permaculture is a well-thought out and experimented design of human settlements and perennial agricultural systems that mimic the relationshiops found in natural ecologies. Permaculture means *permanent agriculture and permanent culture* welded together; it is a core set of design principles that allow individual communities to design their own environments and build increasingly self-sufficent human settlements, ones that reduce society's *reliance on industrial systems of production and distribution identified as totally destroying Earth's ecosystems.* Two Australians, Bill Mollison and David Holmgren,[171] designed premaculture's principles, which have expanded to include alternative cultures and rewilding the human species. Permaculture draws heavily on the practical application of ecological theory to analyse the characteristics and potential relationships between design elements: each element of a design is carefully analysed in terms of needs, outputs and properties *synergically* to feed and benefit the needs of all involved to the disadvantage of none. No waste of nature's resources or their pollution or contamination is allowed in permaculture.

2.3.4 Permaculture, Polyculture, Nature Preservation

PERMACULTURE, POLYCULTURE, and the use of renewable energies – *mimic nature to preserve nature*, in the interest of nature and those that dwell in it - rather than *unmimic nature and destroy nature* in disfavor of nature and of those that dwell in it. Permaculture is designed to tool-look at a whole system of nature or problem; observe how the plant relates; plan to mend sick systems by applying ideas learnt from long-term sustainable systems; fix the damaged landscapes of human agricultural and city systems; see connections between key parts; its core values are Earth- and People-care, and Fare-share or placing limits on consumption and ensuring that Earth's limited resources are used in ways that are equitable and wise. One of permaculture's designs is to appreciate the efficiency and productivity of natural ecosystems, to use natural energies like *wind, gravity, solar, fire, wave* and more and seek to apply these to the way human needs for food and shelter are met.

Polyculture uses multiple crops in the same space, in imitation of the diversity of natural ecosystems; it avoids large stands of single crops, or *monoculture.* Multiple cropping includes *crop rotation, multi-cropping* and *inter-cropping* or *alley-cropping* – a simplification of the layered system which typically uses just two layers, with alternative rows of trees and smaller plants. Crop rotation or *crop sequencing* is the practice of growing a series of dissimilar types of crops in the same area in sequential seasons for various benefits. Benefits of multiple cropping include: avoiding the buildup of *pathogens* and pests that often occurs when one species is continuously cropped; balance the fertility demands of various crops to avoid excessive depletion of nutrients; replenishment of nitrogen through the use of *green manure* in sequence with cereals and other crops; improve soil structure and fertility by alternating deep-rooted and shallow-rooted plants. *Green manure* is a type of cover crop grown primarily to add nutrients and

---

[171] Bill Mollison & David Holmgren: *Bill Mollison:* Born in Tasmania, Australia 1928; researcher, teacher, scientist, naturalist, and is considered to be the Father of Permaculture – an integrated system of design, co-developed with David Holmgren – that encompasses agriculture, horticulture, architecture, ecology, economic systems, land access strategies and legal systems for businesses and communities; *David Holmgren:* Born 1955 in Western Australia; ecologist; ecological designer engineer; writer; co-originator of the permaculture concept with Bill Mollison; through the spread of permaculture around the world, David's environmental principles have extended a global influence.

organic matter to the soil; it is grown for a specific period and then plouged under and incorporated into the soil. Green manures perform multiple functions that include *soil improvement* and *soil-protection,* as follows:

- *Leguminous* green manures such as *clover* and *vetch* contain nitrogen-fixing symbiotic bacteria in root noddles that fix atmospheric nitrogen in a form that plants can use.

- *Green manure* increases the percentage of organic matter or *biomass* in the soil, thereby improving water retention, aeration, and other soil characteristics.

- *The root-system* of some varieties of green manure grow deep in the soil and bring up nutrient resources unavailable to shallow-rooted crops.

- *Common cover* crop functions of weed suppression and prevention of soil erosion and compaction are often also taken into account, when selecting and using green manures.

- *Some green* manure crops, when allowed to flower, provide forage for pollinating insects. Green manure crops include: *Winter cover crops,* such as oats or rye; *Fava beans; Mustard; Fenugreek; Lupin; Sun hemp; Winter tares* or vetch; *Winter field beans; Alfalfa; Tyfon; Buckwheat; Ferns* of the genus Azolla; *Velvet bean* or Mucuna prurien.

Multi-cropping, a form of polyculture, is growing two or more crops in the same space during a single growing season, and can take the form of *double-cropping;* a double-cropping allows a second crop to be planted after the first has been harvested; or *relay-cropping,* in which the second crop is started amidst the first crop before it has been harvested. There is also *companion-planting,* oftimes used in gardening and intensive cultivation of vegetables and fruits. *Inter-cropping* involves cultivating two or more crops in the same space at the same time, often associated with sustainable agriculture and organic farming.

2.3.5 Permaculture Design Principles, Zones

PERMACULTURE DESIGN functions on laid down principles and zones. Zoning is the method of ensuring that elements are currectly placed, and are numbered 0-5, in concentric rings moving outward from the centre point. The centre point is where human activity and need for attention is most concentrated, where there is no need for intervention at all.

The principles of design and zoning are presented by Holmgren's 12 design principles:[172]

---

[172] David Holmgren: *"Permaculture: Principles and Pathways Beyond Sustainability."* Holmgren Design Services, 16 Fourtheenth ST, Hepburn Victoria, 3461, Australia; info@holmgren.com.au; http://www.holmgren.com.au.

- *. Observe and Interact:* By taking the time to engage with nature we can design solutions that suit our particular situation; *2. Catch and Store Energy:* By developing systems that collect resources when they are abundant, we can use them in times of need; *3. Obtain Yield:* Ensure that you are getting truly useful rewards as part of the work that you are doing; *4. Apply Yourself-regulation and Accept Feedback:* We need to discourage inappropriate activity to ensure that systems can continue to function well; *5. Use and Value Renewable Resources and Services:* Make the best use of *nature's abundance* to reduce our consumptive behaviour and dependence on *non-renewable* resources; *6. Produce No Waste:* By valuing and making use of all the resources that are available to us, nothing goes to waste, *7. Design from Patterns to Details:* By stepping back, we can observe patterns in nature and society. These can form the backbone of our designs, with the details in as we go; *8. Integrate Rather than Segregate:* By putting the right things in the right place, relationships develop between those things and they work together to support each other; *9. Use Small and Slow Solutions:* Small and slow systems are easier to maintain than big ones, making better use of local resources and producing more sustainable outcomes; *10. Use and Value Diversity:* Diversity reduces vulnerability to a variety of threats and takes advantage of the unique nature of the environment in which it resides; *11. Use Edges and Value the Marginal:* The interface between things is where the most interesting events take place. There are often the most valuable, diverse and productive elements in the system; *12. Creativity Use and Respond to Change:* We can have a positive impact on inevitable change by carefully observing, and then intervening at the right time.

- *Zone 0: The House or Home Center:* Here permaculture principles would be applied in terms of aiming to reduce energy and water needs, harnessing natural reources such as *sunlight*, and generally creating a harmonious, sustainable environment in which to live, work and relax.

- *Zone ONE: The Zone Nearest the House:* The location for those elements in the system that require frequent attention, or that need to be visited often, such as salad crops, herb plants, soft fruit like strawberries, raspberries, greenhouse and cold frames, propagation area, worm compost bin for kitchen wastes, and so on.

- *Zone TWO: Area for Perennial Plants:* This area is used for siting perennial plants that require less frequent maintenance, such as occasional weed control (preferably through natural methods such as spot-mulching) or pruning, including currant bushes and orchards. This would be a good place for beehives, larger home composting bins, and so on.

- *Zone THREE: The Area for Main Crops:* The area where maincrops are grown, both for domestic use and for trade purposes. After establishment, care and maintenance required are fairly minimised (provided mulches or protective cover placed over soil), such as water or weed control once a week or so.

- *Zone FOUR: A Semi-Wild Area:* This zone is mainly used for forage and collecting wild food as well as timber production. An example might be a *coppice*-managed woodland. A coppiced-woodland is a traditional method of woodland management in which young tree stems are repeatedly cut down to near ground level. In subsequent growth years, many new shoots will emerge, and after a number of years, the coppiced tree, or stool, is ready to be harvested, and the cycle begins again. A coppiced woodland is harvested in sections or coups on a rotation. In this way, a crop is available each year somewhere in the woodland. Coppicing has the effect of producing a rich variety of habitats, as the woodland always has a range of different-aged coppice growing in it, which is beneficial for biodiversity. The cycle length depends upon the species cut, the local custom, and the use to which the product is put. Birch can be coppiced for faggots on a three- or four-year cycle, whereas oak can be coppiced over a fifty-year cycle for poles or firewood.

- *Zone FIVE: A Wild Area:* There is no human intervention in zone five apart from the observation of natural ecosystems and cycles. Here is where the most important lessons of the first permaculture principle of working with, rather than against, nature are learned.

- *Patterns:* The use of patterns both in nature and reusable patterns from other sites is often key to permaculture design used in architecure: *"all things, even the wind, the waves and the earth on its axis, moving around the Sun, from patterns."* (Christopher Alexander); Permaculture designs are encouraged to develop awareness of the patterns that exist in nature (and how they function); and application of pattern on sites in order to satisfy specific design needs.

The principles and application of permaculture, so inclusive and so diverse, wise, economic and substainable in the use and preservation of nature and natural systems – provide a powerful weapon for poverty minimisation and less inequality multiplication.

2.4 Poverty Minimisation, Inequality Multiplication

THE WORLD is going through the roughest times economically, socially, and environmentally. The highest commodities and energy price rises, spiralling unemployment, job layoffs and bankcruptcies are prevailing. Deforestation, monocultural agriculture and factoring farming predominate. Poverty minimisation has deteriorated and inequality has multiplied. Poverty, whether relative or absolute, is brought about by economic conditions, and factors such as *precarious livelihoods; excluded locations; physical limitations; lack of security; abuse of power by ruling elites; limited capacities* and *lack of opportunities* such as *education, healthcare and employment; weak community organisations;* and *environmental factors.* Poverty may affect individuals or groups in both developed and developing economies.

Pauperism, especially in the developing or underdeveloped countries, lounges *two steps forward* and *one step backwards.* Extreme poverty, defined by the World Bank in its *December 2006 Report[173]" as* "living on less than US$1.00 per day" or *moderate poverty* ("living on less than US$2.00 per day") has prevailed. In 2001, up to 1.1 billion people had consumption levels below US$1 a day, and 2.7 billion lived on less than US$2.00 a day. The proportion of the developing world's population living in *extreme economic poverty* fell from 28% in 1990 to 21% in 2001; but between 1981-2001, the world's largest population lived on less than US$1 per day. The report predicts that "in 2030, the number living on less than the equivalent of $1 a day will fall by half, to about 550 million...But much of Africa will have difficulty keeping pace with the rest of the developing world even if conditions there improve in absolute terms...and Africa in 2030 will be home to a larger proportion of the world's poorest people than it is today."

[173] World Bank: *"World Bank 2007 Global Economic Prospects: Managing the Next Wave of Globalization."* Press Release No. 2007/159/DEC, 13 December 2006. Merrell Tuck; mtuckprimdahl@worldbank.org; http://www.worldbank.org/globaloutlook; http://worldbank.org/gep2007.

The products of poverty are many and varied. One of the most odious is *Street Children.* About 100 – 150 million street children world wide live on the streets of India (11 million); Brazil 1-10 million; Russia 1-3 million; Egypt 200,000 – 1 million; Kenya 250,000-300,000; Philippines 200,000; Morocco 30,000; Vietnam 23,000; Jamaica 6,500; Mongolia 3,700-4,000; Uruguay 3,000; Germany 10,000; U.S., 750,000 – 1 million - according to UN and UNICEF statistics. Stereotypal names-calling for legions of street children around the world include: *Street kid, Urchin, Enfants des rues, Saligoman, Garmins, Poussins, Moustiques; Ninos dela calle, Pajaro frutero, Piranitas, Polillas, Resistoleros, Chinches; Scugnizza; Batang Lansangan; Bui Boi; Pivetes, Meninos da rua; Strassenkinder; Balados.*

Characteristics of street children may include:

- They are aged between five and 18 years, and 70% are boys; they are children not taken care of by their parents or other protective guardians; they live in abandoned buildings, cardboard boxes, parks or on the street itself; they spend some time in the streets and sleep in a house with ill-prepared adults or live entirely in the streets and have no adult supervisor or care; those who live in the street engage in some kind of economic activity ranging from begging to vending; most go home at the end of the day and contribute their earnings to their family; they may be attending school and retain a sense of belonging to a family, but because of the economic fragility of the family, these children eventually may opt for a permanent life on the streets; they exist in major cities, especially in developing countries.

- Street Children maybe the subject of abuse, neglect, exploitation, or even in extreme cases, murder by "clean up death squads" hired by local businesses; a common element of abandonment of their children by poor families is because they are unable to feed all their children or the AIDS/HIV infection. Causes for street childrening, according to a World Health Organisation (WHO) 1993 report, is: family breakdowns; armed conflict; natural disasters; poverty and famine; dislocation through migration, urbanisation and over-crowding, and acculturation; physical, commercial and sexual abuse and exploitation by adults.

- Responses to the Street Children problem include: feeding programmes; medical and financial services; legal assistance; street education, and out-reach programmes designed to bring the children into contact with the agency, conscientisation-changing street childrens' attitudes to their circumstances; family re-unification; and drop-in centres and night shelters.

Another totally different group of Street Children, are those children subject to commercial sexual exploitation (CSEC). This involves the vicious cycle of sexual exploitation, HIV/AIDS, and violation of childrens' human rights. Forms of CSEC include: sex tourism and child trafficking; child prostitution and pornography; child marriages and child soldiers.

Commodity and energy price hikes stimulate street childrening, poverty and extreme poverty. The World Bank's updated global poverty level,[174] provides solid data, thus:

---

[174] The World Bank: *"Poverty At A Glance: New Estimates."* 1990-2009. The World Bank, 1818 H Street NW, Washington DC 204433, U.S.

- The poverty headcount rate [in East and South Asia] at the $2-a-day levelis estimated to have fallen to about 27% (2007), down from 29.5% in 2006 and 69% in 1990...In Sub-Saharan Africa, extreme poverty went from 41% in 1981 to 46% in 2001, which combined with growing population increased the number of people in poverty from 231 million to 318 million...In the early 1990s some of the transition economies of Eastern European and Central Asia experienced a sharp drop in income. The collapse of the Soviet Union resulted in large declines in GDP per capita, of about 30-35% between 1990-1998.

- New World Bank estimates show about 1.4 billion people living below the international poverty line of US$1.25 a day in 2005 – equivalent to over one-fourth of the developing world's population. These [estimates] replace previous estimates of poverty at "a dollar a day," and are based on the new threshold for *extreme poverty* derived from the 2005 International Comparison Programme (ICP), which provides vastly improved and expanded data on *global purchasing* power parities.

- The reduction in poverty over time remains as *significant* to the new poverty line as with the previous one. Poverty incidence has declined from 52% of the global population in 1981 to 42% in 1990 and 26% in 2005. However, lags in survey data availability mean that these new estimates *do not include* the effects of the recent *sharp rise in food and fuel prices.* If the recent trends continue, the first Millennium Development Goal (MDG) of reducing extreme poverty by 50% from its 1990 level by 2015 will be achieved. However, one billion people will still live on $1.25 a day or less in 2015 at the current rate of progress.

- Using the new *international poverty line* of $1.25 a day in 2005 prices, the decline in poverty continues to vary considerably across regions. Led by China, the East Asia and the Pacific Region has made dramatic progress, with poverty incidence dropping from 80% to 18% between 1991-2005. At the other extreme is Sub-Saharan Africa (SSA) where poverty rate remained at 50% in 2005 – no lower than in 1981, although with more encouraging recent signs of progress. The $1.25 poverty rate fell from 58% in 1996 to 50% in 2005, though this was not sufficient to bring down the number of people living in poverty.

- Developing and developed countries need to anchor efforts to reduce poverty and achieve the MDGs in country-led development strategies, improve the environment for growth, scale up human development and infrastructure provision, and enhance mechanisms for smoothing the effects of economic shocks.

Poverty-reduction measures adopted both by the World Bank, governments, UN and other institutions and policy-makers, include: through increase in average (per capita) incomes; progressive distributional change; identical growth rates but lower levels of income inequalities that leads to a more substantial reduction in poverty rates due to economic growth; good governance; international debt relief; import substitution and export industries (i.e., discourage imported goods, increase and encourage exports); land net distribution that reduces the inequality in land ownership and create small farms; and micro-loans[175] to poor farmers.

One prominent micro-loans scheme that has made a great impression on the world is the *Grameen Bank*[176] of Bangladesh. Grameen has created a new class

---

[175] Micro-loans: Are small amounts of money loaned to farmers or villages so these can obtain the things they need to increase their economic rewards.

of women entrepreneurs who have raised themselves from poverty. It has improved the livelihoods of farmers and others who are provided access to critical market information and lifeline communications previously unattainable in 28,000 villages in Bangladesh. The Grameen Bank, owned by its poor borrowers (most of whom are women who own 94%, and the remaining 6% by the government of Bangladesh), has since grown into two dozen enterprises, among them: Grameen Trust Fund; Grameen Communications; Grameen Shakti (Energy); Grameen Telecom; Grameen Shikkha (Education); Grameen Motsho (Fisheries); Grameen Baybosa Bikash (Business Development); Grameen Phone; Grameen Software Limited; Grameen CyberNet Limited; Grameen Knitwear Limited; and Grameen Uddog (owner of the brand Grameen "Keck"), represented by the Grameen Family Enterprises. In 2003, Grameen Bank started a new interest-free, arbitrarily-long payment period, credit loan distribution programme targeted at Bangladeshi street beggars. The beggar-borrower is covered under a free-cost life insurance, and the loans are used to generate income by selling low-priced items. About 73,000 beggars have taken loans of Tk 58.32 million ($833,150) and repaid Tk 34.78 million ($496,900).

2.5. Big-scale Farmers, Small-scale Farmers, Farm Subsidies

ECONOMIC SHOCKS and rising commodity and oil price wars are the crude facts that small-scale and big-scale farmers are faced with daily.

Big-scale farmers or the agribusiness industry, is involved with food production, including *farming, corporate farming* (large-scale vertically integrated food production), *seed supply, farm machinery, wholesale* and *distribution, processing, marketing, rental sales, trade associations, publications* and *contract farming.* Contract farming (or *Outgrower schemes)* is based on contractual procedures between a buyer and a farmer which establishes conditions for the production and marketing of farm produce. The farmer (usually a small-scale farmer) agrees to provide established quantities of a specific agricultural product, meet the quality standards and delivery schedule set by the purchaser. The purchaser (usually a big-scale farmer complex) commits to buy the product at a pre-determined price, and in some cases, the purchaser also commits to support production, for example, THROUGH supplying farm inputs, land preparation, technical advice, or arranging transport of produce to the buyer's premises.

---

[176] Prof. Muhammad Yunus, Nobel Peace Prize Laureate 2006, launched Grameen Bank in October 1983 to examine the possibility of designing a credit delivery system to provide banking services targeted to the rural poor, under its *"16 Decisions Set of Values":* 1. We shall follow and advance the four principles of Grameen Bank: Discipline, Unity, Courage and Hardwork – in all walks of life; 2. Prosperity we shall bring to our families; 3. We shall not live in dilapidated homes. We shall repair our houses at the earliest; 4. We shall grow vegetables all the year round. We shall eat plenty of them and sell the surplus; 5. During the plantation seasons, we shall plant as many seedlings as possible; 6. We shall plan to keep our families small. We shall minimise our expenditures. We shall look after our health; 7. We shall educate our children and ensure that they can earn to pay for their education; 8. We shall always keep our children and the environment clean; 9. We shall build and use pitlatrines; 10. We shall drink water from tubewells. If it is not available, we shall boil water or use alum; 11. We shall not take any dowry at our sons' weddings, neither shall we give any dowry at our daughters' wedding. We shall keep our centre free from the curse of dowry. We shall not practise child marriage; 12. We shall not inflict any injustice on anyone, neither shall we allow anyone to do so; 13. We shall collectively undertake bigger investments for higher incomes; 14. We shall always be ready to help each other.If anyone is in difficulty, we shall all help him or her; 15. If we come to know of any breach of discipline in any centre, we shall all go there and help restore discipline; 16. We shall take part in all social activities collectively. [Grameen Bank is criticised for higher interest rates compared to those charged by traditional banks.]

Contract farming or *Outgrowth schemes,* link farmers with a large-scale farm or processing plant which supports production planning, input supply, extension advice, transport and a wide variety of agricultural products. Contract farming supplies agri-food chains.

Small-scale farmers who enter into contract farming encounter contractual problems such as:

- Selling to different competing buyers (side selling or extra-contractual marketing).

- A buyer company's refusal to purchase products at the agreed prices.

- Down-grading of product quality by the buyer.

- Contractors defaulting to pay agreed prices or buying less than the pre-agreed quantities.

- Potential buyers take advantage of farmers, being more powerful than the farmers, and use their bargaining clout to their financial advantage: if farmers are not well-organised or where there are few alternative buyers for crop or it is not easy to change crop, there is always that danger that farmers may have an unfair deal.

- Tactics sometimes used by buyers are: changing pre-agreed standards, down-grading crops on delivery so offering lower prices, or over-pricing for imports and transport provided. Exploitation of farmers by unscrupulous buyers can be ameloirated through: strengthening farmer organisations to better access appropriate services such as credit, extension services and market information; improving their contract negotiation skills to redress the issue of exploitation of farmers and poorly-formulated contracts and their enforcement.

Small-scale farming is always equated with backwardness, as non-productive, non-commercial, subsistence agriculture, prevalent in underdeveloped, rural regions of mostly Third World countries. The rural poor and small-scale farmers and farming-families with little or no access to markets, credit, education, energy, medicine and sanitation, live in these impoverished regions. In 1977, the UN specialised agency, the International Fund for Agricultural Development (IFAD)[177] was established with the mission to: *"enable poor rural people to overcome poverty, dedicate to eradicating rural poverty in developing countries, where about 75% or 1.05 billion live in rural areas and depend on agriculture and related activities for their livelihoods..."* IFAD's Strategic Framework for 2007-2010 goal was to empower poor rural women and men in developing countries to achieve higher incomes and improved food security; its objective to ensure that poor rural people have better access to, and the skills and organisation they need to take advantage of, were to be met through:

---

[177] UN International Fund for Agricultural Development (IFAD): *"IFAD's Strategic Framework for 2007-2010: Enabling the Rural Poor To Overcome Poverty."*

- Natural resources, especially secure access to land and water; improved natural resource management and conservation practices; and improved agricultural technologies and effective production services.

- A broad range of financial services; transparent and competitive markets for agricultural inputs and produce; opportunities for rural off-farm employment and enterprise development; and local and national policy and programming process.

Since 1978, IFAD has offered low-interest loans and grants to governments to develop and finance programmes and projects that enable rural poor to overcome poverty. It has invested $10.6 billion in 796 projects and programmes that have reached more than 330 million poor rural people; other financing sources in recipient countries, including project participants have contributed $15.3 billion; and multilateral, bilateral and other donors have provided another $9.5 billion in co-financing – altogether representing a total of approx. $24.8 billion. IFAD tackles poverty not as a lender, but as an advocate for rural people; it provides a base for a normal global platform to discuss important policy issues that influence the lives of rural poor people; and it draws attention to the central theme of rural development to meet the UN Millennium Development Goals (MDGs).

In September 2008, the UN World Food Programme (WFP), broke its taboo and announced that it will now begin to purchase its surplus supplies food from small-scale farmers as well. According to *Guardian.co.uk*,[178] more than 350,000 small-scale farmers in Africa and Central America will soon begin selling produce to the WFP in an initiative that could transform the way food aid is purchased. Traditionally, the WFP has bought about 80% of its stocks virtually from traders and large-scale farmers in the developing countries. 2007 food supplies totalled $612 million and fed 86 million people. WFP executive director, Ms Jossette Sheeran, said the new project initiative was made possible by charitable foundations established by Bill and Melinda Gates, Microsoft Co-founder; and Howard Buffet, son of Billionaire investor Warren Buffet. The purchase project will target some of the world's poorest countries including Sierra Leone, Malawi, Ethiopia and El Salvador.

Some details of the small-farmer-purchases deal include:

- It is expected that 40,000 metric tons of food – enough to feed 250,000 people for a year – will be purchased from small-scale farmers in the first 12 months. The farmers who sign up will be required to form into local collectives, and to set up a bank account in the group's name. The usual UN requirements for the growers to provide surety bonds, transport and packaging materials, will be relaxed or waived.

- By selling directly to the WFP, rather than to middlepeople, it is expected that small-scale farmers will receive higher-than-normal prices for their produce. This will also ensure that

---

[178] Guardian.co.uk: *"UN to buy surplus crops from small-scale farmers."* Xan Rice, Nairobi, Wednesday 24 September 2008.

local markets are not distorted by only purchasing from farmers with surplus crops. There are also plans to negotiate seasonal contracts with the small-holder collectives to give them additional security; small-scale farmers may also in the future be connected to other local and regional markets.

But the majority of small-scale farmers and the rural poor continue to wallow in exacerbating poverty, hunger and malnutrition. Nearly a billion people worldwide are starving, as the UN Food and Agricultural Organisation (FAO) warned in December 2008. "Almost a billion people go hungry each day after food price rises pushed 40 million more people around the world into the ranks of the undernourished," wrote Borger and Jowitt[179] in December 2008. Food prices halved their historic peaks a few months ago, but the cost of basic staples measured by an FAO index is still high: *28% higher on average than two years ago.* This has led to an increase in the number of people unable to afford to eat enough calories to lead a normal active life. There are now estimated to be 963 million people, 14% of the world's population – going hungry in 2008 – up by 40 million from 2007.

The FAO's *"State of Food Security in the World 2008"* report found that:

- The majority of the hungry live in the developing world, 65% of them in just seven countries: India, China, DR Congo, Bangladesh, Indonesia, Pakistan, and Ethiopia. The worst affected are landless families, particularly households headed by women. "For millions of people in the developing countries, eating the minimum amount of food everyday to live an active and healthy life is a distant dream...The structural problems of hunger, like the access to land, credit and employment, continued while high food prices remain a dire reality," said the FAO's assistant director-general, Haez Ghanem.

- Farmers in the developed world have been able to respond to higher prices by raising production, increasing cereal output by 10%. But farmers in poorer countries have not had access to the fertilisers, seeds, water and markets necessary to capitalise on the price rises.

Farm or *agricultural subsidies* impact on both big- and small-scale farming. Subsidies are paid by governments to farmers and agro-businesses to supplement the latters' income, manage the supply of agricultural commodities, and thus influence the cost and supply of such commodities. *Wheat, feed grains* (for fodder such as maize, sorghum, barley, oats) *cotton, milk, rice, peanuts, sugar, tobacco,* and *oilseed* (such as soybeans) subsidies – are intended to encourage food self-sufficiency; farm subsidies are also meant to guarantee high quality standards, positive impact on society, preserve cultural heritage, landscape management, generation of biodiversity, and agro-tourism, etc.

The advantages and disadvantages of farm or agricultural subsidies include:

---

[179] Julian Borger / Juliette Jowitt: *"Nearly a billion people worldwide are starving, UN agency warns; Rising prices mean 14% now undernourished; Urgency over food crisis lost amid credit crunch."* The Guardian, London, Wednesady 10 December 2008.

- *Advantages of Agricultural Subsidies In Developed Industrial Countries:* Agricultural or farm subsidies in developed industrial nations make agricultural commodities prices like sugar higher; They cause an oversupply of agricultural commodities that are sold on the world market at very low price and can harm local farmers in less developed countries who rely on stable prices for their products; Agricultural subsidies in developed industrial countries promote poverty in developing countries by artificially driving down world crop prices; Low world food prices encourage developing countries to be dependent buyers of food from wealthy countries, instead of local farmers improving the agricultural and economic self-sufficiency of their home country; Since agribusiness becomes more and more industrialised in advanced nations, the market is dominated by a few corporations that get the major part of the European Union (EU) agricultural subsidies; Benefiting from reduced competion, these firms set higher prices on the EU market than people would have to pay at "fair" competitiin; Consumers shoulder the main burden of agricultural subsidies; Biofarmers manage to be competitive without huge amounts of subsidies.

- *Advantages of Agricultural Subsidies In Developing Countries:* Agricultural or farm subsidies in developing poor countries result in increased agricultural production and driving down domestic food prices. Consumers pay less for their food because compared with wealthier individuals, poor people generally pay a smaller proportion of their income in taxes, and they generally spend a larger portion of their incomes on food; Lower food prices financed through tax revenues, will provide larger benefits for the poor than for the wealthy, thus indirectly transferring wealth to lower income individuals; Subsidies that result in lower food prices or domestic overproduction, can provide benefits for the poor in many ways, while subsidies export drive down the price of commodities, which in turn provides cheap food for consumers in developing countries; Without farm subsidies, domestic farmers would not be able to compete with an influx of foreign cheap exports, and removing the subsidies would therefore, drive domestic farmers out of business, thus leaving the country with a much smaller (or possibly nonexistent) agricultural industry.

A country that is unable to domestically produce enough food to feed its people is at the *mercy of the world market,* and is more vulnerable to *trade pressures* and *global food shortages* and *price shocks.* The loss of the domestic farming industry is undesirable, as it feeds unemployment and the loss of traditional cultural way of life. Farm or agricultural subsidies in developing countries are a *necessary evil* that is needed for the survival of developing world economies, but which developed economies can do without. The excruciating World Trade Organisation (WTO) Doha Round negotiations confirm this.

Undoubtedly, climate change and the resulting global warming, have bedevilled and rendered redundant all efforts to eradicate, eliminate, reduce or minimise poverty and destitution, hunger and malnutrition, disease and bacterial-resistance to medication. It has enhanced the rapid destruction of our civilisation. Global warming is still raging and making its indelible and irreversible imprimatur on our environments, lives, cultures, economies and morals. Farming practices, particularly large-scale agribusiness farming, have contributed to global warming. They have destroyed rural communities. Intercontinental food transport, *intensive monoculture,* land and forest destruction, the use of chemical inputs in agriculture, and transforming agriculture into an energy-consumer, instead of an energy-producer, have destroyed ecosystems and biodiversity. The World Trade Organisation (WTO), World Bank and International Monetary (IMF) have imposed these policies whereby food is produced with *oil-based pesticides* and *fertlizers* which have contributed to global warming.

La Via Campesina or International Peasant Movement[180] believes that industrial agriculture is a major contributor to global warming and climate change, and therefore, rising prices, hunger, poverty, malnutrition, and diseases, etc., as the organisation explains below:

- *By Transporting Food All Around The World:* Fresh and packaged food is *unnecessarily* travelling around the world, while simultaneously, local farmers are denied appropriate access to local and national markets. In Europe and the USA, for example, it is now common to find fruits, vegetables, meat or wine from Africa, South America or Oceanis; and we find Asian rice in the Americas or in Africa. *Fossil fuel* used for food transport is releasing tons of CO^2 into the atmosphere. The Swiss peasants association UNITERRC calculated that one kilo of asparagus imported from Mexico needs *5 litres of oil* to travel by plane (11,800 km) to Switzerland, while a kilo of asparagus produced in Switzerland only needs *0.3 litres of oil* to reach the consumer.

- *By Imposing Industrial Forms Of Production (Mechanisation, Intensification, Use Of Agrochemicals, Monoculture...):* The so-called "modernised" agriculture, especially *industrial monoculture,* is destroying the natural soil processes which lead to the storing of carbon in soil organic matter, and replaces them with *chemical processes based on chemical fertilisers, intensive agriculture* and *animal production monocultures* which produce important quantities of *nitrous oxides* (NO^2) – the third most significant greenhouse gas responsible for global warming. In Europe, 40% of the energy consumed on the farm is due to the production of nitrogen fertilizers. Moreover, industrial agriculture production consumes much more energy (and releases much more carbon dioxide) to run its giant tractors to harrow and ploy the land to process food.

- *By Destroying Biodiversity And Its Capacity To Capture Carbon:* Carbon is naturally captured from the air by plants and it is stocked in wood and organic matter in the soil. Some ecosystems such as native forests, peatlands and meadows stock more carbon than others. This carbon cycle has been part of the climate balance for thousands of years. Corporate agribusiness has now *shattered* this balance by imposing widespread chemical agriculture (with massive use of oil-based pesticides and fertilisers), by burning forests for monoculture plantations, and by destroying peatlands and biodiversity.

- *By Converting Land And Forest Into Non-agricultural Areas:* Forests, pastures and cultivated lands are rapidly converted into industrial complexes, big houses, large infrastructure projects or tourist resorts. This in turn causes *massive carbon releases* and reduces the capacity of the environment to absorb the carbon released into the atmosphere.

- *By Transforming Agriculture From An Energy Producer Into An Energy Consumer:* On the energy level, the first role of plants and agriculture is to *transform* solar energy into energy in the form of *sugars* and *cellulose* that can be directly absorbed in food or transformed by animals into animal products. This is a natural process which brings energy into the food chain. However, the *industrialisation process* of agriculture over the last two centuries has led to an agriculture which *consumes energy* (fertilisers, use of tractors, oil-based agrochemicals, etc.).

- *False Solutions:* Agrofuels (fuels produced from plants, agriculture and forestry) are often presented as one of the solutions to the current energy crisis. Under the Kyoto Protocol, 20% of the global energy consumption should come from *renewable energy sources by 2020;* this includes agrofuels. However, leaving aside the *insanity* of *producing food to feed cars* while so

---

[180] La Via Campesina (International Peasant Movement): *"Small-scale sustainable farmers are cooling down the earth."* International Operative Secretariat (IOS) JS Mampang Prapatan XIV No. 5, Jakarta Selatan, DKI Jakarta, Indonesia 12790.

many people are starving, industrial agrofuel production will actually increase global warming *instead of reducing it.* Agrofuel production will revive colonial plantation systems, bring back slave work and seriously increase the *use of petrochemicals,* as well as *contribute* to *deforestation* and *biodiversity destruction.* Intensive agrofuel production is not a solution to global warming; neither will it solve the global crises in the agricultural sector. The impacts will again be felt most seriously in developing countries, as industrialised countries will not be able to cover their agrofuel demand and will need to import huge amounts from the South.

- *Carbon Trading:* Under the Kyoto Protocol and other international schemes "carbon trading" is presented as a solution for global warming. It is a *privatisation of carbon after the privatisation of land, air, seeds, water and other resources.* It allows governments to allocate permits to big industrial pollutants so they can trade "rights to pollute" among themselves. Some other programmes encourage industrial countries to finance cheap carbon dumps such as large-scale plantations in the South as a way to avoid reducing their own emissions. This allows companies to make a double profit while claiming falsely that they contribute to carbon sequestration. On the other hand, natural areas in Asia, Africa and Latin America are being treated as carbon sinks and privatised through the so-called sale of environmental services, thus kicking communities out of their lands and reducing their right to access to their own forests, fields and rivers.

- *Genetically-Modified Crops And Trees:* Genetically modified trees and crops are now being developed for agrofuel production. Genetically modified organisms will *not solve* any environmental crisis as they themselves pose a risk to the environment as well as to health and safety. Moreover, they increase corporate control over seeds, depriving farmers of their right to grow, develop, select, diversify and exchange their own seeds. These GM trees and crops are part of the "second generation" of agrofuels based on cellulose while the first generation is based on the different forms of sugar crops. Even when it doesn't use genetically modified varieties, this "second generation" raises similar concerns.

- *Correct Solutions:* The true and complete solution is to provide food sovereignty as the key to provide livelihoods to millions and protect life on earth. Solutions to the current crisis have emerged from *organised social actors* that are developing modes of production, trade and consumption based on justice, solidarity and healthy communities. No *technical fix* will solve the current global environmental and social disaster, but a set of solutions which include *sustainable small-scale farming,* which is labour-intensive and requires little energy use – can actually contribute to stop and reverse the effects of climate change, thus: *by storing more carbon dioxide in soil organic matter through sustainable production; by replacing nitrogen fertilisers with organic agricultural or/and cultivating nitrogen-fixing plants which capture nitrogen directly from the air; by making possible the decentralisation production, collection and use of energy.* A true argrarian reform, that strengthens small-scale farming, promotes the production of food as the primary use of lands, and regards food as a basic human right that should *not be treated* as a commodity. Local food production will stop the unnecessary transportation of food and ensure that what reacheas our tables is *safe, fresh* and *nutritious.*[181]

---

[181] Foods To Avoid: *"10 Foods To Never Eat Again: While most foods can be enjoyed in moderation, there are some things worth cutting out of your diet entirely when trying to lose weight. ":* White Flour Products: Cut out simple carbohydrates, such as white bread, white pasta and white rice, and replace them with *complex carbohrates* which release their energy more slowly, keeping you fuller for longer and helping you to avoid blood sugar highs and lows; Healthier Option: *Wholegrain bread, brown rice, brown pasta;* Cinema Popcorn With Butter: A large bucket of butter popcorn can contain as many as 1500 calories and 70 grammes of fat. Studies show that we eat much more when distracted (when watching a film, for example); another good reason not to eat in the cinem! Healthier Option: *Go without. If the film is any good, you'll soon lose interest in snacking;* Processed Cheese: Processed cheese that comes in thin slices is high in calories, saturated fat and trans-fat, and is packed full of unhealthy additives and preservatives. Healthier Option: *Low-fat spreading soft cheese, low-fate cottage cheese, reduced-fat hand cheese;* Tinned Soup: Tinned soups are one of the worst culprits for hidden salt. Some brands have as much as half the recommended daily salt intake per serving. Salt not only increases high blood pressure, but it also increases the risk of stomach cancer. Healthier Option: *Homemade vegetable soup;* Alcohol: Do you find that one glass of wine leads to another? If you do, it's best to go *teetotol.* Wine contains approx. 120 calories per glass and, what's worse, it is terrible for weakening willpower. Healthier Option: *Weight Watchers reduced-calorie wine contains 1 point or 80 calories per glass at 90% alcohol. If you can't stop at one glass, drink only water at meal times;* Mc Fattening Whoppers: Fast food buyers are full of saturated fat, trans-fat and cholesterol. The average Whopper with cheese contains 720 calories, while a medium portion of French fries contain more than 600 calories. Health campaigners warn that some family meals at the 'big four' restaurants – Pizza Hut, KFC, McDonald's and Burger King – are almost as salty as sea water. Healthier Option: *Homemade burgers made with lean mince;* Hot Dogs: Most processed meat, such as hot dogs, pies, and meat used in canned soup, contain sodium nitrate preservatives which have been linked to chemicals

Moral and social responsibility expects and presses us to change consumption and production patterns which promote *waste* and *unnecessary consumption* by a minority of humankind, while hundreds of millions of humankind still suffer hunger and deprivation. Fair and just distribution of food and necessary goods, as well as reducing unnecessary consumption should be core aspects of new development patterns. *Industry should not be allowed to impose unnecessary consumption and waste by means of increasing disposable products or by artificially shortening their lives.* Research and implementation of diverse and decentralised systems, based upon local resources and technologies that do not harm the environment or take land from food production.

Applying only the correct solutions for food sovereignty, not *artificial* ones, will save our Earth from the pending doomsday. This includes fish production. Fish farming like food farming, thrives on biodiversity; fishing in waterways is similarly roped in destructive and greedy farming methods, climate change and global warming that threaten our lives and civilisation.

2.6 Fish Rogueries

FISH-FARMING on the global high- and low-seas, has seen a combination of cultural, economic, and existential tug-of-war involving acts of piracy, roguery and pollution of waterways in and around developing or underdeveloped countries. Well-equipped and technically developed fishing trawlers, ships and boats from developed nations raid fishing grounds in and around southern hemispheric seas and waterways, unhindered or unperturbed by any existing laws and regulations. Poorly-equipped domestic fish farmers provide no match to compete with or fend off powerful thieving fish invaders; fish reserves in the victimised waters are receding fast; the unemployed fishermen who depend on fish-farming frantically seek other ways of supplementing their livelihoods. Many unemployed former fishermen, like along the coasts of Somalia, Africa – have turned to piracy as a survival strategy. Another reason for Somali piracy is that foreign fishing vessels dump industrial garbage on their waters and coasts.

The Regional Fisheries Management Organisations (RFMOs) formed by the 57-Member UN Fish Stocks Agreement (UNFSA) in 2001, continue to register gross violations in regards to over-capacity and over-fishing continuing to undermine efforts to achieve long-term sustainanblity of many Straddling Fish Stocks (SFS) and Highly Migratory Fish Stocks (HMS); the provisions of the Agreement with

---

that accelerate the formation of cancer cells in the body. Healthier Option: *Quorn vegetarian sausages. These contain 48 calories and 1.6 g of fat each;* Oily Salad Dressings: Don't ruin an otherwise healthy salad by drenching it with creamy mayonnaise or oily vinaigrette. Watch out for potted potato salads and coleslaw which are laden with fat. Healthier Option: *Balsamic vinegar, low-fat salad dressing;* Sausage Rolls: Man-made hydrogenated or trans-fats used in pastry and biscuits are known to increase the risk of colonary heart disease by raising levels of bad cholesterol. Processed meat products are also often high in saturated salt – making Sausage rolls and pies doubly bad. Healthier Option: *Grilled chicken sandwich on wholemeal bread;* Fizzy Drinks: Sugary fizzy drinks are full of emptycalories. While artificially sweetened fizzy drinks can help you cut down, it's healthier to switch to sugar-free cordials and water. Healthier Option: *Naturally flavoured waters, sugar-free cordials, herbal teas.* [AOL United Kingdom "Living", 20 March 2009.]

respect to compatibility have not been fully applied in some areas of the oceans for some fisheries; RFMO implementation of long-term sustainability measures, States' efforts to address fisheries not regulated by an RFMO are proceeding unevenly; States have begun to apply the precautionary approach to fisheries management, but practical implementation varies widely; and additional work is needed to advance the Agreement's implementation through RFMOs. During the UNFSA's last meeting in 2006[182] the following actions were reviewed and recommended:

- *Reviewed Actions:* The *extent to which* the UNFSA provisions have been incorporated into national laws and regulations, as well as in the charters and/or measures of regional fisheries management organisations (RFMOs); these provisions are actually being implemented in practice, and elements for reviewing and assessing the adequacy of the UNFSA provisions; conservation and management stocks, including adoption of measures; overfishing, and capacity management; effects of fishing on the marine environment; fisheries not regulated by RFMOs; data-collection and sharing; mechanism for international cooperation, and non-members, including integrity of RFMO regimes, fishing activity by non-members, functioning of RFMOs, and participatory rights; monitoring, control and surveillance, and compliance and enforcement, including implementation of flag State duties, investigation and penalisation for violations, and international cooperation; developing States and non-parties, including recognition of special requirements, provision of assistance and capacity-building; increasing adherence to the UNFSA.

- *Recommended Actions for States Individually and Collectively Through RFMOs:* Strengthen their commitment to adopt and fully implement conservation and management measures for Straddling Fish Stocks (SFS) and Highly Migratory Fish Stocks (HMS), including currently regulated stocks, in accordance with best available scientific information and the Agreement's provisions on the precautionary approach; improve cooperation between flag States on the high seas and coastal States to ensure achievement of compatibility of measures in accordance with UNFSA Art 7 (Compatibility of Conservation and management measures); establish new RFMOs for SFS, HMS and high seas discrete stocks, where needed, and agree on interim measures until such RFMOs are established; enhance understanding of ecosystem approaches, commit to incorporating ecosystem considerations in fisheries management, and request the FAO to continue its work on the subject, as appropriate; develop management tools, including closed areas, marine protected areas and marine reserves, to effectively conserve and manage SFS, HMS and high sea discrete stocks and protect habitats, marine biodiversity and vulnerable marine ecosystems, on a case-by-case basis, and in accordance with the best available scientific information, the precautionary approach and international law; commit to urgently reducing capacity of the world's fishing fleets to levels commensurate with the sustainability of fish stocks, while recognising the *legitimate rights of developing States* to develop their fisheries; eliminate subsidies that contribute to Illegal Unreported and Unregulated (IUU) fishing, over-fishing and fishing over-capacity, while complementing the efforts undertaken at the World Trade Organisation (WTO) in accordance with the 2001 Doha Ministerial Declaration to clarify and improve its disciplines on fisheries subsidies; enhance efforts to address and mitigate the incidence and impacts of all kinds of lost or abandoned gear, establish mechanisms for the regular retrieval of derelict gear, and adopt mechanisms to monitor and reduce discards; provide required catch and effort data and fishery-related information, and to develop processes to strengthen data collection and reporting by RFMO members, including through regular audits of member compliance with such obligations; cooperate with the FAO in implementing and further developing the Fisheries Resources Monitoring System (FRMS) initiative; commit to submit information on deep sea fish catches and contribute to FAO's work regarding the collection and collation of information concerning past and present deep water fishing activities.

---

[182] Earth Negotiations Bulletin – A Reporting Service for Environment & Development Negotiations: *"UNFSA Review Highlights."* Published by the International Institute for Sustainable Development (IISD), Vol. No. 59, Thursday 25 May 2006; edited/written by Nienke Beintema/Andrew Brooks/Reem Hajjar/Ekisa Morgera: enb@iisd.org; Editor: Pamela S. Chasek, PhD, pam@iisd.org; Dir., IISD Reporting Services: Langsto James "Kimo" Gorea VI, kimo@iisd.org.

Strong measures against Illegal Unreported and Unregulated (IUU) fishing activity by non-members, was urgently recommended. Suggested measures included:

- Lists of authorised and unauthorised vessels; vessel monitoring system (VMS); observers; and implementation of systems and procedures for vessel inspectors; Canada encouraged non-members fishing in an RFMO area to either join the RFMO, obey the management rules, or abstain from fishing; Australia called for better harmonisation of compliance measures across RFMOs, including measures targeted at non-RFMO members; Mexico emphasised the need to move from single-species to ecosystem-based management, while Canada cautioned against competition between sectoral and integrative ecosystem-based approaches to oceans management; regarding participatory rights, many called for more equitable, science-based and sustainable quota systems, noting that quotas based on historical catch data disadvantaged developing countries and that unfair allocation of participatory rights is a disincentive for non-members to join the UNFSA; Japan suggested using incentives, such as quota allocations and market-based incentives, to encourage non-parties to join RFMOs; the U.S., emphasised that allocation of fishing opprtunities should be based on scientific advice rather than solely on economic concerns.

Greenpeace International keeps a Blacklist Database[183] of IUU vessels and their companies. The purpose of such a database is: "...*to publicly expose irresponsible*

---

[183] Greenpeace International:"*Greenpeace Blacklist: Official Blacklist & Official Blacklist of Companies.*" Blacklist: Total Vessels Recorded: **73** – *Adelita*, American Samoa; *Ahmed Helmi*, Tunisia; *Ahmed Khalil*, Tunisia; *Aladin*, Tunisia; *Antares*, Russian Federation; *Athena*, Italy; *Aveirense*, Portugal; *Biagio Anna*, Italy; *Brites*, Portugal; *Caribe*, Cuba; *Chang Jaan No. 1*, Vanuatu; *Chin Yuh*, Taiwan; *Chun Ying 212*, Vanuatu; *Chung Ying No. 777*, Vanuatu; *Dang Yang II*, China; *Da Yang 18*, China; *Diomede II*, Italy; *Dong Won 117*, South Korea; *Drennec*, Ecuador; *El Jazira*, Tunisia; *Feng Rong Shen*, Taiwan; *Fong Kuo No. 136*, Taiwan; *Fong Kuo No. 3*, Vanuatu; *Guayatuna Dos*, Ecuador; *Huang Shin*, Taiwan; *Joana Pricesa*, Portugal; *Kasei Maru 53*, Japan; *Kenken 888*, Philippines; *Ligny Primo*, Italy; *Lootus II*, Estonia; *Luca Maria*, Italy; *Luna Rossa*, Italy; *Lung Soon 666*, Taiwan; *Lung Yurin*, Panama; *Lutador*, Portugal; *Madins*, Estonia; *Mahkota Abadi 196*, Estonia; *Maria Antonietta*, Italy; *Marshall's 201*, Marshall Islands; *Molka*, Tunisia; *Montelucia*, El Salvador; *Mumrinskiy*, Russian federation; *Odissea*, Italy; *Orsa Maggiore*, Italy; *Oryong 316*, South Korea; *Polestar*, Panama; *Ponoy*, Russian Federation; *Queen Evelyn 168*, Philippines; *Sadik*, Tunisia; *Sajo Familia*, South Korea; *San Andres*, Ecuador; *San Sheng 168*, Taiwan; *Seed Leaf*, Marshall Islands; *Shang Jen 168*, Taiwan; *Shilla Harvester*, South Korea; *Shin Yeou 6*, Taiwan; *Slebech*, Sierra Leone; *Solomboloa*, Russian Federation; *Suruga No. 1*, Panama; *ISP Unidentified Pole and Line*, Philippines; *Tai Sheng*, China; *Tangoroa*, Vanuatu; *Vincente F*, Panama; *Wakaba Maru 8*, Japan; *Win Trend No. 136*, Taiwan; *Yang Szu No. 666*, Taiwan; *Yaroslavets*, Russian Federation; *Ying Jen 636*, Taiwan; *Yuh Yeou 236*,Taiwan; *Yusei Maru 8*, Japan; Official Blacklist: Total : **101** – *Abdi Baba 1*, Bolivia; *Acros No. 2*, Guinea; *Acros No. 3*, Guinea; *Alboran II*, (Total Loss), Panama: *Aldaba*, Togo; *Alfa* (Total Loss), Georgia; *Amorinn*, Togo; *Aquamarine-II*, Togo; *Athena F*, Venezuela; *Avior* (Broken up), Georgia; *Bhaskara No. 10*, Colombia; *Bhaskara No. 9*, Indonesia; *Bhineka*, Indonesia; *Bigaro*, Togo; *Bigeye*, Sierra Leone; *Bravo*, Sierra Leone; *Camelot*, Sierra Leone; *Carmen* (Broken up), Georgia; *Cefey* (Broken up 2007), Georgia; *Cevahir*, Bolivia; *Chi Hao No. 66*, Bolivia; *Comet*, Togo; *Contant*, Equatorial Guinea; *Daniaa*, Guinea; *Daniela F*, Venezuela; *Dolphine*, Russian Fedration; *Dragon 18*, Russian federation, *Dragon III*, Cambodia; *Duero*, Panama; *East Ocean*, China; *Eros Dos*, Panama; *Eva* (Broken up), Georgia; *Gala I*, Bolivia; *Galaxy*, Unknown; *Gold Dragon*, Equatorial Guinea; *Gorilero*, Sierra Leone; *Gunuar Melyan 21*, Sierra Leone; *Hiroyoshi 17*, Indonesia; *Iannis I*, Panama; *Isabella* (Broken up), Georgia; *JINN FENG TSAIR No. 1*, Taiwan, *Jimmy Wijaya XXXV*, Indonesia; *Juanita* (Broken up), Georgia; *Jyi Lih 88*, Georgia; *Liberty*, Unknown; *Lila NO. 10*, Unknown; *Limpopo*, Unknown; *Madua 2*, Unknown; *Madua 3*, Unknown; *Maine*, Guinea; *Maria*, Sierra Leone; *Maria grazia Genovese*, Afghanistan; *Marta Lucia R*, Colombia; *Mary Lynn*, Colombia; *Maya V*, Panama; *Melilla No. 101* (= Dong Won No. 630), South Korea; *Ming Yu Sheng 8*, South Korea; *Mumrinskiy*, Russian Federation; *Murtosa*, Togo; *Nemanskiy*, Togo; *Nicolay Chudotvorets*, Russian federation; *No. 2 Choyu*, Russian Federation; *No. 101 Gloria*, Russian federation; *No. 3 Choyu*, Russian Federation; *North Ocean*, China; *Ocean diamond*, China; *Orca*, China; *Oriente No. 7*, China; *Paloma V*, Unknown; *Pavlovsk*, Panama; *Permata, Permata I, Permata 102, Permata 138, Permata 2, Permata 6, Permata 8*, Indonesia; *Perseverance*, Equatorial Guinea; *Polestar*, Panama; *Red* (Broken up), Panama; *Rex*, Togo; *Reymar 6*, Togo; *Rosita* (Broken up), Georgia; *Sharon I*, Bolivia; *Sibley* (Sunk), Panama, *South Ocean*, Taiwan; *Southern Star 136*, Taiwan; *Sunny Jane*, Taiwan; *TRITON-I*, Sierra Leone; *TaFu I*, Sierra Leone; *Taruman*, Cambodia; *Thor 33*, North Korea; *Tonina No. 5* (ex Tonina V), South Korea; *Toto*, South Korea; *Ulla* (Broken up), Georgia; *Viarsa I* (scrapped), Uruguay; *Wen Teng No. 688*, Uruguay; *West Ocean*, China; *Yu Maan Wong*, Georgia; Official Blacklist of Companies: Total Company Records: **32** – *AVS Co. Ltd* (000'ABC'), 5316428.0; *Agricola Providencia CA*, 5201219.0; *Alpha Camara; Cecibell Securities; China National Fisheries Corporation*, 1318220.0; *DN Juan A Arqibay Perez*, 5302450; *Dong Won Fisheries Co Ltd*, 200181; *Dongwan Industries Co Ltd*, 200203.0; *Grupo Oya Perez SL*, 5049799.0; *Grupo Segade*, 1314920.0; *Grupo Silva Viera; Hayama Senpaku KK* (Hayama Shipping Ltd), 941637.0; *Hong, Cinq-Jin; Iannis Corp*, 5065171.0, *Infitco Ltd; J.L. JALABERT-S; PEREZ; Jose Ariqbay Perez*, 5216542.0; *Jose Manuel Salqueiro*, 5055815.0; *Mabenal SA*, 4091086; *Manarat Al Sahil Fishing Company; Manuel Barros Rivadulla*, 5126014.0; *Muner SA*, 51594290.0; *Nor Russ Trading AS; Pesquera Viba SL*, 5158099.0; *Pionerskiy Ocean Fishing; Marine Center* (Pionerskaya Baza Okeanichenskogo Rybolovnogo Flota (BOFA), 1554080.0; *Punta Brava Fishing SA*, 5288557.0; *Rajan Corp*, 9042001; *Redfin Investments SA*, 5335436.0, *Sharks Investments AW*, 5204638.0; *Smart Shipping C0*,

fishing operators and the companies behind them. It provides a convenient tool for national fisheries administrators, and others quickly to check on the compliance status of a foreign vessel trying to unload its catch in port, seeking services in port, seeking fishing license or to register or flag in a country. Greenpeace also encourages retailers and suppliers to use the database to ensure the fish they source do not come from pirate fishing vessels or from companies involved in such activities."

Illegal Unreported and Unregulated (IUU) fishers are the scorge of the oceans, marine life and the environment: they destroy and degradate the environment and ruin the lives of the already poor inhabitants who depend on fish for their livelihoods. Armed-and-masked fish pirates aboard their pirate-vessels roam the high seas en route to their million-dollar fish-poaching business sprees. Their targets: *Islands of the Pacific and coastal waters of West Africa.* They are responsible to no one and are a law unto themselves. Greenpeace estimates that there were about 1,300 industrial scale pirate fishing ships at sea in 2001. The UN estimates that Somalia and Guinea lose US$300 million and US$100 million each yearly, respectively, and globally more than US$4 billions are lost yearly to fish-pirate fishermen.

Fishing techniques used by fish-pirates destroy ocean life, as tuna stocks around Tanzania, Somalia, Papua New Guinea and Tuvalu are targeted each year with giant nets. These unscrupulous nets scoop up entire shoals, including the young fish vital for breeding and future stock growth. More than half of the world's tuna supply, about 2 million tonnes each year, comes from the Western and Central Pacific Ocean, home to over 20 island nations. This region has become the key target being increasingly preyed upon by distant nations and illegal, unreported and unregulated (IUU) pirate-fishing boats. Shrimp trawling, known as "bycatch from longlining" accounts for 3-4% of the world's fishing industry and is responsible for over 27% of the unnecessary destruction of marine life. While a fleet of locally-based vessels (owned by foreign and local companies) catch about 200,000 tonnes or 10% of the total catch of tuna a year, *increasing numbers of industrial distant water fishing boats are moving into the Pacific, taking about 1,800,000 tonnes or 90% of the total catch.* Economic return from access fees and licenses to the region is a mere 5% or less of the estimated US$2 billion the smuggled fish is worth on the market.

Fish-pirate-vessels hide their identities and origins, ignore rules and often fly the flags of countries that ask no questions about their fishing activities; and pirate fishing returns are *non-existent.* Many so-called "flags of convenience" used by fish-pirate-ships can be bought online for as little as US$500 from island nations like Malta, Panama, Belize, Honduras and St. Vincent & Grenadines. About 80 different countries, including the European Union (EU), Taiwan, Panama, Belize, Honduras, China, Korea, Taiwan, Japan, and the U.S., host owners of these pirate-ships and trawlers. Enforcement of the UNFSA's RFMOs rules and regulations are lax, especially in the victimised fish-pirated seas, oceans and waterways.

1748880.0; *Topaz Fishing Inc; Transglove Investments,* 5056209.0. [Greenpeace International, Ottho Heldringstraat 5, 1066 AZ Amsterdam, The Netherlands; support.services@int.greenpeace.org.

Government policing the rogue fish traders and merchants around the ports where the fish loot enters the markets is as good as nothing and dead as a dodo.

The Northwest Atlantic Fisheries Organisation (NAFO), established in 1979, regulates international waters between Canada and Greenland and manages the fish stocks, except for salmon, tuna, marlins, whales and sedentary species. The European Fisheries Fund (EFF) which replaced the Financial Instrument for Fisheries Guidance (FIFG) on 1 January 2007 – is designed to secure a sustainable European fishing and aquaculture industry within the Common Fisheries Policy (CFP). EFF policy focuses on the exploitation of fisheries resources and achieving a stable balance between these resources and the capacity of Community fishing fleet; strengthening the competitiveness and viability of operators in the sectors, promoting environmentally-friendly fishing and production methods; providing adequate support to people employed in the sector; and fostering the sustainable development of fisheries areas. *The EFF and CFP neither control illegal fishing outside EU seas, oceans and waterways, nor the illegal fish loot landing on EU ports through fishing-pirate-vessels.*

2.7 Economic Downturns, Natural Disasters

UNRELENTING FISH thievery by pirate fisherpeople is one of those frustrating incidents that contributed to price hiking of commodities. A wide-ranging global economic downturn or chronic recession was the end-result of the high oil and food prices crisis, coupled with deregulated banking and financial practices. Frequent occurrences of natural climatic disasters added to the chain of woes attending the high food and energy prices debacle. Natural climatic disasters have been itched on by climatic changes and global warming. Climate change and global warming are the direct consequences of our obsessive use of fossil fuels in all sectors of our industrial and economic lives, wrong agricultural production methods, and continued gassing of the atmosphere with increasing volumes of greenhouse gases. Greenhouse gases result from our use of fossil fuels, petrochemicals, unabated destruction of forests, dilution of food with chemicals and genetic modified crops, and seemingly irreversible trade practices and consumption imbalances.

World economies – developed and developing alike – experienced recessions or economic downturns in one form or other, during the era of spiralling oil and food prices 2006-2009. There was general slowdown in economic activities; gross domestic products (GDPs) or gross domestic incomes (GDIs) which monitor economic performance, employment, investment spending, capacity of utilisation, household incomes and business profits – all fell collosally; mass bankcruptcies, credit crunches, inflation, deflation and disinflation, foreclosures and unemployment took control of economies. Governments had no alternative but to respond by adopting expansionary macroeconomic policies, such as *increasing the money supply and government spending,* and regulate banking and financial institutions. A summary of the inconveniences, actors and counter-measures implemented, brought on by the high oil and food prices crises follows:[184]

At the height of the oil and food prices in 2008, natural disasters claimed over 220,000 human lives. This was one of the most devasting years on record, and the most destructive in terms of the number of victims and financial cost of the damage done – laid to *debit account* of climate change. Many families were unable to bear the costs of burial of their family members, relatives or friends. "This continues the long-term trend we have been observing. Climate change has already started and is probably contributing to increasingly frequent weather extremes and ensuring natural catastrophes," said *Munich, Re* Insurer board member Torsten Jeworrek.[185] The catalogue of human and economic catastrophes and woes caused by climate-changing disasters in 2008 alone, included:

- Most devastating in terms of human fatalities was Cyclone *Nargis*, which lashed Myanmar on 2-3 May, killed more than 135,000 people and left more than one million homeless; days later, an earthquake shook China's Sichuan province, leaving 70,000 dead, 18,000 missing, almost five million homeless, and caused around US$85 billion worth of damages.

- Around 1,000 people died in a severe cold snap in January in Afghanistan, Kyrgystan and Tajikistan; 635 perished in August and September in floods in India, Nepal and Bangladesh. Typhoon *Fengshen* killed 557 people in China and the Philippines in June, while earthquakes in Pakistan in October left 300 dead. Six tropical typhoons also slammed into the southern U.S.,including *Ike* causing insured losses of US$10 million – the insurance industry's costliest catastrophe of the year. In Europe, low-pressure storm *Emma* in March caused a US$2 billion

---

[184] 2008 G-20 Washington Summit/APEC Peru 2008/2009 G-20 London Summit: *"Financial Crisis of 2007-2009:Late 2000s Recession."* : Specific Issues: U.S., marketing house correction; World food crisis; Energy crisis (Central Asia); Subprime mortgage crisis (timelines, List of writedowns); Global financial crisis; Automative industry crisis; Future of newspapers; List of entities involved (Bankcrupt or acquired banks; Bankcrupt retailers); Effects upon museums; Banking revelations in Ireland; Resurgence of Keynesianism; By Country (or Region): Belgium; Iceland; Ireland; Latvia; Russia; Spain; Ukraine (Europe, Africa, Asia, Australia); Legislation and policy responses: (Banking and finance stability and reform): Banking (special Provisions) Act 2008; Commercial Paper Funding Facility; Emergency Economic Stabilisation Act of 2008; Troubled Assets Relief Programme; Term Asset-Backed securities Loan Facility; Temporary Liquidity Guarantee Programme; 2008 United Kingdom bank rescue package; 2008 East Asian meetings; Anglo Irish Bank Corporation Bill 2008; 2009 G-20 London summit; Irish emergency budget, 2009; Natural Assets Management Agency National fiscal policy response to the late 2000s recession; Housing and Economic Recovery Act of 2008; Economic Recovery Stimulus Act 2008; 2008 Chinese economic stimulus plan; 2008 European Union stimulus plan; American Recovery and Reinvestment Act of 2009; Green New Deal; Companies and banking institutions: Companies in bankruptcy, administration, or other insolvency proceedings; or in the future: New Century Financial Corporation; American Freedom Mortgage; American Home Mortgage; Bernard L. Madoff Investment Securities LLC; Charter Communications; Lehman Brothers (bankcruptcy); Linens'n Things; Mervyns; Netbank; Terra Securities (scandal); Sentinel Management Group; Washington Mutual; Icesave; Kaupthing Singer & Friedlauder; Yamato Life; Circuit City; Allco Finance Group; Waterford Wedgwood; Saab Automobile; BearingPoint; Tweeter; Baccock & Brown; Silicon Graphics; Conquest Vacations; General Growth Properties; Chrysler; Thornburg Mortgage; Government bailouts and takeovers: Northern Rock (nationalisation); IndyMac Federal Bank; Fannie Mae (takeover); Freddie Mac (takeover); AIG; Bradford & Bingley; Fortis; Glinir; Hypo Real Estate; Dexia; C L Financial; Landsbanki; Kaupthing; Straumur; ING Group; Citigroup; General Motors; Chrysler; Bank of America; Anglo Irish Bank (nationalisation); Bank of Antigua; ACC Capital Holdings (re-organisation); U.S. Central Credit Union; Bank of Freland; Allied Irish Bank; Company Acquisitions: Ameriquest Mortgage; Countrywide Financial; Bear Stearns; Alliance of Leicester; Merrill Lynch; Washington Mutual; Derbyshire Building Society; Cheshire Building Society; HBOS; Wachovia; Sovereign Bank; Barnsley Building Society; Scarborough Building Society; national City Corp; Dunfermline Building Society; Other topics: Alleged frauds and fraudsters: Stanford Financial Group (Allen Stanford); Fairfield Greenwich Group; UBS AG; Sean Fitzpatrick (Anglo Irish Bank); Kazutsugi Nami (Enten controversy); Nicholas Cosmo; Arthur Nadel; Marc Dreier; Joseph S. Forte; Paul Greenwood; Stephen Walsh; Proven or admitted frauds and fraudsters: Bernard Madoff (Ponzi scheme); Satyam Computer Services (accounting scandal; Ramalinga Raju); Related entities: Federal Deposit Insurance Corporation; Federal Reserve System; Federal Housing Administration; Federal Housing Finance Agency; Federal Housing Finance Board; Government National Mortgage Association; Office of Federal Housing Enterprise Oversight; Office of Financial Stability; UK Financial Investment Limited; Federal Home Loan Banks; Securities involved and financial markets: Auction rate securities; Collateralised debt obligations; Collateralised mortgage obligations; Credit default swaps; Mortgage-backed securities; Secondary mortgage market, Related topics: Bailout; Bank run; Credit crunch; Economic bubble; Error account; Financial contagion; Financial crisis; Interbank lending market; Liquidity crisis; 'Tea Party' protests. [Reproduced from Wikipedia Foundation, Inc., from page last modified on 29 April 2009.]

[185] Associated Press (AP): *"Natural disasters 'killed over 220,000' in 2008."* 29 December, 2008.

worth of damage, and storm *Hilal* in late May and early June inflicted US$1.1 billion catastrophe.

- In 2005, a large number of hurricanes visitation in the southern U.S., caused US$200 billion worth of damages. The Kolbe earthquake in Japan in 1995, wreaked more destruction since records began in 1900.

- Estimates of the World Meteorological Organisation (WMO): 2008 was the 10th warmest year since the beginning of routine meausrements and the 8th warmest in the northern hemisphere; the number of tropical cyclones in the North Atlantic in 2008 was much higher than the long-term average, and in terms of both the total number of storms and the number of major hurricanes, 2008 was the fourth most severe hurricane season since reliable data have been available.

2.8 Conclusion

RENEWABLE ENERGIES' production and use offer the best available solution to cap and hold down any future occurrences of price hikes premeditated by speculated high oil prices. In contrast to fossil fuels, renewable energies present vast-ranging remedies against chemical pollution of food production and consumption, environmental degradation, atmospheric pollution and water contamination, and preservatrion of ecosystems and their biodiversities.

Chapter 3

RENEWABLE ENERGIES AND FOOD PRICES

3.1 Introduction: Unlimited Alternative Energies

UNLIMITED ALTERNATIVE, non-polluting energies abound around us. They are waiting to be tapped, harnessed and consumed alongside multiple food production methods much more sustainably, economically and efficiently. Renewable energies are generated from natural resources and are naturally replenishable, non-polluting and recyclable. In 2006, about 80% of global final energy consumption came from renewables with 13% coming from traditional biomass. On 16 June 2009, German energy companies announced plans to begin a 400 billion euros Solar Energy project in the Sahara Desert to harness solar energy for Europe by 2020. Financial-intensive fossil-fuel technologies can and should be replaced with less financial-intensive renewable energies' technologies. All that is lacking or needed, is the political will, commonsense urgency, and financial investment to embrace renewable energies' technologies. Existing fossil-fuel technologies can be used or remodelled to suit renewables or new renewables' technologies using *nano, hydrogen, fuel cell, piezoelectricity,* and the already functioning *wind, biomass, photovoltaic, thermal, geothermal, water* and *solar* technologies.

3.2 Solar Technologies

SOLAR ENERGY or solar radiation from the Sun called *sunlight*, is the super energy that suffuses all Nature and gives life to all living and non-living creatures and things; sun-energy also drives the Earth's climate and weather. The Sun's surface consists of about *74% of its mass or 92% of its volume* of *Hydrogen, 24% of its mass or 7% volume* of *Helium,* and trace quantities of ofther elements, including *Iron, Nickel, Oxygen, Silicon, Sulphur, Magnesium, Carbon, Neon, Calcium, Chromium,* etc. All renewable energies, other than geothermal and tidal, derive their energy from the Sun. Solar technologies, including (existing remodelled fossil-fuel plants) which capture, tap, convert and harness solar energy for use in our homes, industry and socety in the forms of *electricity, heat* and *power* are classed either as *active* or *passive: Active* solar techniques use *photovoltaic* panels, pumps, and fans to convert sunlight into useful inputs, and increase the supply of energy. *Passive* solar techniques select materials with favorable thermal properties, design spaces that naturally circulate air, reference the position of a building to the Sun, and reduce the need for alternative resources.

Solar energy and solar technologies impact and influence a great number of our activities, among them: *Agriculture* and *Horticulture; Architecture* and *Urban Planning; Solar Thermal, Water* and *Process Heating; Cooling, Ventilation* and *Cooking; Electrical Generation* and *Thermogenerators; Solar Chemical; Solar Vehicles* and *Energy Storage,* as follows:

- *Agriculture and Horticulture:* The art of successful agriculture to ensure high productivity depends on the maximum capture of sunlight for plants. Planting and sowing techniques such as timed *planting cycles, tailored row orientation, staggered heights* between rows and the *mixing of plant varieties,* do improve crop yields. *Greenhouses* convert solar light to heat and thus helps maintain a year-round production and the growth in enclosed environments of specialty crops and other plants not naturally suited to the local climate; Greenhouses remain an important part of horticulture; plastic transparent materials made of *polyethylene* are used to grow plants that require a higher temperature or humidity in polytunnels; lightweight polyethylene or spunbonded polyester synthetic materials are used to row-cover or provide protective covering or shield vegetable plants from undesirable effects of cold, wind and insect damage; an extremely lightweight row cover, called a *cloche,* or belljar covering made of glass or plastic floating row cover – are placed directly over plants.

- *Water Treatment:* Solar energy can be used in water stabilisation ponds, treat waste water without chemicals or electricity, or distil water used to make saline or brackish water portable, and detoxification of contaminated water via photolysis. Solar stabilised ponds allow algae to grow in them and consume carbon dioxide in photosynthesis. In solar desalination, individual *still* designs include single-slope, double-slope or greenhouse type, vertical, conical, inverted absorbers, multi-wick, and multiple effect. These stills can operate in *passive, active,* or *hybrid* modes; double-slopes stills are the most economical for decentralised domestic purposes, while active multiple effect units are more suitable for large-scale applications; solar water disinfection (SODIS) involves exposing water-filled plastic polyethylene terephthalate (PET) bottles to sunlight for several hours, with exposure times varying depending on weather and climate, from a minimum of six hours to two days during overcast conditions.

- *Architecture and Urban Planning:* Building design is greatly influenced by sunlight. For example, advanced *solar architecture* and urban planning methods orient buildings toward the south to provide light and warmth *(passive solar),* compact proportion or a *(low surface area to volume ratio),* selective shading or *(overhangs)* and thermal mass or heat capacity. The most recent solar design approaches use computer modelling which tie together solar lighting-heating- ventilation systems in an integrated solar design package. *Active solar* equipment such as pumps, fans and switchable windows can complement *passive* design and improve system performamance. In metropolitan areas known as *urban heat islands* (UHI), temperatures are higher than temperatures of the surrounding environment, due to increased absorption of the solar light by urban materials such as asphalt and concrete; UHI areas have lower *albedos* or the extent to which an object diffusely reflects light from the Sun, and higher heat- or thermal heat capacities, or the capacity of a body to store heat – than those in the natural environment. The UHI-effect can be counteracted by painting buildings and roads white and planting trees. Thermal mass materials store absorbed sunlight heat, among common ones being: stone, cement and water – used both in arid climates, temperate regions, and cold temperate areas, to cool, warm, heat and provide light.

- *Solar Thermal, Water and Process Heating; Cooling, Cooking, Ventilation* and *Lighting:* Water heating, Space heating and cooling, Process heat generation, Cooling, Cooking, Lighting and Ventilation, are done more economically and sustainably with solar energy. Solar hot water systems use sunlight to heat water, where the most common types of solar water heaters for domestic hot water are *evacuated tube collectors* and glazed *flat plate collectors,* and *unglazed plastic collectors* for swimming pools. A total of approx. 154 gigawatts (GW) of solar hot water systems were installed worldwide as of 2007, with China installing 70 GW of these, and to install 210 GW by 2020. While the U.S., and Australia heat their swimming pools with 18 GW installed solar power, Israel and Cyprus lead in the use of solar hot water systems; heating, ventilation and air conditioning (HVAC) systems account for 30% of the everyday use of energy in commercial buildings, and nearly 50% in residential buildings in the U.S. In solar lighting, *natural lighting* replaces *artificial lighting,* using daylight techniques and hybrid solar lighting that reduce energy consumption. *Hybrid solar lighting* (HSL) is an active solar method of providing interior lillumination: HSL systems collect sunlight using mirrors that track the sun and use optical fibres to transmit it inside the building to supplement conventional lighting. Daylight systems collect and distribute sunlight for interior illumination, and their designs imply careful selection of window types, sizes and orientation, exterior shading devices may considered as well; individual features, including sawtooth roofs, clerestory, clearstorey, overstorey windows or domes placed at the top of buildings, etc. A solar or thermal chimney is a passive solar ventilation system composed of a vertical shaftconnecting the interior and the exterior of a building. As the chimney warms, the air inside is heated causing an updratf that pulls air through the building. *Green building* design encompasses passive solar and increasing the efficiency of resource use – energy, water, and materials, while reducing building impacts on human health and the environment during the building's lifecycle. Natural lighting offsets energy use, indirectly offsets non-solar energy use by reducing the need for air-conditioning, and offers physiological and psychological benefits (compared to artificial lighting). Artificial lighting represents a major component of energy consumption, and accounts for a significant part of all energy consumed worldwide through use of *electric lights, gas lighting, candles,* and *oil lamps.* A variety of adverse health effects may be caused by *light pollution* or excess light exposure; over-illumination or improper spectral composition of light may include: increased headache incidence, worker fatigue, medically defined stress – a state of alarm and adrenaline production, short-term resistance as a coping mechanism, and exhaustion, with common stress symptoms being irritability, muscular tension, inability to concentrate and a variety of physical reactions, such as headaches and elated heart rate.), decrease in sexual function and increase in anxiety. A list of worldwide solar thermal power stations include:[186] Solar cookers use sunlight for cooking, drying and pasteurization, and can be grouped into: *box*, *panel* and *reflector* cookers. The simplest solar box cooker consists of an insulated container with a transparent lid, and can be used effectively with partially overcast skies and will typically reach temperatures of 90-1500°C. Panel cookers use a reflective panel to direct sunlight onto an insulated container and reach temperatures comparable to box cookers; reflector cookers use various concentrating geometries – *dish, trough, fresnel mirrors* - to focus light on a cooking container, with temperatures of 315°C and above, and require direct light to function properly; the solar bowl is a concentrating technology employed by the Solar Kitchen in Auroville, Tamil Nadu State, India.

- *Electrical Generation* and *Thermogenerators:* Sunlight is converted into electricity by using *Photovoltaic* (PV), *Concentrating solar power* (CSP), and various experimental technologies. PV is applied to power small and medium-sized applications, such as the *calculator*–powered by a single solar cell, *off-grid* homes powered by a PV array (linked collection of PV nodules, PV panels – a packaged interconnected asssembly of PV cells - or solar cells). CSP is used in large-scale applications, such as CSP plants *Solar Energy Generating Systems* (SEGS) which consists of nine solar power plants in California's Mojave Desert, where *insolation* (i.e., a measure of solar radiation energy received on a given surface area in a given time) is among the best in the U.S. The electricity produced by PV plants is expressed as average irradiation in *watts per sq metre* (W/m^2) or *kilowatt-hours per sq metre per day* (kW-h (m^2.day) or

---

[186] List of Worldwide Solar Thermal Power Stations: <u>Operational:</u> Solar Energy Generating Systems, Mojave Desert, California, U.S. / Parabolic trough, Collection of 9 units, 354 MW; Nevada Solar One, Las Vegas, U.S./ Parabolic trough , 64 MW; Andasok I, Granada, Spain / Parabolic trough, 50 MW completed Nov., 2008; Energia Solar De Puertollano, Ciudad Real, Spain / Parabolic trough, 50MW, completed Apr., 2009; PS20 solar power tower, Seville, Spain / Solar power tower, 20 MW, completed Apr., 2009; PS10 solar power tower, Seville, Spain / solar power tower, 11 MW, Europe's first commercial solar tower; Kimberlina Solar Thermal Energy Plant, Bakersfield, California, U.S. / fresnel reflector, 5 MW, Austre demonstration plant; Liddell Power Station Solar Steam Generator, New South Wales, Australia / fresnel reflector, 2 MW, electrical equivalent steam boost for coal station, Jülich Solar Tower, Jülich, Germany / solar power tower, 1.5 MW, completed Dec., 2009; Puerto Errada I, Murcia, Spain / fresnel reflector, 1.4 MW, completed Apr., 2009; Keahole Solar Power, Hawaii, U.S., parabolic trough, 1 MW; <u>Under Construction:</u> Martin Next GenerationSolar Energy Centre, Florida, U.S., 75 MW steam input into a combined cycle, parabolic trough design; Andasol 2 solar power station, Granada, Spain, 50 MW with heat storage, Parabolic trough design; Andasol 3 solar power station, Granada, Spain, 50 MW with heat storage, parabolic trough design; Alvarado / La Risca I solar power station, Badajoz, Spain, 50 MW, parabolic trough design; Solnova I solar power station, Spain, 50 MW, parabolic trough design; Solnova 3 solar power station, Spain, 50 MW, parabolic trough design; Extresol I solar power station, Spain, 50 MW, parabolic trough design; Kuraymat Plant, Egypt, 40 MW steam input for a gas powered plant, parabolic trough design; Hassi R'mel integrated solar combined cycle power station, Algeria, 20 MW steam input for a gas powered plant, parabolic trough design, Beni Mathar plant, Morocco, 20 MW for hybrid power plant, technology unknown; Solar Tres Power Station, Spain, 17 Mwwith heat storage, power tower design; eSolar demonstration plant, 5 MW, Lancaster, California, U.S.; <u>Announced Thermal Power Stations, U.S.:</u> Stirling Emergency Systems (SES) Solar One, California, 850 MW Stirling engine; SES Solar Two, California, 750 MW Stirling engine; Mojave Solar Park, California, 553 MW parabolic trough; Fort Irwin, California, 500 MW unnamed solar thermal technology, military; Hualapai Valley Project, Arizona, 340 MW parabolic trough; TBN, Florida, 300 MW fresnel reflector; Starwood Solar, Arizona, 290 MW parabolic trough; Solana Generating Station, Arizona, 280 MW parabolic trough; Beacon Solar Energy Project, California, 250 MW parabolic trough; Harper Lake Solar, California, 250 MW solar trough; TBNKingman solar project, Arizona, 200 MW parabolic trough; Carrizo Energy Solar Farm, California, 177 MW fresnel reflector; Ivanpah Solar Electric Generating System, California, 110 MW power tower; San Joaquin Solar 1 and 2, California, 107 MW parabolic trough hybrid with biomass; eSolar 1,California, 84 MW solar power; eSolar 2, California, 66 MW solar tower; City of Palmadale Hybrid Power Project, California, 62 MW parabolic trough steam input for hybrid gas plant; TBN, California, 59 MW parabolic trough with heat storage; Victorville 2 Hybrid Power Project, California, 50 MW parabolic trough steam input for hybrid gas plant; <u>Announced Thermal Power Stations, Spain:</u> Extresol 2-3, Badajoz, 50 MW each with heat storage, parabolic trough design; Andasol 4-7, Granada, 50 MW each with heat storage, parabolic trough design; Manchasol 1-2, Ciudad Real, 50 MW each with heat storage, parabolic trough design; Solnova 2, 4-5, Sevilla, 50 MW each with heat storage, parabolic trough design; Ecija 1-2, Ecija, 50 MW each with heat storage, parabolic trough design; Helios 1-2 Ciudad Real, 50 MW each with heat storage, parabolic trough design; Termesol 50, Sevilla, 50 MW each with heat storage, parabolic trough design; Arcosol 50, Cadiz, 50 MW each with heat, parabolic trough design; Ibersol Badajoz, Fuente de Cantos, 50 MW parabolic trough design; Ibersol Valdecaballeros 1-2,Valdecaballeros, 50 MW,parabolic trough designs; Ibersol Sevilla-Aznalcollar, 50 MW, parabolic trough design; Ibersol Almeria, Tabernas, 50 MW, parabolic troug design; Ibersol Albacete, Lorca, 50 MW, parabolic trough design; Ibersol Murcia, Cubillos, 50 MW, parabolic trough design; Ibersol Zamora, Cubillos, 50 MW, parabolic trough design; Enerstar Villena Power Plant, Villera, 50 MW, parabolic trough design; Aste 1A, 1B, 3, 4, Alcazar de San Juan (Ciudad Real), 50 Mweach, parabolic trough design; Astexol 1-2, Extremadura, 50 MW each, parabolic trough design; Palma del Rio 1-2, Cordoba, 50 MW each, parabolic trough design; AZ 20, Sevilla, 20 MW, power tower design; Almaden Plant, Albacete, 10 MW, linear fresnel design; Gotasol, Gotarrendura, 10 MW, linear fresnel design; Aznalcollar TH, Sevilla, 80 MW, dish Stirling design; <u>Announced Thermal Power Stations,Miscellanoeus:</u> Negev Desert, Israel, 250 MW, design TBA; Upington, South Africa, 100 MW, power tower design; Shams, Abu Dhabi Madinat Zayad, 100 MW, parabolic trough design; Yazd Plant, Iran, 67 MW steam input for hybrid gas plant, technology TBA; Archmede near Siracusa, Sicily, Italy, 28.1 MW with heat storage; Solenha-Aspres sur Buech, France, 12 MW with heat storage, parabolic trough design; Cloncurry solar power station, Australia, 10 MW with heat storage, power tower design. [Source: Wikipedia, the free encyclopedia.]

hours/day) or *multi-megawatt* (MW). As an intermittent power source, solar power requires a backup supply which can partially complement wind power; local backup is usually done by batteries, and utilities use pumped-hydro storage in the form of water. A list of worldwide largest and selected smaller Photovoltaic (PV) power stations and plants, include:[187] Thermogenerators (TEG) are devices which convert heat temperature differences directly into electrical energy. *Thermoelectric effect* is the direct conversion of temperature differences to electric voltage and vice versa. A thermoelectric device creates a voltage when there is a different temperature on each side, and conversely, a voltage applied to it, creates a temperature difference. Thermoelectric effect can also be used to measure temperature, cool, cook or heat objects.

- *Solar, Photovoltaic Cells:* A solar or photovoltaic cell is a semiconductor device that converts light directly into electricity by the *photovoltaic effect.* Assemblies of cells are used to make *solar panels, solar modules,* or *photovoltaic[188] arrays.* Photovoltaic effect or *photoelectric effect* involves the creation of a voltage (or a corresponding electric current) in a material upon exposure to electron magnetic radiation. A photovoltaic *module* or *panel,* is a packaged interconnected assembly of *photovoltaic cells* or *solar cells.* An illumination of photovoltaic modules or panels is called a *photovoltaic array.* A photovoltaic installation includes any array of photovoltaic modules or panels, an inverter, batteries and interconnection wiring. Photovoltaic systems are used for either on- or off-grid applications, and for solar panels on spacecraft. Solar panels must withstand heat, cold, rain and hail for many years, and many crystalline silicon module manufacturers offer warranties that guarantee electrical production for 10 years at 90% of rated power output and 25 years at 80%. There are three generations of solar cells: *1st generation* consists of a large area, high quality and single-junction devices, and involves high energy and labor inputs with high costs comparable to fossil fuel energy generation; *2nd generation* are developed to address energy developments and costs, and techniques such as vapor deposition, electroplating, and Ultrasonic Nozzles spray that uses high 20–50 kHz frequency vibrations to produce nearly narrow drop size distribution and low velocity spray from a low viscosity liquid, are of advantage to high temperature processing. Materials used include cadmium telleruride (CdTe), copper indium gallium selenide, amorphous silicon, and micromorphus silicon applied in *thin-film* to support *substrates* such as glass or ceramics for reduced material mass and costs; *3rd generation* aims to enhance poor electrical performance of the 2nd generation thin-film technology.

- *Solar Chemical:* Solar induced chemical reactions - *thermochemical* and *photochemical* -

---

[187] List of Worldwide Largest and Smaller Selected Photovoltaic (PV) Power Stations/Plants by Country: Germany: Waldpolenz Solar Park, 40 MW, completed Dec., 2008; Koethen, 14.75 MW, completed Dec., 2008; Erlasee Solar Park, 12 MW; Pocking Solar Park, 10 MW; Gottelborn Solar Park, 8.4 MW; Bavaria Solar Park in Muhlhausen, 6.3 MW; Rote Jahne Solar Park, 6 MW; Japan: Kameyama, 5.2 MW; South Korea: SinAn power plant, 24 MW, completed Oct., 2008; Gochang power plant, 15 MW, completed Oct., 2008; Portugal: Moura photovoltaic power station, 46 MW, completed Dec., 2008; Serpa solar power plant, 11 MW; Spain: Olmedilla Photovoltaic Park, 60 MW, completed Sept., 2008; Puertollano Photovoltaic Park, 50 MW, 2008; Arnedo Solar Plant, 34 MW, completed Oct., 2008; Merida/Don Alvaro Solar Park, 30 MW, completed Sept., 2008; Planta Solar La Magascona & La Magasquila, 30 MW; Planta Solar Ose de la Vega, 30 MW, Planta Fotovoltaico Casas de Los Pinos, 28 MW; Planta fotovoltaica de Lucainena de las Torres, 23.2 MW; Parque Fotovoltaico Abertura Solar, 23.1 MW; Parque Solar Hoya de Los Vincentes, 23 MW; Huerta Solar Almaraz, 22.1 MW, completed Sept., 2008; Parque Solar El Coronil I, 21.4 MW; Solarpark Calveron, 21.2 MW; Huerta Solar Almaraz, 20MW, completed Sept., 2008; Planta solar fotovoltaico Calasparra, 20MW; Beneixama photovoltaic power plant, 20 MW; Parque Solar Olivenza, 18 MW; Parque Solar Bonillo, 18 MW, completed Oct., 2008; Planta Sola Calzada de Oropesa, 15 MW; Planta de energia solar Mahora, 15 MW, completed Sept., 2008; Planta Solar de Salamanca, 13.8 MW; Parque Solar Guadarranque, 13.6 MW; Parque Solar El Realengo, 13.2 MW; Parque Fotovoltaico Solten I, 13 MW; Lobosillo Solar Park, 12.7 MW; Parque Solar Hinojosa del Valle, 12 MW; Parque Solar Fotovoltaico Villafranca, 12 MW; Monte Alto photovoltaic power plant, 9.5 MW; Viana Solar Park, 8.7 MW; Huerta solar de Aldea del Conde, 6.3 MW; Huerta Solar Crevillent, 6 MW; Huerta Solar de Olmedilla, 6 MW; Darro Solar Park, 5.8 MW; U.S.: Nellis Solar Power Plant, 14.02 MW; Sempra Generation PV Plant, 12.6 MW,completed Dec., 2008; Alamosa photovoltaic power plant, 8.2 MW;. [Large PV Systems in Planning or Under Construction: Rancho Cielo Solar Farm, U.S., 600 MW, thin film silicon from Signet; Solar Topaz Solar Farm, U.S., 550 MW, thin film silicon from Optisolar; High Plains Ranch, U.S., 250 MW, monocrystalline silicon from SunPower with tracking; Mildura Solar concentrator power station, Australia, 154 MW, Heliostat concentrator using GaAs cells from Spectro-lab; KCRD Solar Farm, U.S., 80 MW, scheduled to be completed in 2012; DeSoto County Florida, U.S., 25 MW, to be constructed by SunPower for FPL Energy, completion date 2009; Davidson County farm, U.S., 21.5 MW, 36 individual structures; Cadiz solar power plant, Spain, 20.1 MW; Kennedy Space Center, Florida, U.S., 10 MW, to be constructed by SunPowerfor FPL Energy, completion date 2010.]

[188] Photovoltaics: Deals with the technology and research to all applications of solar cells in producing electricity for practical use, and such energy produced in this way is referred to as solar power. The photovoltaic effect and photoelectric effect are different: in photovoltaic effect, electrons are ejected from a materials surface upon exposure to radiation of sufficient energy; in photoelectric effect, the generated electrons are transferred from one material to another resulting in the buildup of a voltage difference between two electrons.

offset energy that would otherwise come from an alternative source, and convert solar energy into storable and transportable fuels. Hydrogen production technologies split water at high temperatures (2300-2600°C) into oxygen and water to produce electric power; electrolysis driven by *photovoltaic* or *photochemical cells* and solar concentrators drive the steam from re-formation of natural gas in order to increase the overall hydrogen yield, as compared to conventional methods; thermochemical cycles characterised by the decomposition and regeneration of reactants produce hydrogen. Sandia National Labs' (U.S.) Sunshine-to-Petrol (S2P) technology, uses the high temperatures generated by concentrating sunlight along with zirconica and ferrite catalyst to break down atmospheric carbon dioxide into oxygen and carbon monoxide (CO); the CO can then be used to synthetize conventional fuels such as methanol, gasoline and jet fuel. Energy-rich chemical intermediaries like *ferric-thionine* cell solution or equivalent, can potentially be stored and subsequently reacted at the electrodes, using a *photogalvanic* device, to produce electrical potential. Photoelectrochemical cells (PEC) - consisting of a semiconductor (typically titanium dioxide or related titinates) - produce an electrical potential when illuminated and immersed in an electrolyte. Photoelectrochemical cells are of two types: PECs that convert light into electricity and PECs that use light to drive chemical reactions such as electrolysis.

- *Solar Vehicles & Energy Storage:* Solar-driven cars are being developed since the 1990s, although many pre-1990s vehicles were equipped with solar panels for auxiliary power for air-conditioning and reduced fuel consumption, solar-driven and PV-powered boats, and solar sails for aircraft. A black solar-balloon, for example, is filled with ordinary air, and as sunlight shines on it, the air inside is heated and expands, causing an upward buoyancy for force, much like an artificially heated hot air balloon. Some solar ballons have enough space for human flight. Solar sails require no fuel like rockets do, but use large membrane mirrors to exploit solar radiation pressure from the Sun. The High-altitude Airship (HAA) is an unmanned, long-duration, lighter-than-air vehicle using helium gas for lift, and then film solar cells for power. The U.S. Missile Defense Agency's *Ballistic Missile Defense System* (BMDS) Airships do not require power to remain aloft because an airship's envelope presents a large area to the Sun.

- *Solar Extracting & Solar Ponds:* Solar extracting techniques *parabolic trough* or *dish, reflectors* and *fresnel,* can provide processed heat for commercial and industrial applications. A solar pond is a pool of salt water (usually 1-2 m deep) that collects and stores solar energy, through preventing convection[189] currents or movement of molecules within fluids. A prototype solar pond was constructed in 1958 on the shores of the Dead Sea near Jerusalem, Israel, and consisted of layers of water that successfully increased from a weak salt solution at the top to a high salt solution at the bottom. Evaporation ponds are shallow pools that concentrate dissolved solids through evaporation, and their use to obtain salt from sea water, is one of the oldest applications of solar energy. Modern uses include: concentrating brine solutions used in leach mining and removing dissolved solids from waste streams. Clothes' lines, clotheshorses, and clothes rackets dry clothes through evaporation by wind and sunlight without consuming electricity or gas. *Unglazed* transpired collectors (UTC) are perforated sun-facing walls for preheating ventilation air. UTCs can raise the incoming air temperature up to 22°C and deliver outlet temperatures of 45-60°C. The short payback period of transpired collectors (3-12 years) makes them a more cost-effective alternative than *glazed* collection systems. Over 80 systems with a combined collector area of 35,000 square metres (sq m) had been installed as of 2003 worldwide. These include 860 sq m collector in Costa Rica used for drying coffee beans and a 1,300 sq m collector in Coimbatore, India, used for drying marigolds.

The pre-eminence of solar energy power and technologies, reduces the need for continued production and consumption of fossil fuels inexcusable.

---

[189] Convection: Is one of the major modes of heat transfer and mass transfer. In fluids, convection heat and mass transfer take place through both diffusion and by *advection* in which matter or heat is transported by motion of individual particles in the fluid. Liquids, gases, and *rheids.* Rheids are solid materials that deform by vicous flow; almost any tpye of rock can behave like a rheid under appropriate conditions of temperature and pressure.

3.3 Wind Technologies

WIND ENERGY or power and technologies is the conversion of wind motion into a useful form, such as electricity, using turbines. Wind is the flow of air or other gases that compose principally the atmosphere of the Earth and the atmosphere of the universe; wind is air molecules in motion on Earth. In outer space, the *Solar wind* is the movement of gases or charged particles from the Sun through space. *Planetary wind* is the outgassing of elements from a planet's atmosphere into space. Solar wind is a stream of charged particles or a plasma ejected from the upper atmosphere of the Sun, which consists mostly of electrons and protons, with energies of about 1 kW. The stream of particles varies in temperature and speed with the passage of times, and are able to escape the Sun's gravity. An *Air mass* is a large volume of air defined by its temperature and water vapor content. Air masses cover many hundreds or thousands of square miles, adopting the characteristics of the surface below them, and classified according to latitude and their continental or maritime source regions. Difference in heating between the poles and the equator lead to the development of jet streams. *Jet streams* are fast flowing, narrow air currents found at the *Tropopause,* the transition between the *Troposphere* where the temperature decreases with height, and the *Stratosphere* where the temperature increases with height, and are located 10-15 km above the Earth's surface. Jet streams form near boundaries of adjacent air masses with significant differences in the temperature, such as the *Polar* region and the warmer air to the south.

Wind – one of the standard atmospheric and weather phenomena - assumes many forms and sizes; it pervades and controls the entire length and breadth of the Earth, weather and weather patterns, lives, economies, agriculture, and industry. Wind power is so pervasive and all-inclusive in humankind's everyday living that it shapes landforms, and occurs on a wide range of scales: from local *breezes* generated by heating of land surfaces; global winds resulting from difference in absorption of solar energy between climate zones; differential heating between the equator and the poles; and Earth's motion on its axis and orbit around the Sun.

Wind permeates and manipulates all of the following natural or common phenomena:

•
   Weather, Climate: Weather is the natural phenomena that is happening in and around the Earth's atmosphere's lowest *troposphere* layer. The troposphere contains approx. 75% of the atmosphere's mass and 99% of its water vapor and aerosol or particulate matter (PM) or fine particles, (i.e., tiny particles of solid or liquid suspended in atmospheric gases and liquids), which can be natural or man-made. Weather occurs because of *density* or strong temperature and moisture differences between *polar-* and *tropical-air* between one place and another, caused by the Sun's angle being at any particular spot, which varies by latitude from the tropics. *Climate* is average atmospheric conditions over longer periods of time. Long-term

climate is affected by changes brought about by solar energy distribution by the Earth in its rotation duties. In the middle altitudes (away from the polar and tropical latitudes), weather systems, such as *extratropical* cyclones, are the result of instabilities of the *jet stream* flow – between the *troposphere* and the *stratosphere,* the upper layer of the atmosphere. Surface temperature differences cause *pressure differences:* higher altitudes are cooler than lower altitudes due to *compressional heating.*

- *Clouds, Precipitation:* The formation of clouds in the troposphere are associated with rain or liquid precipitation, heat and mass exchange and wind action. The various stages of cloud formation, from simple to dangerous, that leads both to good and tempestuous weather, rain, floods, ocean and sea waves and tides; wind executes clouds' formation, during and after suspended liquids and gases are changing their forms: *Cumulonimbus* (CB) cloud, which forms from *Cumulus congestus* (CC) or towering cloud, is a type of tall, dense, cloud involved in thunderstorms and other intense weather or weather instability. CBs can appear alone, in clusters, or along a cold front in a *squall line,* can further develop into a *supercell* (i.e., a severe thunderstorm with special features), and create lightning through the heart of the cloud. CC cloud foreshadows characteristics of unstable areas of the atmosphere which are undergoing *convection* - the movement of molecules within fluids including heat and mass transfers. CC clouds, characterised by sharp outlines and great vertical development, are produced by strong *updrafts* typically taller than CC is wide; CCs are formed by the development of *Cumulus mediocris* (CM) from the cumulus family, slightly larger in vertical development than *Cumulus* humilis (CH) or fair weather cloud, which is hot and appears in hot countries or in the tropics and over mountainous terrain. CHs do not produce precipitation but may further advance into clouds such as CC and CB which do, but can also be formed from *Altocumulus castellanos* (ACCAS); ACCASs are evidence of mid-atmospheric instability and a high-mid-altitude lapse rate, and typically accompanied by moderate turbulence as well as potential icing conditions. ACCASs are harbingers of bad weather, and if surface-based convection can connect to the mid-tropospheric unstable layer, their continued development can produce CBs; the appearance of ACCASs early in a sunny day may indicate a high probability of the formation of thunderstorms in the afternoon. CCs may develop into *Cumulonimbus capillatus-fibrous-top* or *Cumulus incus-anvil-top* (CCCFT–CCCIAT); *Cumulonimbus calvus* (CCC) develops from CC, and its further development can result to a CCCFT-CCCIAT.

- *Hierarchy of Prevailing Winds:* Differential heating between the Equator and the Poles which causes the *Jet stream,* associated climatological mid-latitudes, and the Earth's rotation motion of air around areas of high and low pressures – are the major driving forces of large-scale atmospheric wind circulation. This has established major wind hierarchies called *Westerlies, Polar easterlies* and *Trade winds,* which dominate and affect any region's prevailing and dominant winds. The Westerlies flow in and across the mid-latitudes and are at the mercy of polar cyclones; the Easterlies flow in and across low and high-latitudes. The Trade winds or *trades* are the prevailing pattern of easterly surface winds dominant in the tropics near the Earth's Equator; trades blow predominantly from the northeast in the *Northern Hemisphere,* and from the southeast in the *Southern Hemisphere;* they act as the steering flow for tropical storms that form over the Atlantic, Pacific, and Indian Oceans, that make landfall in North America, Southeast Asia, and India, respectively. Trade winds also steer Saharan Desert dust westward across the Atlantic Ocean into the Caribbean Sea, and positions of southeast North America; In areas where the wind tends to be light, the *Sea breeze* and *Land breeze cycles* are the prevailing winds; highly elevated surfaces can induce thermal low to augment the environmental wind; *mountain* and *valley breezes* dominate the wind pattern in variable terrains, mountains and valleys, while highly elevated surfaces, such as near the *Sonoran Desert, Mexican Plateau, Sahara Desert, South America* over northwest Argentina, *Iberian Peninsula,* and *Tibetan Plateau,* can induce a thermal low because of intensive heating. Prevailing winds in large deserts move dust large distances from their source regions; *Monsoons* or seasonal rain-bringing prevailing winds supply rain to vast regions of the world considered to receive their majority rainfall during a particular season, classified into: North American Monsoon; South American Monsoon; Indian Sub-Continent Monsoon; Sub-Saharan Africa Monsoon; Australian Monsoon; East and South Asian Monsoons.

- *Storm Species:* Storms, snowstorms and gusts are the Black Sheep of the wind family, as they leave total anihilation after their violent visitations. In fact, a common *storm* means any disturbed state of an astronomical body's atmosphere that implies severe weather, marked by strong winds, thunder and lightning, heavy precipitation, ice, and uprooting, transportation and deposition of substances through the wind. The storm species includes: storms; snowstorms, snow; ice storms, winterstorms, blizzards or winter hurricanes; thunderstorms; squalls, squall-lines, supercells, multicells, single-cells, gusts, dust- or sandstorms; hailstorms; Northeastern storm; gales, updrafts, downdrafts, downbursts, microbursts, straight-line winds; and thunder and lightning.

- A *Snowstorm* or *thundersnow storm* or *winter thunderstorm* or *blizzard,* is a rare thunderstorm with snow falling as the primary precipitation, instead of rain. Regions with strong upward motion within the cold sector of extratropical cyclones between autumn and spring, experience snow and snowstorms; a *Blizzard* is a severe winter storm condition characterised by low temperatures, strong winds, and heavy blowing snow, formed when a high pressure system or ridge interacts with a low pressure system; *Thunderstorm* or *electrical storm* or *lightning storm* or *storm* is a form of weather characterised by the presence of lightning and its effects of thunder, usually accompanied by a heavy rainfall and sometimes with snow, hail, or no precipitation at all. Thunderstorms may live up in a series, and strong or severe thunderstorms may rotate. Thunderstorms are classified into four types: *supercells, squalls, multicells,* and *single-cells. Thunder* is the sound made by lightning, whose sudden increase in presence and temperature produces rapid expansion of air surrounding and within a bolt of lightning, and expansion of air that creates sonic shockwaves; A *Hailstorm* is a form of precipitation which consists of balls or irregular lumps of ice hailstones, and consist mostly of water ice, with the largest stones coming from severe and dangerous thunderstorms. Hailstorms sometimes occur during, but not always from, thunderstorms, and can occur within any thunderstorm; An *Updraft, Downdraft,* or *Downburst,* is a movement of warm, cold, or significantly rain-cooled air pockets *vertically* or *horizontally,* and includes *dry-, wet-downbursts* and macro-, heat-bursts associated with thunderstorms with very little or heavy rain. An Updraft or Downdraft is less dense than the surrounding region, and will rise or descend until it reaches air that is either warmer, colder or less dense than itself, and create clouds, which then mainly cause thunderstorms.

- A *Squall* is a sudden sharp increase in wind speed which is usually associated with active weather, such as rain showers, thunderstorms, heavy snow, hurricanes and cyclones. A squall line is an organised line of thunderstorms which are classified as a multicell cluster. Independently formed squalls occur along front lines and may contain heavy precipitation, hail, frequent lightning and dangerous straightline winds – and possibly, funnel clouds, tornadoes and watersports.

- A *Supercell* is a thunderstorm characterised by the presence of a *mesocyclone* or a deep continuously rotating updraft. A supercell is the largest of the four types of thunderstorms, [supercells, squalls, multicells, single-cells] and have the potential to be the most severe, are often isolated from other thunderstorms, and can dominate the local climate up to 20 miles (32 km) away.

- A *Multicell cluster* or *Multicellular thunderstorm cluster* is a thunderstorm composed of multiple cells, each being at a different stage in the life cycle of a thunderstorm. A cell is a complete updraft, downdraft, downburst, or microburst, and each cell takes turn as dominant cell within the centre of the thunderstorm, while new cells form within the centre by upward (*western* or *southwestern*) or downward (*eastern* or northeastern) movement of the wind; a microburst is a very localised column of sinking air that produces damaging divergent and straight-line winds at the surface that are similar to, but not distinguishable from tornadoes; a *Straight-line wind* or *thundergust* or *hurricane of the prairie,* is a very strong wind that produces damage, demonstrates a lack of rotational pattern but associated with cylonic storms, tropical cyclones and tornadoes.

- A *Single-cell* or *Single-cell thunderstorm* or *Pulse storm,* is usually mild, producing severe weather only for short periods of time, but upgrades itself, if provoked by a strong updraft winds; a *Gust* is wind, air or other gases flow of shorter duration, in contrast to intermidiate-duration winds such as squalls, and long-term duration winds such as storms, gales, hurricanes, and typhoons.

- A *Dust- sandstorm* is wind action common in arid and semi-arid regions arising when a gust front passes or when the wind force exceeds the threshold value where loose sand and dust are removed from the dry surface. Particles thus transported by saltation and suspension, cause soil erosion from one place and deposition in another, such as in the Sahara and drylands around the Arabian Peninsula.

- A *Northeastern storm* or *nor' easter* is a low-pressure area storm prevalent along East Coast of the U.S., and Atlantic Canada, whose centre of rotation is just off the East Coast, and causes coastal flooding, coastal erosion, hurricane force winds and heavy snowfall. Its precipitation pattern is similar to other extratropical storms; a *Gale* is a very strong wind with a speed between 28-55 knots (32-63 mph).

- *Cyclonic Dynasty:* The autocratic and destructive dynasty of cyclones is long, and includes: cyclones, tropical cyclones, subtropical cyclones, extratropical cyclones, mesocyclones, typhoons or hurricanes, tornadoes, European windstorm, and anticyclones. A *Cylone* is an area of closed, circular fluid motion rotating in the same direction as the Earth's rotation around the Sun, usually charaterised by inward spiralling winds that rotate *counter-clockwise* in the *Northern Hemisphere* (NH), and *clockwise* in the *Southern Hemisphere* (SH). An *Anticylone* is the opposite of the cyclone, and involves the descending movement of air, along with surface systems lighter than average atmospheric pressure over the part of the planet's surface. Anticyclones clear the skies, introduce cooler air, and fogs and hazes can form overnight within a region of higher pressue. Large cyclonic circulations are almost always centred on areas of low atmospheric pressures with cold-core polar that produce cyclones, extratropical cylones, and mesocyclones, which always occur together with updrafts in supercells, where tornadoes may form. *Anticyclones along with extratropical cyclones drive the weather over much of the Earth, producing cloudiness, heavy gales and thunderstorms;* a *European windstorm* is a severe cyclonic storm associated with areas of low-pressure that track across the North Atlantic towards Northwestern Europe, and are most in the winter months.

- A *Tropical cyclone* is tropical depressions in the tropics that produce extremely powerful winds and torrential rain, high waves and damaging storm surge that spawns tornadoes. Tropical cyclones develop over large bodies of warm water, and lose their strength if they move over land, and coastal regions receive significant damage, while inland regions are relatively safe from receiving strong winds. Heavy rains, however, produce formidable flooding up to 40 km (25 miles) from the coastline. The devastating damage inflicted by tropical cyclones are enormous on population and property; a *Subtropical cyclone* takes off from tropical cyclones and forms into extratropical cyclones which have colder temperatures. Subtropical cyclones are also likely to form outside the traditional bounds of the hurricane season.

- An *Extratropical cyclone* or *Mid-latitude cyclone* or *wave cyclone* is a group of cyclones with synoptic scale low pressure weather systems that occur in the middle latitudes outside the tropics, and has neither tropical nor polar characteristics. It is connected with *fronts* or boundaries separating two masses of air of different densities, and horizontal gradients in temperature and dew point or *baroclinic zones. Extratropical cyclones, along with anticyclones, drive the everyday weather over much of the earth, producing cloudiness, mild showers, heavy gales and thunderstorms;* a *Mesocyclone* is a *vortex* or often turbulent, flow of air within a convective storm (i.e., air that rises and rotates around a verical axis, usually in the same direction) in a given hemisphere. It is most often cyclonic, associated with a localised low-pressure region within a severe storm that often occurs together with updrafts in supercells, where tornadoes may form.

- A *Typhoon* or *hurricane* or *tropical cyclone* is a storm system with a large low-pressure centre and numerous thunderstorms that produce strong winds and heavy rain. It feeds on heat released when moist air rises, resulting in condensation of water vapor contained in the moist air, and fuelled by a different heat mechanism than other cyclonic windstorms, such as nor'easters, European windstorms, and polar lows. It originates in the *doldrums* about 10° away from the Equator.

- A *Tornado* is a violent, dangerous, rotating column of air which is in contact both with the surface of the earth and a cumulonimbus cloud (CC), and in some cases, forms the basis of a cumulonimbus cloud. Tornadoes come in many sizes, but typically, in the form of a visible condensation *funnel* or funnel-shaped cloud of condensed water droplets. The rotating column of wind can extend from the base of a cloud or a towering cloud but not reaching the ground or water surface. Funnel-shaped clouds form most frequently in association with supercell thunderstorms.

So, wind's ubiquitous energy power, is everywhere in abundance and super-abundance. In fact, *wind energy and its total amount of economically extractable windpower is considerably more than present human power use from all sources.* An estimated 72 *terrawatt* (TW), where each terrawatt is equal to a *1 Trillion watts* of windpower on Earth, is potentially viable, compared to about 15 TW average global power consumption *from all sources in 2005.* Also, not all the energy of the wind flowing past a given point can be recovered. Windpower potential in the world is much greater than current energy consumption. According to a study by Cristina L. Archer and Mark Z. Jacobson,[190] global wind power potential for the year 2000 was estimated to be ~72TW (or ~54,000 million tons of oil equivalent (MToe) per year. As such, sufficient wind exists to supply all the world's energy needs (i.e., 6995-10177 MToe), "although many practical barriers need to be overcome to realise this potential." The potential takes into account only locations with mean annual wind speed 6.9 m/s at 80m, and assumes that six turbines per sq km for 77 m diameter, 1.5MW turbines on roughly 13% of the total global land area (though that land would also be available for other compatible uses, such as farming).

Commercial wind power production and consumption for electricity, heat and industrial applications is borne on the shoulders of *Wind turbine* technology. Wind power produce is expressed in *Megawatts* (MW) when it is pumped out by generators stuffed by turbine rotation. A wind turbine is a machine that calculates the *kinetic* energy in wind into mechanical energy. Worldwide as of 2008, wind power electricity produce was 1.5%, (a 3-year double from 2005), is upward-trending, and 80 countries, as of May 2009, are commercial wind power producers and users. Large-scale windfarms,[191] a group of wind turbines in the

---

[190] Cristina L. Archer/Mark Z. Jacobson: *"Evaluation of global windpower."* Abstarct. Journal of Geophysical Research-Atmospheres in 2005; stanford.edu; lozej@stanford.edu.
[191] Worldwide Installed Windpower Capacity & List of Turbine Manufacturers: U.S.: 2008: 25,170 MW; 2007: 22,247 MW; 2006: 20,622 MW; 2005: 18,415 MW; Germany: 2008: 23,903 MW; 2007: 22,247 MW; 2006: 20,622 MW; 2005: 18,415; Spain: 2008: 16,740 MW; 2007: 15,145 MW; 2006: 11,615 MW; 2005: 10,028; China: 12,210 MW; 2007: 6,050 MW; 2006: 2,604 MW; 2005: 1,260 MW; India: 2008: 9,587 MW; 2007: 8,000 MW; 2006: 6,270 MW; 2005: 4,430 MW; Italy: 2008: 3,736 MW; 2007: 2,726 MW; 2006: 2,123 MW; 2005: 1,718; France: 2008: 3,404 MW; 2007: 2,454 MW; 2006: 1,963 MW; 2005: 757 MW; UK: 2008: 3,288 MW; 2007: 2,389 MW; 2006: 1,963 MW; 2005: 1,332 MW; Denmark/Faeroe Is: 2008: 3,160 MW; 2007: 3,129 MW; 2006: 3,140 MW; 2005: 3,136 MW; Portugal: 2008: 2,862 MW; 2007: 2,150 MW; 2006: 1,716 MW; 2005: 1,022 MW; Canada: 2008: 2,369 MW; 2007: 1,856 MW; 2006: 1,459 MW; 2005: 683 MW; Netherlands: 2008: 2,225 MW; 2007: 1,747 MW; 2006:

same location, generate electric power connected to the local electricity power transmission network; smaller turbines generate electricity for isolated locations, and the surplus electricity so generated by small domestic turbines is bought back by utility companies.

Wind turbines are considered dangerous to bats and birds. This danger is negligible when compared to their deaths caused by human activities and fossil fuel contamination: offshore wind sites 10 km or more from shore do not interfere with bats lives, compared with agricultural and other activities which do. Meanwhile, producing and consuming wind energy means: *Zero pollution; Extra income for rural farmers* by renting land for turbines, about $200-400 per year/turbine; *Creates more jobs* per gigawatt of electricity generated than coal power stations; *Renewable* and *sustainable source of electricity.* Generally, wind turbines perform highest in higher wind speeds, and lowest in lower wind speeds, so the linking of widely distributed windfarms, and hydro storage of wind energy, can maintain the regular supply of windpower.

Windpower-grid management essentials to efficient, sustainable and cost-effective wind energy farming, include the following factors:

- *Grid Management, Environmental Effects:* The success, sustainability and renewalbility of windpower production and consumption relies on good, efficient and cost-effective grid management, and its effect on the environment. As the electricity is generated by a windfarm

---

1,560 MW; 2005: 1,219 MW; Japan: 2008:1,880 MW; 2007: 1,538 MW; 2006: 1,394 MW; 2005: 1,061 MW; Australia: 2008: 1,494 MW; 2007: 824 MW; 2006: 817 MW; 2005: 708 MW; Sweden: 2008: 1,067 MW; 2007: 788 MW; 2006: 527 MW; 2005: 510 MW; Ireland: 2008: 1,245 MW; 2007: 805 MW; 2006: 745 MW; 2005: 496 MW; Austria: 2008: 995 MW; 2007: 982 MW; 2006: 965 MW; 2005: 819 MW; Greece: 2008: 990 MW; 2007: 871 MW; 2006: 746 MW; 2005: 573 MW; Poland: 472 MW; 2007: 276 MW; 2006: 153 MW; 2005: 83 MW; Turkey: 2008: 333 MW; 2007: 146 MW; 2006: 51 MW; 2005: 20 MW; Norway: 2008: 428 MW; 2007: 333 MW; 2006: 314 MW; 2005: 267 MW; Belgium: 2008: 384 MW; 2007: 287 MW; 2006: 193 MW; 2005. 167 MW; Egypt: 2008: 390 MW; 2007: 310 MW; 2006: 230 MW; 2005: 145 MW; Taiwan: 2008: 350 MW; 2007: 282 MW; 2006: 188 MW; 2005: 104 MW; Brazil: 2008: 338 MW; 2007: 247MW; 2006: 237 MW; 2005: 29 MW; New Zealand: 2008: 325 MW; 2007: 322 MW; 2006: 171 MW; 2005: 169 MW; South Korea: 2008: 278 MW; 2007: 191 MW; 2006: 171 MW; 2005: 98 MW; Bulgaria: 2008: 158 MW; 2007: 35 MW; 2006: 20 MW; 2005: 6 MW; Czech Rep.: 2008: 150 MW; 2007: 116 MW; 2006: 50 MW; 2005: 28 MW; Finland: 2008: 140 MW; 2007: 110 MW; 2006: 86 MW; 2005: 83 MW; Morocco: 2008: 125 MW; 2007: 114 MW; 2006: 124 MW; 2005: 64 MW; Hungary: 2008: 127 MW; 2007: 65 MW; 2006: 61 MW; 2005: 18 MW; Ukraine: 2008: 90 MW; 2007: 89 MW; 2006: 86 MW; 2005: 77 MW; Mexico: 2008: 85 MW; 2007: 87 MW; 2006: 88 MW; 2005: 3 MW; Iran: 2008: 82 MW; 2007: 66 MW; 2006: 48 MW; 2005: 23 MW; Costa Rica: 2008: 74 MW; 2007: 74 MW; 2006: 74 MW; 2005: 71 MW; Rest of Europe: 2006: 163 MW; 2005: 129; Rest of Americas: 2006: 109 MW; 2005: 109 MW; Rest of Asia: 2006: 38 MW; 2005: 38 MW; Rest of Africa/Middle East: 2006: 31 MW; 2005: 31 MW; Rest of Oceania: 2006: 12 MW; 2005: 12 MW; World Total MW: 2008: 121,188 MW; 2007: 93,849 MW; 2006: 74,223 MW; 2005: 59,091 MW; List of Turbine Manufacturers, Alphabetically: AAER Systems, Canada; Accona, Spain; AeroCity Wind Power, U.S.; Aerostar Wind Turbines, U.S.; AN Winenergie, Germany; A. Ayvazian & Associates, Iran; Bard Engineering, Germany; Bergey, U.S.; Bonus Energy, Denmark; Clipper windpower, U.S.; darwin, Netherlands; DeWind, Germany; Doosan, South Korea; DyoCore Inc, U.S.; Dragonfly Industries, Inc., U.S.; Energya Wind Technologies, Netherlands; Ecotecnia, Spain; Enerco, Germany; Entegrity Wind Systems, Canada; Ecotechnia, Eozen, Spain; Fhrländer, Germany; Gamesa Eolica, Spain; General Electric,U.S.; Goldwind, China; GreenWave, U.S.; Hara Kosan, Netherlands/Japan; Heartland Energy Solutions, U.S.; HelixWind,U.S.; Hush Wind Power Limited, Australia; Hyosung, South Korea; Hyundai Heavy Industries, South Korea; Impsa, Argentina;Jacobs/Wind Turbine Industries,U.S.; Leitwind,Italy; Liquid Wind, U.S.; Kenersys, India; Kestrel Wind Turbines, South Africa; Mitsubishi Heavy industries, Japan; Multibird GmnH, Germany; Mtorres, Spain; Neo-Aerodynamic,U.S.; Nordex, Germany;Norwin, Denmark; Nordic Windpower, U.S.; Northern Power Systems,U.S.; PacWind, U.S.; Power Wind GmbH, Germany; Proven Energy Wind Turbines, UK; Repower, Germany; Samsung Heavy Industries,South Korea; Scanwind, Norway; Shanghai Aeolus Windpower (SAWT), China; Siemens, Denmark/Germany; Sinovel, China; Southwest Windpower, U.S.; SRC Vertical, Russia; Suzlon, India; TMA; Turbotricity; Turbowinds; Unison; Vensys, Germany; Vestas, Denmark; Vergnet, France; WES Canada, Canada; Wind Energy Solutions, Netherlands; Windflow, New Zealand; WindWind, Finland; Sub-providers of Wind Turbines: ABB Generators; Bosch Rexroth Gearboxes, Germany; DMI Industries, Towers, U.S.; DONGKUK S&C, South Korea; Hansen Transmissions Int., Gearboxes, Belgium; LM Glasfiber, Denmark; Marmen inc., Towers, Finland; Movemtas Gearboxes, Finland; Polymarin Composites Blades,South Korea; PSM, South Korea; TAEWOONG Fasteners, South Korea; Wind-Fix Fasteners, Netherlands; Minenergy Gearboxes, Germany, Gexpro Services, U.S./Europe/Asia; Sadad Machinesco, Iran; Satellite Communications Source Inc., Canada.

or a combination of them, it is fed into the national electric power transmission network. Individual turbines are *interconnected* with a medium voltage power collection system and communications network; the medium-voltage electrical current is increased in voltage at a substation with a *transformer* for connection to the higher voltage transmission system; *surplus* power produced by small-scale or *microgenerators,* can be fedback into the network and sold to the utility company, thus producing a retail credit for the consumer to offset their energy costs. The environmental impacts of windfarming are negligible, compared to fossil fuel production and consumption: the energy consumed to manufacture, transport and building a windfarm or a wind power plant, is equal to the new energy produced by the plant within a few months of operation.

●

Offshore Windfarms, Small-scale Windpower: Offshore windpower farms are most favoured due to their proximity to strong sea and ocean wind resources, shallow waters that lead to seas, and to easy transporation of barges by water. Due to limitations on suitable locations on land because of dense populations, offshore wind farming has boomed in Europe along the shallow waters of the North and Baltic Seas. Transportation of large wind turbine components, such as tower sections, nacelles, and blades, is much easier over water than on land; ships and barges can handle large loads easily than trucks and lorries or cars. The first wind offshore farms were installed in Denmark and since then offshore and onshore windfarms have sprung up worldwide, as shown here:[192] Small-scale wind power generation affects both offshore and onshore windfarms with a capacity to produce 50 kW or less of electrical power. Isolated communities that otherwise rely on diesel generators, may use wind

---

[192] List of Onshore Wind Farms that are Operating or Under Construction Worldwide: U.S.: Adair Wind Farm, 175 MW; Altamont Pass Wind Farm, 596 MW; Ashtabula Wind Farm, 196 MW; Barton Wind Farm, 160 MW; Barton Chapel Wind Farm, 120 MW; Beech Ridge Wind Farm, 186 MW; Benton County Wind Farm, 130 MW; Big Horn Wind Farm, 130 MW; Biglow Canyon Wind Farm, 125 MW; Black Law Wind Farm 124 MW; Bliss Wind Farm, 100 MW; Blue Canyon, 225 MW; Blue Sky Green Field, 145 MW; Brazos Wind Ranch, 160 MW; Buffalo Gap, 523 MW; Blue Creek Divide, 180 MW; Callahan Divide, 114 MW; Camp Grove, 150 MW; Camp Springs, 130 MW; Capricorn Ridge, 662 MW; Carroll, 150 MW; Cedar Creek, 300 MW; Century, 150 MW; Champion, 126 MW; Colorado Green, 162 MW; Crystal Lake, 350 MW; Desert Sky, 160 MW; Dutch Hill/Cohocton, 125 MW; Elbow Creek, 150 MW; Elk River, 150 MW; Fenton, 206 MW; Forest Creek, 124 MW; Forward Wind Energy Center, 129 MW; Fowler Ridge,750 MW; Glenrock, 118.5 MW; Goodland I, 130 MW; Gray County, 102 MW; Green Mt. Energy, 160 MW; Gulf, 283 MW; Hackberry, 165 MW; High Winds, 162 MW; Horse Hollow Wind Energy Center, 736 MW; Intrepid, 160 MW; Judith Gap, 160 MW;Kibby Wind Power Projects, 132 MW; King Mountain, 281 MW; Klondike, 400 MW; Langdon Wind Energy Center, 159 MW; Lone Star, 400 MW; Maple Ridge, 322 MW; Marengo, 140 MW; McAdoo, 150 MW; Milford Wind Corridor Project, 203 MW; Mount Storm, 264 MW; NedPower Mount Storm I, 164 MW; Nw Mexico Wind Energy Center, 204 MW; Noble Chateaugay wind park, 106 MW; Noble Weathersfileld Windpark, 126 MW; Pauther Creek, 257 MW; Peetz, 400 MW; Penascal, 202 MW; Pine Tree, 120 MW; Pioneer Prairie, 293 MW; Pomeroy, 196 MW; Prairie Star, 100 MW; Red Hills, 105 MW; Roscoe, 781 MW; San Gorgonio Pass, 619 MW; Sherbino, 750 MW; Shiloh, 300 MW; Smoky Hills, 249 MW; Stanton Energy Center, 120 MW; Stateline Wind Project, 300 MW; Story Country, 150 MW; Sweetwater, 585 MW; Tatanka, 180 MW; Tehachapi Pass, 685 MW; Trent, 150 MW; Turkey track, 169 MW; Twin Groves, 396 MW; Walnut, 153 MW; Wethersfield Wind Park, 124 MW; White Creek Wind Power Project, 204 MW; Whitelee, 322 MW; Wild Horse, 229 MW; Wildorado Wind Ranch, 161 MW; Woodward, 159 MW; Canada: Anse-a-Valleau Wind Farm, 100 MW; Centennnial Wind Power Facility, 150 MW; Enbridge Ontario, 181 MW; Melancthon EcoPower Centre, 199 MW; Port Alma, 101 MW; St Leon Project, 104 MW; Prince Township, 189 MW; Wolfe Island Wind Park, 197 MW; Mexico: Eurus, 250 MW; Albania: Kryevidhi, 150 MW; Vlore, 500 MW; Belgium: Thorntonbank, 300 MW; Portugal: Alto Minho Wind Farm, 240 MW; Arada-Montemuro Wind Farm, 112 MW; Gardunha, 106 MW; Pinhal Interior, 144 MW; Spain: Ventominho, 240 MW; El Marquesado, 198 MW; Higuerruela, 161 MW; Maranchon, 208 MW; Sisante, 198 MW; Romania: CEZ Fantanela, 600 MW; EDP Dobrogea, 266 MW; Sabloal Valea Dacilor, 147MW; Tomis Team Dobrogea, 600 MW; Norway: Smola, 170 MW; UK: Crystal Rig, 180 MW; Hardy Yard Hill, 120MW; Thanet Offshore Wind Project, 300 MW; Australia: Lake Bonney, 239 MW; Portland Wind Project, 195 MW; Snow Town, 170 MW; Waubra, 192 MW; Woolnorth, 140 MW; India: Vankusawade Wind Park, 201 MW; New Zealand: Project West Wind, 143 MW; Tararua, 161 MW; Proposed Onshore Wind Farms: Bald Fields, Australia, 104 MW; Beech Ridge, U.S., 186 MW; Belwind, Belgium, 330 MW; Blackspring Ridge, Canada, 300 MW; Clyde Wind,UK, 548 MW; Crows Nest, Australia, 124 MW; Dobrich, Bulgaria, 200 MW; Enel Agichiol, Romania, 210 MW; Eolica Baia, Romania, 126 MW; Eolica Beidaud, Romania, 128 MW; Eolica Casimcea, Romania, 244 MW; Eolica Cogealac, Romania, 448 MW; Eolica Sacele, Romania, 252 MW; Franklin County, U.S., 200-300 MW; Golden Hills, U.S., 400 MW; Griffin,UK, 204 MW; Halkirk, Canada, 150 MW; Hartland, U.S., 500-1000 MW; Hauauru ma raki, New Zealand, 540 MW; Hawke's Bay, New Zealand, 225 MW; High Country Energy, U.S., 300 MW; Kavarna, Bulgaria, 250 MW; Kaiwera Downs, New Zealand, 240 MW; Koekenaap, South Africa, 100 MW; Kruger Energy Port Alua, Canada, 101 MW; Lac-Alfred, Canada, 300 MW; Lincoln Gap, Australia, 124 MW; Macarthur, Australia, 329 MW; Mahinerangi, New Zealand, 200 MW; Mariselu, Romania, 300 MW; Markbygden, Sweden, ca. 4000 MW at most; Massif-du-Sud, Canada, 150 MW; McAdoo, U.S., 150 MW; McCormick Ranch Wind Park, U.S., 300 MW; Mount Mercer, Australia, 131 MW; Motorimbu, New Zealand, 108 MW; Musselroe, Australia, 140 MW; Owerinny, Ireland, 320 MW; Pampa Wind Wind Project, U.S., 320 MW; Pine Canyon,U.S., 150 MW; Pine Tree Wind Project,U.S., 120 MW;Plambeck Bulgaria, Bulgaria, 250 MW; Project Hayes, New Zealand, 630 MW; Project Central Wind, New Zealand, 130 MW; Puketiro, New Zealand, 130 MW; Riviere-du-Moulin, Canada, 350 MW; Rototuna, New Zealand, 500 MW; Sherphards Flat, U.S., 909 MW; Silverton, Australia, 1000 most MW; Sinus Holding, Romania, 700 MW; Slopedown, New Zealand, 150 at most MW; Taharoa, New Zealand, 100 MW; Tangier, Morocco, 140 MW; Taralga, Australia, 105 MW; Te Waka, New Zealand, 156 MW; Tehachapi Renewal Project, U.S., 4500 MW; Titan Wind Project, U.S., 5050 MW; Tiritea, New Zealand, 300 at least MW; Underwood, Canada, 199.7 MW; Valley City/Lake Ashtabula, U.S., 200 MW; Waitahora, New Zealand, 177 MW; Waverley, New Zealand, 135 MW.

turbines to displace diesel fuel consumption. *Capacity Factor & Fuel Saving* aims at preserving constant wind power for regular supply of electricity, even at periods when speeds are low, at cost-effective levels. This can be achieved through a combination of windfarms connected which pool their services together to yield year-round results.

- *Turbine Placement:* Winfarm designs use specialised wind energy software applications to evaluate the impact of these issues on a given wind farm design. Good selection of a wind turbine site, availability of transmission lines, value of energy to be produced, cost of land acquisition, land use considerations, and environmental impact of the construction and operations, are basic to starting a windfarming enterprise. Offshore location costs which are higher than onshore location costs, may be offset by higher annual local factors, thereby reducing energy producing costs. Moreover, *Penetration Limits & Intermittency* show the highly variable different timetables, from hour to hour, and seasonally.

The eternal existence of wind energy power and solar energy power, along with other renewable energy sources and their technologeies, are the one of the best reasons why fossil fuel and petrochemical energies must be replaced.

3.4 Thermal & Geothermal Technologies

THERMAL AND geothermal energy production and use rely on the Sun, centre of our planet Earth, and part of our vast *Milkyway Galaxy* and its vast *Sun Systems*. Radiation from the sun or heat *in transfer* is thermal energy. Thermal energy is part of the overall *internal energy* of a system, which comes from the movement of atoms and molecules in matter, as a form of kinetic energy produced by the random movements of those molecules. Thermal energy can be increased or decreased, like temperatures, for example. The transformation or tranfer of thermal energy from one form to another, is channelled via *conduction, convection, radiation* or by a combination of all three. To change from solid to liquid, or from liquid to gas states, chemical substances release or absorb *latent heat.* Living organisms like green plants, rely on energy *radiation* from the Sun; and animals need *chemical* energy to be able to grow and reproduce. A thermal power station is primarily driven by steam from heated water, to spin steam turbine. The spin turbine drives an electrical generator or propels a ship or boat. Steam is usually recyclable.

Geothermal energy exists in, and extracted from the original radioactive heat of decayed animals and solar energy absorbed at the surface which is stored in the earth from the origin formation of the Earth. Its location is limited to areas near *tectonic plate* boundaries[193] which lateral movement typically is at speeds of 50-

---

[193] Tectonic Plate Boundaries: Denote those areas in and around the large motions of Earth's *lthosphere,* or outermost part of the Earth's interior, made up of two layers – *lithosphere* consisting of the crust and rigid uppermost part of the mantle; and *asthenosphere* below the lithosphere, which although is solid, has relatively low viscosity and shear strength and can flow like a liquid on geological time scales. The deeper mantle below the astheniosphere is more rigid due to the higher pressure. The *mantle* is a highly viscous layer between the *crust* or outermost solid shell of a rocky planet or moon, which is chemically distinct from the underlying mantle, and the *outer core*or a liquid layer about 2260 km thick composed of iron and nickel which lies above the Earth's solid *inner core* below its mantle. The lithosphere is broken up into about 8 major *tectonic plates* and many minor *lithospheric plates* which ride on the *asthenosphere,* and move in relation to one another at one of three types of plate boundaries: *convergent* or collisional boundaries; *divergent* or spreading centres boundaries; and *transform* boundaries; earthquakes, volcanic activity,mountain building, and oceanic trench formation, occur along plate boundaries.

100 mm annually. Besides the heat emanating from deep within the Earth, the top 10 m of the ground accumulates solar energy during the summer, and releases that energy during the winters. *Geothermal power has the potential to help mitigate global warming if used wisely, even though geothermal wells tend to release greenhouse gases trapped deep within the earth;* trapped greenhouse gases deep within the earth are much lower than those produced by fossil fuels. Geothermal energy is the most reliable, economically-friendly, and cost-effective renewable energy source. About 10 GW of geothermal power electricity or 0.3% of global electricity, is installed around the world as of 2007, used for heating, bathing and generating electricity. For district- and space-heating, spas, industrial processes, desalination and agricultural applications – an additional 28 GW of direct geothermal capacity has been installed worldwide. A large geothermal plant can power entire cities, while smaller ones can supply rural villages. Compared with the production and consumption of fossil fuels, geothermal energy production and consumption is *safer, requires no fuel and therefore immune to fluctuations, and generates less greenhouse gases than fossil fuels.*

The environmental impacts likely to emerge with geothermal source exploitation may include:

- *Dangerous and Noxious Gases:* Dangerous and noxious gases or a mixture of them, like *carbon dioxide* and *hydrogen sulphate* are exposed to the environment with the sourcing and resourcing of geothermal energy, and contribute to *acid rain* precipitation. Aquatic animals, plants, and animals and infrastructure are adversely affected. The use of sulphur and nitrogen compounds frequently causes acid rain when these react in the atmosphere. Noxious smells are daily fixtures around or in the vicinity of geothermal electric plants, which additionally, emit an average of 122 kg of carbon dioxide per MWh of electricity. Dissolved gases and hot water from geothermal sources may contain trace amounts of dangerous elements such as *mercury, arsenic* and *antimony,* which if disposed of into rivers, can render water unsafe for human use.

- *Hornets Nest of Destabilisation of Nature*: The construction of geothermal power plants can adversely affect land stability and trigger off earthquakes.

The art of geothermalling energy production begins with drilling, which accounts for most of the costs of electrical plants, and exploration of deep resources with very high financial risks. A geothermal electric plant and well costs about EUR 2-5 million per MW of capacity to realise. Operational costs are EUR 0.04-0.10 per kWh, with or without government subsidies, either for electricity production or direct use. Until recently, geothermal plants have been built only on the edges of tectonic plates. With the use of *binary cycle* technology's *enhanced geothermal systems* (EGSs), heat pumps now extract enough heat from shallow ground anywhere in the world to provide wintertime home heating. Wells are rarely more than 3-10 km deep. Water may be piped directly into radiators where natural hot springs are available.

Natural *hot springs* are produced by the emergence of geothermal heated groundwater from the earth's crust available on every continent and even under the oceans and seas. In shallow, hot but dry ground, earth tubes or downhole heat exchangers may be used without a heat pump. Also, in areas where shallow ground is too cold and disappear completely below 10 m of depth, the heat is extracted with a heat pump. Around the world, geothermal electricity is generated in 24 countries.[194] Estimates of the electricity generating potential vary from 35-2000 MW.

3.5 Biomass Technologies

BIOMASS ENERGY is a renewable energy source which is also a direct beneficiary of solar and wind power energies. It is available in its various forms, such as: trees and branches; trash, yard clippings and wood chips; plant or animal matter; biodegradable-, food-, paper-, and slaughterhouse-wastes; manure, and sewage. Industrial biomass can also be grown from numerous types of plants, such as miscanthus, switchgrass, hemp, corn, poplar, willow, sorghum, sugarcane; and a variety of tree species like eucalyptus, oil palm or palm oil. *Fossil fuels* are not considered biomass because they contain carbon dioxide that have been "outed" of the *carbon cycle[195]* for a long time. Biomass is part of the carbon cycle, and is one of the *four* reservoirs of carbon interconnected by pathways of exchange.

The carbon in biomass, which is ca. 50% of its dry matter content, is already part of the atmospheric carbon cycle; biomass absorbs CO_2 from the atmosphere during its growing lifetime, which reverts to the atmosphere as a mixture of CO_2 and methane (CH_4), depending on the ultimate fate of the biomass material. When biomass is used as a fuel, it still puts the same amount of CO_2 into the atmosphere. *But when biomass is used to deforestate or urbanise green sites, it contributes to global warming.* The disposal of biomass residues from a mixture of CO_2 and CH_4, including feces, displaces the production of an equivalent amount of fossil fuels, as well as shifts the composition of the recycled carbon emissions. Rotting biomass produces a mixture of up to 50% CH_4, biomass burning up to 5-10% CH_4, and biomass-controlled combustion in a power plant converts almost all of the carbon to CO_2; CH_4 converts to CO_2 in the atmosphere, completing the cycle, and thus significantly reduces greenhouse gas emissions because CH_4 is much stronger than CO_2.

The miscreant prodigal carbon or *carbon dioxide "outed"* from the carbon cycle is one of the major *greenhouse gases;* its volume is increased by the CO_2 produced from human or *anthropogenic* activities. Anthropogenic Effects'[196] processes produce gases and materials dangerous and hazardous to the environment, as

[194] Installed Geothermal Electric Capacities, 2007: U.S.: 2685 MW; Philippines: 1969.7 MW; Indonesia:992 MW; Mexico: 953 MW; Italy: 810.5 MW; Japan:535.2 MW; New Zealand: 471.6 MW; Iceland: 421.2 MW; El Salvador: 2042 MW; Costa Rica: 162.5 MW; Kenya: 128.8 MW; Nicaragua: 87.4 MW; Russia: 79 MW; Papua New Guinea: 56 MW; Guatemala: 53 MW; Turkey: 38 MW; China: 27.8 MW; Portugal: 23 MW; France: 14.7 MW; Germany: 8.4 MW; Ethiopia: 7.3 MW; Austria: 1.1 MW; Thailand:0.3 MW; Australia: 0.2 MW; Total World: 9731.9 MW.

[195] Carbon Cycle: Carbon exists in the Earth's terrestial biosphere atmosphere (0.04%) primarily as the gas Carbon dioxide (CO_2), and in other gases like methane and chlorofuorocarbons produced by human activities. Carbon also exists in plants (86% above-ground, 73% in the soil), oceans and sediments of fossil fuels.

[196] Anthropogenic Effects: Are processes or materials produced from Industry, Agriculture, Botany, Mining, Construction, and Habitation.

opposed to those that occur naturally without human influence, thus: *Industry* releases gases and dust into the atmosphere and disposes waste improperly that lead to global warming, air and water contamination and pollution; *Agriculture* converts woodlands into fields and pastures, contaminates and pollutes air, water and soil from fertilizer and pesticide use; *Botany* alters plants by breeding, selection, genetic engineering and tissue-fusion; *Mining* removes topsoil, and diverts and pollutes groundwater; *Construction* destroys natural habitats, diverts groundwater, and fills in marshes, bays, swamps, ponds and streams; *Habitation* concentrates on human activities, waste products, sewage and debris in discrete zones.

Biomass power energy for heating, lighting and electricity production that can be weaned from raw biomass, is divided into *First-generation biofuels* or *agrofuels* and *Second-generation biofuels* or *agrofuels*. First-generation biofuels are those made from *sugar, starch, vegetable oil* or *animal fats*, with the basic feedstocks for their production often being *seeds, grains, crops* or *grasses* such as *wheat, corn* or *maize, soybean, rapeseed, sugar beet, sugarcane, cassava, palm oil, potatoes* and *sunflower seeds* which yield starch and liquids. The starch is then fermented into *bioethanol* and the liquid is pressed into *vegetable oil* and then used as *biodiesel*. The fermenting of plant-derived sugars is also used to produce other alcohols such as wine and beer. First-generation feedstocks are also the basic foodstuffs consumed by vast populations of human beings and animals. Accordingly, the biofuel industry has been roundly criticised for diverting food crops to produce biofuels to feed vehicles. The production of biofuel feedstocks requires higher agricultural inputs in the form of *fertilisers, pesticides,* and *herbicides* which thus limits the reduction of *greenhouse gases,* while the fertilisers, pesticides and herbicides end up in the food chain and polluting groundwater. The most common first-generation biofuels are *Vegetable oil; Bioalcohols, Bioethers, Biogas, Syngas,* and *Solid biofuels.*

The production of second-generation biofuels is a response to the rebukes and accusations levelled against the bioenergy industry for *robbing humans and animals to feed vehicles and machines.* Thus, second-generation biofuels are those made from *non-food* crop items or from *wastes,* and *environmentally-friendly-and-degradable* substances; such second-generation non-food crops include *switchgrass, miscanthus, straw, timber, woodchips, skins* and *pulps* from fruit pressing, *manure, sewage, husks, foodwastes, algae, buchloe dactyloides, jatropha, pongamia pinnate, hemp, linseed, flax, borage, cannabis sativa, echinaceae artemisa, tobacco, lavender, oilseed rape* used to produce *bioethanol, biobutanol, biodiesel* or *agrofuel, syngas, bioelectricity, hemp-lime-* and *straw-building materials, insulation paints, varnishes, drugs, botanical and herbal medicines, nutritional supplements, plant-made pharmaceuticals, plastics and packing, essential oils, printing inks,* and *paper coatings.*

Woody or *fibrous* biomass provides feedstocks with useful sugars locked in by *lingin* and *cellulose* or *complex carbohydrates* or sugar-based molecules contained in all plants, from which *lignocellulosic ethanol*[197]can be produced

---

[197] Lignocellulosic Ethanol: Is made by freezing the sugar molecules from cellulose using enzymes, steam heating, other pre-treatments. Lingin, the by-product, can be bundled as a carbon neutral fuel to produce heat and power for the processing plant.

from the sugars by fermentation. Lignocellulosic ethanol can reduce greenhouse gas emissions by about 90% when compared with fossil fuels, as evidenced by an IOGEN CORP-run lignocellulosic plant in Canada. Bio-synthetic liquid and biofuels under development include: *Biohydrogen;* BioDimethyl *ether* (DME); *Biomethanol; Butanol* and *isobutanol; 2,5-Dimethylfuran; High Temperature Upgrading* (HTU) *diesel; Mixed alcohols* (mostly ethanol, propanol, and butanol, with some pentanol, hexanol, heptanol, and octanol); *Wood diesel; Fischer-Tropsch* (FT) *fuels.*

Both first- and second-generation biofuels or agrofuels are either liquid or gaceous and are produced by growing crops high in sugar (e.g., sugarcane, sugar beet, sweet sorghum) or starch (e.g., corn, maize), and using yeast to ferment them to *ethyl alcohol* (i.e., ethanol or pure alcohol or grain alcohol or drinking alcohol); or by growing plants that contain high amounts of vegetable oil (e.g., palm oil, soybean, algae, jatropha, pongamia pinnate), which after harvesting are heated, then burned directly in a diesel engine or chemically processed to produce fuels such as biodiesel. *Woodgas, methanol,* or ethanol are biofuels or agrofuels that can be converted from wood and its by-products; *cellulosic ethanol* is made from non-edible plant parts; *pelletised* and burnt wood and dried grass can be used as energy. Biomass technologies include fermentation of sugars into ethanol or alcohol; *anaerobic digestion*[198] breaks down biodegradable microorganisms in landfills, wastewater sludges and other organic wastes in the absence of oxygen, and sets free carbon dioxide (CO^2) and methane (CH4), rich in *biogas*; gasification; direct combustion and co-firing (direct, indirect and parallel). International biomass power commercialisation is coordinated by the UN Global Bioenergy Partnership (GBEP), working from the Food and Agricultural Organisation (FAO) offices in Rome, Italy. The GBEP[199] was launched on 11 May 2006, to support wider, cost-effective biomass and bioenergy development, particularly in developing countries.

Biomass power generating industries exist in North America and Europe. For example, in the U.S., 1,700 MW of operating capacity or about 0.5% of U.S., electricity is supplied. *This avoids an equivalent of 11 million tons of CO² and 2 million tons of CH4 emissions per year from biomass residues that would otherwise be placed in landfills, disposal piles or by ploying agricultural residues.* In Europe, the European Bioenergy Industry Association (EUBIA) launched a campaign from 2005-2008, to raise the public's awareness and promote sustainable energy production use by implementing projects, which included:

-
  BIOPROS 2005: Collective research aimed to improve the biomass production of Short-Rotation-Plantation through the safe application of wastewater and sludge.

-
  AGROBIOGAS: Using agricultural wastes for the production of biogas through the *anaerobic digestion* (without oxygen) process.

---

[198] Anaerobic Digestion (AD): A naturally-occurring process of decomposition of microorganisms in the absence of oxygen; this is similar to fermentation.

[199] UN Global Bioenergy Partnership (GBEP): Was formed under the 2005 Gleneagles Plan of Action by the G8+5 to provide a mechanism for partners to organise, coordinate and implement targeted international research, development and commercial activities to production, delivery, conversion and use of biomass energy, with a focus on developing countries.

- *EMINENT 2 2006:* Encourages early market introduction of new energy technologies in liaison with energy research groups and industry.

- *RESTMAC 2006:* Creating markets for Renewable Energy Technologies – EU RES technology marketing campaign.

- *COMPETE 2007:* Competence Platform in Energy Crop and Agro-forestry Systems for Arid and Semi-Arid Ecosystems-Africa.

- *PELLET@LAS 2007:* Develop and promote transparency in the European fuel pellets market by facilitating pellets trade, removing market barriers, information gaps and local supply bottlenecks.

- *INTAS-COMPETITIVE HYDROGEN FROM AGRO-FORESTRY RESIDUES 2007:* Agro-forestry residues to hydrogen: Development of the second step (carbonisation) of pellets, and the third step (steam reforming of the carbonised pellets) to give the maximum yield of bio-hydrogen.

- *AQUATERRE 2008:* Integrated European Network for Biomass and waste reutilisation of Bio-products.

Widespread uncontrolled biomass burning for energy, including the production and use of bioenergy crops, are not sustainable and efficient uses of this renewable energy source.

3.6 Water Technologies

TECHNOLOGIES that generate electricity through the use of water make ultimate use of the Earth's *Water Cycle* or *Hydrologic Cycle,*[200] which entails the continuous movement of water on, above, and below the surface of the Earth. And since it is truly a cycle, there is no beginning or end, because water can change states among *liquid, vapor* and *ice* at various places in the water cycle. Rain, precipitation, ice, rivers, seas, oceans, tides, clouds, air, winds, sun, vegetation, temperature, seasons, and Earth's rotation – are all complicit to upholding the status quo of the water cycle. Water power is the most widely used form of renewable energy.

---

[200] Water or Hydrological Cycle: The balance of water on Earth that is deposited on Earth remains fairly constant: it occurs when the atmosphere, a large gaseous solution, becomes saturated with *water vapor* and condenses and falls out of solution, (i.e., *precipitates*), and thus cools the air or adds water vapor to the air. For example, *Virga* is an observable streak or shaft of precipitation that falls from a cloud but evaporates before it reaches the ground. At high altitudes, it falls mainly as *ice crystals* before melting and finally evaporating, usually due to compressional heating, because air pressure increases closer to the ground. Virga is very common in deserts and in temperate climates, and can cause very interesting weather effects, because as rain is changed from *liquid* to *vapor* form, it removes heat from the air due to the high heating voporisation of water.

Hydroelectricity is produced by hydropower using the gravitational force of falling or flowing water, to drive a water turbine and generator. The energy extracted from dammed water depends on the volume and on the difference between the source and the water's outflow or the *head*. A very *high head* water is obtained by running the water through a large pipe or *penstock*. Movement of water between reservoirs at different elevations is used to provide pumped storage hydroelectricity for high peak demands, while excess generation capacity is used to pump water into the higher reservoirs at times of lower reservoir, and vice versa. Hydrolelectric power is converted from tidal energy, wave power or wind waves, "rogue" waves, and dams, using standard and modern technolgies. In 2005 worldwide, hydroelectricity supplied some 816 gigawatts of electricity (Gwe), approx. 20% of the world's electricity supply or 88% of electricity from renewable energy sources.

Tidal power or energy generation converts the energy of tides into electricity or other useful forms of power, and is considered more predictable than wind or solar power. This is because tidal rise and fall motions are tied in with the rising of Earth's ocean surface caused by the tidal forces of the Moon and the Sun acting on the oceans. Tides cause changes in the depth of marine and estuarine water bodies and produce oscillating currents or *tidal streams;* tidal streams make prediction of tides in coastal navigation into electricity or other useful forms of power. Tidal forces contribute to ocean currents, which moderate global temperatures by transporting heat energy toward the poles.

Tide or *water mills* are driven by tidal rise and fall generate tide energy. These are usually situated in river estuaries, away from the effects of waves but close enough to the sea in order to have a reasonable tidal range. Tidal range is the vertical difference between the *highest high tide* and *lowest low tide,* and the *most extreme* tidal range, that occur around the time of the full or new moons, when gravity of both the Sun and Moon are pulling the same way (new moon), or exact opposite way (full moon).

Wave power or wind wave, is the transport of energy by ocean surface waves, and the capture of that energy to generate electricity, water desalination, or pumping of water into reservoirs. *Wind waves* or wind-generated waves are surface waves that occur on the free surfaces of oceans, seas, lakes, rivers, canals, small puddles and ponds, and are quite distinct from the daytime flux of tidal power and the steady gyre of ocean currents. They usually result from the wind blowing over a vast enough stretch of fluid surface, and some waves in the oceans can travel thousands of miles before reaching land. These waves range in size from small *ripples* or *capillary* waves to huge *rough* waves. Wind waves are *wind seas* when generated by local winds; they are *swells* when they cease to blow; they are *ocean surface winds* when they occur in the ocean; *tsunamis* are specific *wave winds* not caused by wind but by geological effects, such as underwater earthquakes or landslides and waves generated by underwater explosions or the fall of meteorites; tsunamis are not visible in deep water as they are small in height and very long in wavelength, but grow to devastating proportions at the coast due to reduced depth. *Wave power winds* or wind waves are sometimes polluted through the action of individual *"rogue waves"* or freak-, monster-, or king waves; or sometimes they occur up to heights of 30 metres or higher than wind waves at certain locations and moments. Rogue winds are distinct from tides, and are caused by the Moon and Sun's gravitation pull.

Wind waves are farmed using various technology devices to capture the energy of waves in shoreline, nearshore and offshore locations, such as: absorb or buoy *through* surface following attentuator, terminator, lining perpendicular to wave propagation; oscillating water column; overtopping; and takeoff with hydraulic ram, elastomeric hose pump, pump-to-shore, hydroelectric turbine, air turbine; and linear electrical generator. Some devices incorporate designs such as parabolic reflectors (or dish mirror), parabolic-shaped reflective device used to collect or distribute energy such as light, sound, or radio waves from distant sources and assemble them to a focal point. Wave farming power devices must be *sustainable, cost-effective, environmentally-friendly* and *tough* to resist storms and adverse weather generally, thus:

- *Environmental Impacts:* Impacts on the marine environment and life, such as noise pollution could have negative effects, if not monitored, although the noise and visible impact of each design varies; constructing devices must be able to survive storm damage and saltwater corrosion; likely sources of failure include seized bearings, broken wells, and snapped mooring lines.

- *Costing and Social Factors:* Designs of bad weatherproof prototype devices may raise costs, sour competitiveness and raise electricity costs generated by wave farming. Wave farms could displace fishermen from their traditional fishing grounds occupied by wave energy devices or such devices can prove dangerous to safe shipping.

Among Wave power and Wind wave farming devices used are:

- *Pelamis Wave Energy Convertor:* A surface device which articulates with the movement of the waves, each resisting motion between it and the next section, creating pressurized oil to drive a hydraulic ram with a hydraulic motor. The machine is long and narrow snake-like and points into the waves; it accentuates the waves, gathering more energy than its narrow profile suggests. It's articulating sections drive internal hydraulic generators through the use of pumps and accumulators.

- *Pacific Northwest Generating Cooperative:* This cooperative is funding the building of a commercial wave-pack at Reedsport, Oregon, U.S., aimed at utilising the PowerBuoy Ocean Power Technologies, which consist of modular, ocean-going buoys: the rising and falling of the waves moves the buoy-like structure creating mechanical energy which is converted into electricity and transmitted to shore over a submerged transmission line. A 40 kW buoy has a diameter of 12ft (4m), 52ft (16m) long, with approx. 13ft of the unit rising above the ocean surface. Using the three point mooring system, they are designed to be installed one- to five miles (8 km) offshore in water 100 to 200ft (60m) deep.

- *Wave Dragon Wave Energy Converter:* Has "large" focus waves up into an offshore reservoir, and the water returns to the ocean by the force of gravity via hydroelectric generators.

- *AquaBuOY, SeaRacer & CETO:* This transfers energy by converting the vertical component of wave kinetic energy into pressurized seawater by means of two-stroke hose pumps.

Pressurized seawater is directed into a conversion system consisting of a turbine driving an electrical generator. The power is transmitted to shore by means of a second, undersea transmission line. A commercial wave power production facility utilising this technology is beginning initial construction in Portugal and the U.S. A SeaRacer uses an entirely new pumping technique for gathering the wave energy. A *CETO* consists of a single piston pump attached to the sea floor, with a float tethered to the piston. Deep water wave resources are believed enormous, between 1-10TW, although not all of this can be captured. Locations with the most potential for wave power capturing include the: *western seaboard* of Europe; *northen coast* of the UK and the *Pacific coastlines* of North/South America, Southern Africa, Australia and New Zealand; *north/south temperate zones;* and prevaling *westerlies, horse latitudes* or subtropical high latitudes. Waves cause the float to rise and fall, generating pressurized water which is piped to an offshore facility to drive hydraulic generators or run reverse osmosis water desilanisation. CETO is currently being experimented in Australia. Another type of wave buoys being developed uses special polymers.

-
    Wavebob: This is a leading new wave energy converter technology which has been conducting R&D over a year period in the Atlantic Ocean near Galway, Ireland.

Wave power farming first began with the Agucadora *Wave Park* in Portugal in 2008, with three Pelamis P-750 machines and a total installed 2.25 MW capacity. More 21 MW were to be installed with 25 Pelamis machines. Scotland (UK), approved four million British pounds for a 3 MW 4-Pelamis machines wave farm on 20 February 2007, as part of a 14 million pounds funding package for marine power. Cornwall (UK), also announced funding for a 20 MW Wave Hub wave energy farm to serve 7,500 households, that will potentially save 300,000 carbon dioxide emissions within the next 25 years.

Industrial hydroelectricity plants including small-scale ones (10-30 MW), use *dams* to generate more voltage, to supply public electricity works, serve specific individual enterprises and function as substations. Aluminum electrolytic, aluminum smelting, aluminum, airplane works, irrigation and power to the population, require powerful hydroelectric plants or hydroelectric substations to supply dam-generated electricity. Hydroelectric schemes also include reservoirs which provide facilities for water sports, aquaculture, irrigation support agriculture, which also serve as tourist attractions areas, and flood control.

Dam-generated hydroelectricity plants[201] are nearly immune to increases in the cost of fossil fuels, such as *gas, oil* or *coal,* do not burn fossil fuels, do not emit

---

[201] Worldwide Major Hydroelectric Schemes Under Construction with 2000 MW & Proposed Ones: Under Construction: *Three Gorges Dam,* China: 22,500 MW, largest power plant in the world, first power in July 2003 with 12,600 MW installed by Oct., 2007; *Xiluodu Dam,* China: 12,600 MW, construction once stopped due to lack of environmental impact study; *Siang Upper HE Project,* India: 11,200 MW, multiphase construction over a period of 15 years, construction was delayed due to dispute with China; *Xiangjiaba Dam,* China: 6,400 MW, to be completed 2015; *Longtam Dam,* China: 6,300 MW,to be completed Dec., 2009; *Nuozhadu Dam,* China: 5,800 MW, to be completed 2017; *Jinping 2 Hydropower Station,* China: 4,800 MW, with 23 families and 129 local residents moved to built this dam; *Laxiwa Dam,* China: 4,200 MW, to be completed 2010; *Xiaowan Dam,* China: 4,200 MW, to be completed Dec., 2012; *Jinjing 1 Hydropower Station,* China: 3,600 MW, to be completed 2014; *Pubugou Dam,* China: 3,300 MW, to be completed 2010; *Goupitan Dam,* China: 3,000 MW, to be completed 2011; *Guanyinyan Dam,* China: 3,000 MW, to be completed 2015; *Lianghetou Dam,* China: 3,000 MW, to be completed 2015; *Boguchan Dam,* Russia: 3,000 MW, to be completed 2012; *Chapeton,* Argentina: 3,000 MW; *Dagangshan,* China: 2,600 MW, to be completed 2014; *Jinangiao Dam,* China: 2,400 MW, to be completed 2010; *Guandi Dam,* China: 2,400 MW, to be completed 2012; *Liyuan Dam,* China: 2,400 MW; *Tacoma Dam Bolivar State,* Venezuela: a new power plant to be last developed in the Low Caroni Basin, to bring the total to six plants on the same river, including 10,000 MW *Guvi Dam,* to be completed 2014; *Ludila Dam,* Chaina: 2,100 MW, with River diversion to begin 2009, to be completed 2015; *Bureya Dam,* Russia: 2,010 MW, to be completed 2009; *Shuangjiangkou Dam,* China: 2,000 MW, dam to be 314 m high; *Ahai Dam,* China: 2,000 MW; *Lower Subansiri Dam,* India: 2,000 MW, to be completed 2009; Proposed Major Hydroelectric Projects Greater Than or Equal to 2,000 MW: *RedSea dam,* Middle East: 50,000, still in planning, and would be largest dam in the world; *Grand Inga,* DR Congo: 10,000 MW; *Baichetan Dam,* China: 13,500 MW, still in planning, to be completed 2015; *Wudongde Dam,* China: 7,500 MW, still in planning,to be completed 2015; *Maji Dam,* China: 4,200 MW, to be completed 2013; *Songta Dam,*China: 4,200 MW, to be completed 2013; *Liangjiaren Dam,* China:

CO^2, and no imports of fossil fuels are needed. Some CO^2 is produced during manufacture and construction of the project, which is but a tiny fraction of the operating emissions of equivalent fossil-fuel electricity generation. Operating labor costs are low, and hydroelectric plants have longer economic lives than fuel-fired electric generation. Hydroelectricity eliminates the *flue gas* emissions from fossil-fuel combustion and pollutants such as sulphur dioxide, nitric oxide, carbon monoxide, dust, and mercury in the coal; it avoids the hazards, indirect health effects, and coal emissions of coal mining. Compared to *nuclear power,* hydroelectricity generates no nuclear waste, no nuclear leaks, has none of the dangers associated with uranium mining, and unlike uranium, hydroelectricity is a renewable energy source. Hydroelectricity power plants have a more predictable load factor than wind energy coupled with a strong reservoir. Hydroelectric plants can be easily regulated to follow variations in power demand.

The most glaring environmental *disadvantages* of constructing and operating hydroelectric power plants are:

- *Population Relocation:* Before a hydroelectric dam project can be executed in practice on the ground, the displaced human populations living where the dams and reservoirs are located, are relocated to new lands. This can be very emotional, impacting on personalities, wealth, lifestyles, careers, religions and cultures. As of 2008, an estimated 40-80 million people worldwide had been displaced to make way for a dam project. No amount of compensation, in many instances, can replace lost ancestral and cultural attachments or the spiritual or historical value disrupted or washed away by dams and this relocational movement.

- *Environmental Damage:* The damage to personal, religious, career and cultural icons caused by human relocation to make place for hydroelectric dams, is followed immediately by the programmed damage to the environment. Hydroelectric projects always disrupt and damage surrounding aquatic ecosystems both upstream and downstream of the plant life. Natural fish habitats and migrations are artificially barricaded. Generation of hydroelectricity changes the downstream river environment, while water exiting a turbine usually contains suspended sediment which can lead to scouring of river beds and loss of river banks and sand-bars through erosion.

- *Greenhouse Gas Emissions:* Methane ($CH4$) and Carbon dioxide (CO^2), potent greenhouse gases, are produced in substantial amounts by reservoirs of power in flooded areas in tropical regions decaying in anaerobic environment.

- *Dam Failures:* Dams may fail and some do, like the Banqiao Dam in Southern China that took 171,000 lives, and left millions homeless. Dams may even be subject to terror or sabotage attacks, or bombardment by enemy forces. If dams are created in geologically wrong places, they may cause disasters. The Italian Vajout Dam was one such disaster case, where in 1963, almost 2000 died. Diminished river flow, caused by global warming decreased glacier volume, can lead to power shortages. Hydrogeneration depends on rainfall in the watershed, and may be significantly reduced in years of low rainfall or snowmelt, and long-term energy yield may be affected by climate change vagaries.

4,000 MW, still in planning,to be completed 2015; *Jiran Dam,* Brazil: 3,300 MW, to be completed 2007; *Pati Dam,* Argentina: 3,300 MW; *Santo Antonio Dam,* Brazil: 3,150 MW, to be completed 2007; *Lower Churchill,* Canada: 2.000 MW, to be completed 2014; *HidroAysen,* Chile: 2,700 MW, to be completed 2020; *Lenggu Dam,* China: 2,718 MW; *Subansiiri Upper HE Project,* India; *Changheba Dam,* China: 2,200 MW, to be completed 2015; *Banduo 1 Dam,*China: 2,000 MW, to be completed 2015.

Water power for electricity generation, in spite of some built-in disadvantages, is a priceless renewable energy source, far superior and less costly and less dangerous, compared with fossil-fuels and nuclear energy production and consumption.

3.7 Hydrogen, Fuel Cell, Nano Technologies

HYDROGEN GAS, fuel cells and nano technologies have much in common: they are used to produce, store, and sustainably use energy for electricity generation, heating and other industrial applications and devices. Nano technologies make it possible to manipulate, weld together and produce new fuel modalities with fuel cells, batteries, hydrogen combustibility and usability.

Hydrogen fuel refers to the use of hydrogen for its combustive qualities as fuel and energy carrier, but must first be weaned from its compound form with oxygen as *water* (H_2O). This weaning process uses *electrolysis*; or hydrogen is gathered by other means because hydrogen does not occur naturally by itself, cannot be mined nor drilled (as fossil oils can) and requires more energy input to produce than is generated with its combustion. Solar, wind or oher renewable energy sources friendly to the environment, can also be used to produce hydrogen fuel. The use of hydrogen and its environmental friendliness, can replace fossil-fuel use with its environmental unfriendliness, transported and applied to various industrial and domestic uses: *powering automobiles, boats, airplanes, fuel cell devices,* and as *clean burning energy.* Fuel cells filled with hydrogen can be, and are already, being transported to any location for use and refuelling, as these industrial examples show:

- *The GM 1996 Electrovan:* The General Motors (GM) 1996 Electrovan was the automative industry's first attempt at an automative powered by a hydrogen fuel cell. The Electrovan which weighed more than twice as much as a normal van, could travel up to 70 mph for 30 seconds.

- *Chrysler Natrium 2001:* In 2001, the Chrysler Natrium used its own on-board hydrogen processor to produce hydrogen for fuel cell. This reacted sodium borohydride fuel with Borax, with the resulting hydrogen generating electric power in the fuel cell for near-silent operation and a range of 300 miles without trespassing on passenger space. Chrysler also developed vehicles which separated hydrogen from gasoline in the vehicle, to reduce emissions without relying on a nonexistent hydrogen infrastructure and avoid large storage tanks. Sodium borohydride and Borax occur naturally in great quantities in the U.S.

- *First Public Hydrogen Refuelling Station:* In April 2003, the first public service hydrogen refuelling staion was opened in Reykjavik, Iceland, which serves three DaimlerChrysler buses. The station produces hydrogen it needs by itself with an *electrolyzing unit* produced by Norsk Hydro, and does not need refilling: all that enters is electricity and water. The station has no roof, in order to allow any leaked hydrogen from escape to the atmosphere. Royal Dutch Shell is a partner in the project.

- *Hydrogen Fuel Initiative* (HFI): A 2001 U.S., Bush administration US$1 billion Hydrogen Fuel Initiative (HFT) scheme to produce commercial fuel cell vehicles by 2020, was in May 2009, redirected by the new Obama administration, to fund research related to stationary fuel cell hydrogen vehicles, since other vehicle technologies will lead to quicker reduction in emissions in a shorter time.

- *Hydrogen-Run Motorcycles:* The first hydrogen-run motorcycles or Emission Neutral Vehicle (ENV) were produced in 2005, by Intelligent Energy, a UK-based firm. The ENV motorcycles hold enough fuel to run for four hours, and to travel 100 miles in an urban area, at a top speed of 50 mph; in 2004, Honda also developed a fuel-cell motorcycle which utilised the Honda FC Stack.

- *Fuel Cell Airplane & Submarines:* In February 2008, Europe-wide tests were conducted on Type 212 submarines that used fuel cells to remain submerged for weeks without the need to surface; and manned test airplane or Fuel Cell Demonstrator Airplane (FCDA) powered only by a Proton Exchange Membrane (PEM) fuel cell and lithium-ion and lightweight battery hybrid system - to power an electric motor, which was coupled to a conventional propeller.

- *Revolve Eco-Rally:* In 2007, Britain's Prince Charles launched the Revolve Eco-Rally demonstration of several fuel cell vehicles on British roads, driven by British celebrities and dignitaries from Brighton to London's Trafalgar Square. The fuel cell vehicles (rally was preparatory to the 2008 Formula Zero Championships which began 22 August in Rotterdam) were designed and built by university students from around the world. Fuel cell vehicle designs included: Greenchoice Forze, University of Delfl, Holland and winner of the competition; Element One, Detroit; HerUCLAs, U.S; EURLATecH2, Spain; Imperial Racing Green, UK; and Zero Emission Racing Team, Holland.

- *California and Japan Hydrogen Highway:* The U.S., California Hydrogen Highway (CHH) is an initiative to implement a series of hydrogen refuelling stations along the State, to refuel hydrogen vehicles, such as fuel cell- and hydrogen combustion vehicles. As of July 2007, California had 179 fuel cell vehicles and 25 fuelling stations, with 10 more planned; Japan also has a hydrogen highway as part of its Hydrogen Fuel Cell Project, with 12 hydrogen fuel stations already built in 11 cities so far; Canada, Sweden, Norway and Denmark, have hydrogen highways implemented.

- *Fuel Cell Prototype Cars:* There are numerous prototype or production cars and buses based on fuel cell technology being researched or manufactured by motor or car manufacturers. For example, in 2008, Honda released a hydrogen vehicle, the FCX Clarity; there also exist other examples of bikes and bicycles with hydrogen fuel cell engine; a few companies are conducting hydrogen fuel cell research and practical fuel cell bus trials; Daimler AG, with 36 experimental units powered by Ballard Power Systems fuel cells, in January 2007, completed a successful three-year trial in 11 cities; there are also fuel cell powered buses currently active or in production, such as a fleet of Thor buses with UTC Power fuel cells in California, U.S., operated by SunLine Transit Agency; The FuelBus Club is a good global cooperatrive effort in trial fuel cell buses.

- *First Brazilian Hydrogen Fuel Cell Bus:* The Brazilian Omnibus Brasileiro a Hidrogenio (Brazilian Hydrogen Autobus) project, will begin the first Brazilian hydrogen fuel cell bus prototype in Caxias do Sul; the hydrogen fuel will be produced in Sao Bernardo do Campo, including three additional buses, from water through electrolysis.

A *Fuel Cell* is an electrochemical conversion device. It consists of *electrolyte* or *catalyst, fuel, anode* and *cathode* sides: *electricity* is produced from fuel on the anode side, and an *oxidant* is produced from the cathode side; while the reaction products (electricity and oxidant) flow out of the device, the *electrolyte* or *catalyst* and *fuel* remain within it. Fuel cells can operate virtually continuously as long as the necessary flows are maintained. The electrolyte or catalyst typically comprises a *platinum* group metal or *alloy;* a typical fuel cell produces a voltage between 0.6V-7V, which decreases as the current increases, due to: activation loss or *overpotential; ohmic loss* or drop due to resistance of the cell components and interconnects; *mass transit loss* or depletion of reactants at catalyst sites under high loads that cause rapid loss of voltage. Fuel cells can be combined in series and parallel circuits to produce the desired energy amounts.

The different fuel cell types[202] and hydrogen-oxygen designs include:

- *Proton Exchange Fuel Cells* (PEMFCs) or *Polymer Electrolyte Membrane* (PEM): These are being developed for transport-, stationary-, and portable-fuel cell applications. Their distinguishing features are lower temperature and pressure ranges (50-100°C) and a special polymer electrolyte membrane.

- *Direct Methanol Fuel Cells* (DMFCs), *Indirect-Methanol Fuel Cells* (IMFCs), *Refined Methanol Fuel Cells* (RMFCs): These are subcategory systems of the proton-exchange fuels cells (PEMFCs) or polymer electrolyte membrane (PEM), where the methanol (CH^3OH) is *not reformed* but fed directly to the fuel cell operating ca. 90-120°C (in the case of DMFCs/PEM), and methanol is *reformed* before being fed into the fuel cell (in the case of IMFC/RMFC).

- *Membrane Electrode Assembly* (MEA): This is an assembled stack of proton exchange membranes (PEMs) or semi-permeable membranes generally made from *ionomers* and designed to conduct protons while being impermeable to gases such as oxygen or hydrogen. They separate reactants and transport of protons.

- *Alkali Anion Exchange Membrane* (AAEMs): These are assembled stacks of proton exchange membranes (PEMs) or membrane electrode assembly (MEA), the catalysts and electrodes used in fuel cells, which act as sandwiches between two electrodes which have the catalyst embedded in them.

- *Oxygen Ion Exchange Fuel Cells* (OIEFCs): These are electrochemical conversion devices that produce electricity directly from oxidizing a fuel in a solid oxide fuel cell (SOF). The anode and

---

[202] Types of Fuel Cells: *Metal hydride* / electrolyte: aqueous alkaline solution (e.g., potassium hydroxide); *Electro-galvanic* / electrolyte: aqueous alkaline solution (e.g., potassium hydroxide); *Direct formic acid* (DFAC) / electrolyte: polymer membrane (ionomer); *Zinc-air battery* / electrolyte: aqueous alkaline solution (e.g., potassium hydroxide); *Microbial* / electrolyte: polymer membrane or humic acid (from dark brown major constituents of organic matter humus soil); *Upflow microbial fuel cell* (UMFC) under research; *Regenerative* / electrolyte: polymer membrane (ionomer); *Direct borohydride* / electrolyte: aqueous alkaline solution (e.g., sodium hydroxide); *Alkaline* / electrolyte: aqueous alkaline solution (e.g., potassium hydroxide); *Direct methanol* / electrolyte: polymer membrane (ionomer); *Reformed methanol* / electrolyte: polymer membrane (ionomer); *Direct methanol* / electrolyte: polymer membrane (ionomer); *Direct formic acid* / electrolyte: polymer membrane (ionomer); *Proton exchange membrane* / elctrolyte: polymer membrane (ionomer), e.g., Nafion or Polybenzimidazole fibre); *RFC-Redox* / electrolyte: liquid electrolytes with redox shuttle & polymer membrane (ionomer); *Phosphoric acid* / electrolyte: molten phosphoric acid (H^3PO4); *Molten carbonate* / electrolyte: molten carbonate (e.g., sodium bicarbonate – $NaHCO^3$); *Tubular solid oxide fuel cell* (TSOFC) / electrolyte: O^2-conducting ceramic oxide (e.g., Zirconium dioxide – ZrO^2); *Protonic ceramic* / electrolyte: H+-conducting ceramic oxide; *Direct carbon* / electrolyte: several different; *Planar Solid oxide* / electrolyte: O^2-conducting ceramic oxide (e.g., zirconium oxide – ZrO^2 Lanthanum Nickel Oxide La^2XO4, X=Ni, Co, Cu).

cathode are separated by an electrolyte that is conducive to oxygen ions but not non-conducive to electrons. The electrolyte is usually made from zirconia doped with yttria.

- *Stationary- and Portable-Fuel Cell Applications:* Fuel cell portable and stationary applications are very important for the supply of stored hydrogen energy to generate electricity: portable or movable fuel cell power stations, are used as micropower in consumer electronic devices to provide power or as portable power in emergency power stations for critical areas; stationary or not moving fuel cell connected to the electric grid, to provide supplemental power as emergency power systems for critical areas, or installed as a grid-independent generator for on-site service.

- *Emergency Power Systems:* Stationary and portable fuel cells and batteries service emergency power station systems with the needed electric power. These emergency power systems may include lighting, generators and other apparatus, that provide backup resources in a crisis or when regular systems fail. Locations that need these services include hospitals, scientific laboratories, data centres, telecommunication equipment and modern naval ships. Emergency power stations rely on generators, deep cycle batteries, flywheel emergency storage or hydrogen cells. Some homebrew emergency power systems also use regular lead-acid car batteries, but these do not make a very efficient or reliable system.

A *Battery* is an array of *electrochemical cell* or *thermodynamically* open system for electricity storage or one such cell; contrary to a fuel cell, a battery consumes the *reactant* from an external source, which must be replenished. A battery also stores electrical energy chemically. A battery consists of one or two cells used in a variety of portable and non-portable appliances as a mains power supplies, backups, and sometimes as alternatives to the mains. A *Galvanic* or voltaic cell is part of a battery consisting of an electrochemical cell with two different metals connected by a salt bridge or porous disk between the individual half-cells; a *Flow cell* is a rechargeable battery in which electrolyte containing one or more dissolved electroactive species flows through an electrochemical cell that converts chemical energy directly to electricity. Batteries in common usage - all made of chemicals and chemical combinations - and their common usages, include:[203]

- *Zinc-carbon, Zinc-chloride, Zinc-air, Alkaline,* : Light-drain-heavy applications; hearing aids (Zinc-air).

- *Silver-oxide:* Hearing aids; watches;calculators.

- *Lithium-Thionyl:* Industrial applications; electric meters; other devices that contain volatile memory circuits, wireless gast water meters.

- *Mercury:* Now manufactured only for specialist applications due to mercury's toxicity;

---

[203] Battery Cells: <u>Primary Cells:</u> *Alkaline* or alkaline cell; *Aluminium* or aluminium-air; *Atomic* or nuclear or tritium and radioisotope generator; *Bunsen cell;Chromic acid cell; Clark cell; Daniel cell* or gravity cell or crowfoot cell; *Germ cell; Leclanche cell; Lemon; Lithium; Mercury* or mercuric oxide or mercury cell; *Molten salt; Optoelectric nuclear; Organic radical battery* (ORB); *Oxyride; Paper; Silver-oxide* or Silver-zinc; *Voltaic pile; Western cell; Zinc-air; Zinc-carbon dry cell* or battery; <u>Rechargeable Battery or Secondary Cells:</u> *Alkaline; Lead-acid; Lthium ion polymer* or Polymer lithium ion or Lithium polymer; *Lithium-sulphur* or galvanic cell; *Lithium sulphure* or galvanic cell; *Molten salt;Nickel-cadmium; Nickel-iron; Nickel hydrogen; Nickel metal hybrid; Nickel zinc Rechargeable alkaline; Sodium-sulphur; Super-sulphur;Super-iron;Super Charge ion* (SCIB); *Vanadium redox* and *redox flow; Zinc-bomine flow.*

[mercury was formerly used digital watches, radio communications, portable electric equipment.]

- *Thermal, Water-activated:* High temperature reserve exclusively for military applications; Radiosomes and emergency applications.

- *Nickel oxyhydroxide:* Ideal for applications that use bursts of high currents, such as digital cameras.

- *Paper:* Designed to function as both a lithium-ion battery, and a supercapacitor, using ionic liquids (essentially a salt), as electrolyte. The sheets can be rolled, twisted, folded or cut into numerous shapes with no loss of integrity or or efficiency, or stacked like printer paper or a voltaic pile, to boost total output. It can be made in a variety of sizes from postage stamp to broadsheet. Their lightweight and low cost make them attractive for portable electronics, aircraft and automobiles, while their ability to use electrolytes is blood, make them potentially useful for medical devices such as pacemakers;they are biodegradable.

Nano technologies or nanotech or nanoscience, greatly improve upon battery, hydrogen and fuel cell technologies. Nanotech studies the control of matter on an atomic and molecular scale, and deals with structures of the size 100 *nanometers* or smaller, and involves developing materials or devices within that size. The diverse and extensive range of nanotech includes novel extensions of conventional device physics; completely new approaches based upon molecular self-assembly; developing new materials with dimensions on the *nanoscale;* and speculation on whether matter can be directly controlled on the atomic scale. Nanotech has the potential to create many new materials and devices with wide-ranging applications, such as in medicine, electronics and energy production.

Nanotech has its dark side: concern about toxicity and environmental impact, and potential effects on global economics of *nanomaterials, nanoparticles* and *nanotubes* - all made from naturally-occurring or artificially manufactured elements. Elements from which nanomaterials are made include: *diesel soot; carbon; carbon black; titanium dioxide; alumina; zinc oxide; asbestos; quartz; reactive oxygen species* (ROS); *fullerenes;* and *nanoparticles* or *nanocrystals.* Fullerenes are a class of *carbon allotropes*[204] or graphene sheets, similar to a graphite structure rolled into tubes or spheres which include *carbon nanotubes.*

A *graphite structure* is composed of stacked sheets of linked hexagonal and pentagonal rings. Fullerenes have been researched as able to bind specific antibiotics to the structure of resistant bacteria, can target certain types of cancer, or be used as light-activated antimicrobial agents, heat resistance and superconductors. Commonly, to produce fullerenes, a large current is sent between two nearby graphite electrodes in an inert atmosphere. The resulting *carbon plasma arc* between the electrodes is cooled into *sooty residue* from which many fullerenes can be isolated. Spherical fullerenes are called *buckyballs,* cylindrical ones, *carbon nanotubes* or *buckytubes.*

---

[204] Carbon Allotropes: These are molecules composed entirely of carbon in the form of a hollow sphere, ellipsoid, tube, or plane.

Nanoparticles or *nanocrystals* are made of metals, semiconductors, or oxides, divided into *combustive-derived* (like diesel soot), *manufactured* (like carbon nanotubes), and *naturally-occurring* (like volcanic eruptions, atmospheric chemistry). These have mechanical, electrical, magnetic, optical, chemical and other properties; they have been used in *quantum dots*[205] or *semiconductors* whose excitons are confined in all three spatial dimensions, and as chemical catalysts. They are a bridge between bulk materials and atomic or molecular structures, and exhibit a number of unexpected visual specific properties relative to bulk materials that have constant physical propperties regardless of size.

The marvels and drawbacks of nanotech processes include:

- *Benefits of Nanotechnology:* Improved manufacturing methods, Water purification systems, and Energy systems; Physical enhancement (i.e., an attempt to temporarily or permanently overcome cureent limitations of the human body through natural physical means, or select or alter human characteristics and capabilities, whether or not the alteration results in characteristics and capabilities that lie beyond the existing human reach) - the test being whether the technology is used for non-therapeutic purposes. Some bioethicists restrict the term to the non-therapeutic applications of specific technologies like *neuro-, gene-,* and *nano-technologies,* and *human biology*; Nanomedicine: use of nanomaterials, nanoelectronic biosensors, and even future applications of molecular nanotechnology; Better food production methods and nutrition; Large-scale infrastructure auto-fabrication; Products made with nanotechnology may require little labor, land, or maintenance, be highly productive, low in cost, and have modest requirements for materials and energy.

- *Risks of Nanotoxicology:* The risks of using nanomaterials and their inherent toxicity are huge: *Environmental, health,* and *safety issues* are the consequences, if negative effects of nanoparticles are overlooked before they are released, or if the safety of workers in nanotech industries is insufficient. *Diesel nanoparticles,* for example, have been found to damage the *cardiovascular system* in a mouse model. Because of their extremely small size, nanomaterials can much more readily gain entry into the human body than larger-sized particles do. Nanoparticles are able to cross biological membranes, penetrate the skin and access cells and blood stream following inhalation or injestion, and tissues and organs that larger-sized properties normally cannot. Even larger *microparticles* may penetrate skin when it is flexed; broken skin is an ineffectual and ineffective particle barrier, through which even *acne, eczema, shaving wounds,* or *severe sunburn* may accelerate skin uptake of nanomaterials. Once in the blood stream, nanomaterials can be transported around the body and be taken up by organs and tissues, including the *brain, heart, liver, intestines, testis, kidneys, spleen, lungs, gut, bone marrow, cell cultures* and *nervous system,* resulting in increased oxidative stress, inflamatory cytokine production and cell death; *Nanotoxicity* and *particle toxicology:* the study of nanotoxicity or nanotoxicology, addresses the toxicology of nanoparticles above 100 diameters, which cannot appear to have some unusual toxic effects not seen with larger particles, and tend to be different properties from larger particles that are known to have *pathogenic* effects, like asbestos or quartz. Nanomaterials also influence toxicity: chemical composition, shape, surface structure, surface charge, aggregation and solubility, and the presence or absence of functional groups of other chemicals.

- *Nanotubes, Carbon Nanotubes:* Are members of the fullerene structural family, a class of carbon allotropes, which include the spherical buckyballs. Carbon allotropes are molecules

---

[205] Quantum Dots: Have properties that are between those of bulk semiconductors and those of discrete molecules. They are used in transistors, solar cells, LEDs and diode lasers, or as agents for medical imaging and as *quibits* (quantum binary digits) in computers.

composed entirely of carbon in the form of a hollow sphere, ellipsoid, tube, or plane. Nanotubes are so called because of their size as the diameter of a nanotube is on the order of a few nanometers (i.e., approx. *1/50,000th* of the width of a human hair), and they can be up to several millimeters in length; they have been constructed with length-to-diameter ratio of up to 28,00:1, which is significantly larger than any other material. Nanotubes are classed as *single-walled nanotubes* (SWNTs) and *multi-walled nanotubes* (MWNTs), and their nature of bonding is described by applied quantum chemistry's *orbital hybridisation*. Chemical bonding of nanotubes is by sp bonds similar to those of graphite. Cylindrical carbon molecules have moved properties that make them potentially useful in many applications in nanotechnology, *electronics, optics, architecture* and other fields of materials science. Diamond is one best example of *Carbon Nanotubes* (CNTs) or *buckytubes* whose hardness and high dispersion of light makes it useful for industrial applications and jewellery. Diamond is the hardest known natural mineral, and an excellent abrasive to hold polish and luster extremely well. No known naturally-occurring substance can scratch or cut a cylindrical nanostructured diamond.

- *New Nanomaterials and Device Physics:* These are materials with *morphological* features smaller than a *one-tenth* ($1/10^{th}$ nm) of a micrometer in at least one dimension. *Device physics* produces electronic components and semiconductors that exploit the electronic properties of semiconductor materials, principally *silicon, germanium,* and *gallium arsenide.* Semiconductor devices use electronic conduction in the solid state as opposed to the gaseous state or *thermionic* emission in a high vacuum, and thus effectively replacing *thermionic* devices or *vacuum tubes* in most applications. Semiconductor devices are both *single-* and *integrated circuits* (ICs), which consist of a few to millions of devices manufactured and interconnected on a single semiconductor substrate.

- *Direct Contol Matter on the Atomic Scale :* Molecular nanotechnology (MTN) is concerned with the process by which molecules adopt a defined arrangement without guidance or management from outside sources, quite distinct from nanoscale materials, but uses nanomachines.

Hydrogen fuel cell and nano technologies greatly advance and further the uses of renewable energies, sustainable food production and cost-effective agricultural and transportation systems.

3.8 New Transport, Communication Technologies

IMPROVED AND advanced transportation and communications oil the chain of better food production methods, and environmentally- and health-friendly energy production, management and consumption. Intelligent transportation systems (ITS), Information technology (IT), Information systems (ISs), all combine to make human life and our planet to remain functional and healthier: Transportation infrastructure vehicles, loads, and routes reduce vehicle ware, transportation times and fuel consumption; design, development, and implementation, support or manage computer-based information systems like software application and computer hardware, keeping records and data, and manual and automated processes, to thus quicken the pace of information gathering; recent advances in vehicle electronics have impacted on more computer processors on vehicles; and sensing technologies advance safety capabilities.

Intelligent Transportation Systems (ITS) apply in car navigation; traffic signal control; container management; variable message signs; automatic number recognition or speed monitoring cameras, such as security CCTV-Closed-circuit television for surveillance in areas that may need monitoring, such as banks, casinos, airports, military installations, convenience stores; more advanced applications that integrate live data and feedback from a number of other sources, such as parking guidance and information; weather information; bridge de- or anti-icing frozen contaminant, snow, ice, slush from a surface; electronic toll collection; emergency vehicle notification; autonomous cruise control system (ACC); automatic road zones with congesting pricing; automatic road enforcement; video vehicle detection; inductive loop detection; collision avoidance; floating car- and floating cellular-data; artificial intelligence for machines and computers; artificial passenger; driverless cars, smart cars; advanced modelling and comparison with historical baseline data; wireless communications; and intelligent dynamic traffic light sequence – as briefly described:

- *Computers, Computer Software:* All advanced and intelligent transportation and communication systems rely heavily on computers and computer software programmes. The computer[206] is a machine that manipulates data according to a *set of instructions, using tiny, intergrated circuits.* Embedded computers found in devices like fighter aircraft, industrial robots, digital cameras, toys, airraft, ships, submarines, etc., have the ability to store and execute lists of sophisticated *programmed* software application instructions; these instructions make these concealed computers extremely versatile to carry out their computational tasks. These instructions include model-based process control, artificial intelligence, and ubiquitous computing.

- *Car Navigation, Floating Cellular Data, Video Vehicle Detection:* When mobile phones are built in cars, they routinely transmit their location information to the nextwork, even when no voice connection is established, and are thus used as anonymous traffic probes. Such a car's movement allows the measuring and analysing triangulation network data anonymously, and the data is connected into accurate traffic information. Because no infrastructure is necessary for this to happen, this floating car data technology provides great advantages over existing methods of traffic measurement: it is less expensive than sensor cameras, has more coverage, is faster to set up with less maintenance, and works in all weather conditions, including heavy rainfall. *Video vehicle detection* using cameras also automatically detects traffic flow measurement, and does not require any extra installing of components into the road surface. A move towards vehicle infrastructure integration VII centres around the 5.9 HHz frequency (802.11.x WAVE).

- *Automatic Road Enforcement-, Number Recognition-, Road Cordon-Zones:* A traffic enforcement camera system is used to detect and identify vehicles disobeying a speed limit, some other road legal requirement, or vehicle- and number plate-recognition, and automatically charges ticket offenders through a toll-using electronic toll collection (ETC). Most ETC systems were previously based on using radio devices in vehicles, but ETC protocols have more recently been standardised around the Dedicated Short Range Communications

---

[206] Early Computers: The first electronic computers were the size of a large room and consumed as much power as several hundred modern personal computers (PCs). Computers are programmed to run using various computer languages. Computer langaues are designed to permit no ambiguity but to be precise; they are purely written languages and some often difficult to read aloud, generally translated into *machine language* or *machine code:* a system of instructions and data executed directly by a computer's *central processing unit* (CPU); machine code may be regarded as primitive and combersome or as the lowest-level representation of a compiled and/or assembled computer programme. Commonly used Assembly languages: commonly used computer languages include: ARM – Acorn RISC Machine, (then Advanced RISC) designed by ARM Limited/Holdings; MIPS (originally an acrornym for Microprocessor without Interlocked Pipeline Stages) is a reduced set of computing (RISC) instruction set architecture (ISA) developed by MIPS Computer Systems/Technologies; 86 assembly (x86) architecture; Commonly used High level languages: Ada – a structured, statistically typed, imperative, and object-oriented high-level computer programming language; BASIC or Beginners All-purpose Symbolic Instruction Code; C – a general-purpose computer programming languae; C++ - or C Plus (pronounced six plus); C# - (pronounced C Sharp); COBOL – Common Business-Oriented Language; Fortran – (previously FORTRAN); Java; Lisp or LISP; Pascal; Object Pascal, a derivative of Pascal; Commonly used Scripting languages: Bourne script; Javascript; Python; Ruby;PHP; Perl.

(DSRC) protocols promoted for vehicle safety by the Intelligent Transportation Society of America (ITSA), ERTICO-ITS Europe, and ITS Japan. Traffic tickets are sent by mail. The various cameras used to identify tresspassing vehicles include: *speed cameras* against vehicles travelling over the legal speed limit, many such devices using radar to detect a vehicle speed or electromagnetic loops buried in each lane of the road; *red light cameras* that detect vehicles that cross a stopline or designated stopping place where a red light traffic light is showing; *bus lane cameras* that identify vehicles travelling in lanes reserved for buses; *level crossing cameras* that identify vehicles crossing railways at grade illegally; *double white line cameras* that identify vehicles crossing lines; *high-occupancy vehicle* (HOV) *lane cameras* that identify vehicles violating HOV requirements; *turn cameras* at intersections where specific turns are prohibited on red, mostly used in cities or heavy populated areas. For dynamic traffic light sequence, intelligent RFID (i.e., indestructible infrastructure sensors devices, such as reflectors installed or embedded on the road or surrounding the road), has been developed, to circumvent or avoid the problems that usually arise with systems which use image processing and beam interruption techniques.

- *Bridge De- or Anti-Icing:* A process to remove or protect contaminants, snow, ice, or slush from bridge surfaces, accomplished either mechanically through spraying and pushing, or by use of chemicals such as salts, alcohols, glycols, or by a combination of these both techniques. De-icing fluids are always applied heated and diluted; anti-icing applies a viscous fluid over a surface to absorb the contaminant, and all anti-icing fluids offer only limited protection upon frozen contaminant surfaces depending upon type and precipitation rate.

- *Artificial Intelligence for Machines, Artificial Passenger:* Artificial Intelligence (AI) technologies aim to create AI or intelligent agents (IAs) for machines, which will perceive their environment, and take actions which maximise their chances of success. The *Intelligent Passenger* (IP), is a telematic device developed by IBM to interact verbally to reduce the livelihood of a vehicle operator falling asleep at the controls. The IP is equipped to engage a vehicle operator by carrying on conversations, playing verbal games, controlling the vehicle's stereo system, monitors the driver's speech patterns to detect fatigue, and in response can suggest that the driver take a break or get some sleep. IP is also integrated with wireless services to provide weather and road information, driving directions, etc.

- *Smart-, Driverless-Cars:* Artificial intelligence (AI) technologies in conjunction with *virtual reality* (VR) technologies, have expanded to produce driverless or autonomous vehicles or smart cars that can drive themselves from one point to another without assistance from a driver. VR allows a user to interact with a computer-stimulated environment, whether that environment is a simulation of the real world or an imaginary world. Most current VR environments are primarily visual experiences, displayed either on a computer screen or through special or stereoscopic displays; some simulations are additional sensory information, such as sound through speakers or earphones; some advanced, haptic systems now are tactile information or feedback in medical and gaming applications. Users can interact with a virtual environment or a virtual artificial (VA) either through the use of standard input devices such as a keyboard and mouse, or through multimodial devices such as a wired glove, the Polhemus boom arm, and omnidirectional treadmill. A *smart car* is an automation automobile with AI functionality that uses integrated circuits. An example of a smart car is the TV series *Knight Rider,* where *KITT* (Knight Industries Two Thousand and Knight Industries Three Thousand) car – is an artificial intelligence electronic computer-installed in a highly advanced, very mobile, robot in the form of a 1982 Pontiac Firebird Trans Am. KITT was redesigned in the two-hour pilot movie with a new name, Knight Industries Three Thousand, and revamped even more than the new series aired. KITT is now a 2008 Ford Shelby GT500KR Mustang, equipped with new features. A number of real-life KITT-archetypal real-life vehicles have been designed, built and sold that incorporate AI technology: Mercedes-Benz Robot cars that have led to the development of their S-Class, which is a series of vehicles, with radar-controlled braking system or Brake Assistant Plus and the intelligent cruise control function; and EUREKA Prometheus Project on autononous vehicles (1987-1995); Agetthere passenger vehicles using Netherland's FROG-navigation technology); Italy's ARGO research project; and U.S.'s DARPA Grand Challenge.

- *Autonomous, Adaptive Cruise Control* (ACC): These AI systems of smart and driverless cars use either a *radar* or a *laser* set up to allow the vehicle to slow when approaching another vehicle and accelerate again to the present speed when traffic allows. Examples of ACC vehicles are:[207]

- *Advance Modelling:* This process uses *Computer-aided design* (CAD) or *Computer-aided geometric design* (CAGD) to design objects, real or virtual; models for object shapes; and symbolic information such as materials, processes, dimensions, and tolerances, according to specific applicable and specific conventions, such as curves and figures in two-dimensional ("2D") space; or curves, surfaces, or solids in three-dimensional ("3D") objects. *Geometric models* use applied mathematics and computational geometry or *algorithms* to mathematically describe shapes. Both CAD and CAGD processes are widely applied in automative, shipbuilding, aerospace, architecture and postethetics, movies, advertising, computer graphics, and technical manual industries.

- *Wireless Communications:* Transportation systems use various forms of *wireless* technologies, among them: *short-range* (less than 500 yards) using IEEE 802 II protocols, a set of standards carrying out wireless local area network– WLAN-computer communication in 2.4, 3.6 and 5 GHz frequency bands - promoted by the Dedicated Short Range Communications (DSRC) standard which is being promoted by the Intelligent Transportation Society of America (ITSA) and the U.S., Department of Transportation. The range of these protocols can be extended using Mobile ad-hoc networks (MANET or mobile mesh network, a self-configuring network of devices connected by wireless links), or Mesh networking, a way to route data, voice and instructions between nodes, which allows for continuous connections and reconfiguration around broken or blocked paths by "hopping" from node to node until the destination is reached; *longer range* communications, using infrastructure networks such as WiMAX or World-wide Interoperability for Microwave Access (WIMA), a telecommmunications technology that provides wireless transmission of data using a variety of transmission nodes, from point-to-multipoint links to portable and fully mobile internet. The technology provides up to 3 Mbits/s broadband speed without the need for cables, based on the IEEE 802.16 standard or Broadband Wireless Access (IEEE 802.16), Global System for Mobile Communications (GSM), the most popular standard for mobile phones in the world, or 3G (International Mobile Telecommunications or 3rd Generation – a family of standards for wireless communications defined by the International Communications Union (ITU), which includes EDGE, CDMA2000, the UMTS family and the DECT and WiMAX. Long-range communications using these methods are well established, but unlike the short-range protocols, these methods require extensive and very extensive infrastructure deployment.

Even the improved, versatile and intelligent transportation and communications - cannot save us from our crumbling and poisoned environment; or save us from high food prices or hiked energy prices; or save us from an atmosphere unabatedly blanketed with new greenhouse gases.

3.9 Piezoelectricity Technologies

---

[207] Autonomous or Adaptive Cruise Control (ACC) Vehicles Examples : 2005 Acura RL ; Audi A4, A5, A6, A8, Q7; BMW7 Series, 5 series, 6 series, 3 series (Active Cruise Control); 2004 Cadillac DTS, STS, XLR; 2007 Chrysler 300C; 2006 Ford Mondeo, Taurus, S-Max, Legend; 2003 Honda Inspire Accord, Legend; Hyundai Genesis (Smart Cruise Control delayed); Infinity M, Q45, QX56, G35, FX35/45/50 and G37; 1999 Jaguar XK-R, S-Type, XJ, XF; 2000 Lexus LS430/460 (laser and radar), RX (laser and radar), GS, IS, ES 350, and LX570; Lincoln MKS, MKT; 1998 Nissan Cima, Nissan Primera T-Spec Models (Intelligent Cruise Control); 1998 Mercedes-benz S-Class, E-Class, CLS-Class, SL-Class, M-Class, GL-Class,CLK-Class; Range Rover Sport; Renault Vel Satis; Subaru Legacy & Outback Japan-spec called SI-Cruise; 1997 Toyota Celsior, Sienna (XLE Limited Edition), Avalon, sequoia (Platinum Edition), Prius; Volkswagen Passat, Phaeton, Tuareg; Volvo S80, V70, XC70, XC60.

PIEZOELECTRICITY IS the current harvested or *squeezed* from free power energy which hangs in and around us and especially in crystals, and certain ceramics and bones. Materials that exhibit direct *piezoelectric effect* can also produce piezoelectricity in the reverse direction when stress is applied, as for example: in the *detection of sound; generation of high voltages; electronic frequency generation; microbalances;* and *ultrafine focusing of optical assemblies.* Applying stress on objects with the piezoelectricity effect has also formed the basis of a number of instrumental techniques with *atomic resolution,* and *scanning probe microscopies* such as *STM, AFM, MTA, SNOM,* etc.[208] Direct piezoelectricity is used in applications such as *cigarette lighter ignition; transformers; powered cell phones; sensors; actuators; motors; laser mirror adjustments; accusto-optic modulators; inkjet printers; fuel injectors; X-Ray shutters; XY stages for micro scanning in infrared cameras; quartz clocks,* etc., as follows:

- *Sensors: Sensors* use different modes to load the piezoelectric element to be used in longitudinal, transversal and shear, as well as detection of pressure vibrations in the form of sound, as for examle: in *piezo microphones,* where soundwaves bend the piezoelectric material, creating a changing voltage and piezoelectric picksups for electric guitars; a *piezo sensor* or contact microphone attached to the body of an instrument; *piezo sensors* used with high frequency sound in ultra sound *transducers* for medical imaging and for *industrial non-destructive testing* (NDT); *ultrasonic transducers* that inject ultrasound waves into the body, receive the returned wave, and convert it into an electrical signal or voltage; most medical ultrasound transducers are piezoelectric; piezoelectric elements are used in the detection and generation of *sonar waves;* piezoelectricity is used in *power monitoring in high power applications,* as in medical treatments, *sonochemistry* and industrial processing; piezos are sometimes used in *strain gauges;* piezoelectric transducers are used in *electronic drum pads* to detect the impact of the drum's sticks; *automotive engine management* systems use piezoelectric transducers to detect detonation by sampling the vibrations of the engine block and also to detect the precise moment of fuel injection or *needle lift sensors; ultrasonic piezo sensors* are used in the detection of accoustic emission in accoustic emission testing; *crystal earpieces* are sometimes used in old or low power radios.

- *Actuators:* These very high electric fields *piezo crystals* correspond to only tiny changes in width of the crystal, which width can be changed with better-than-micrometer precision, making them the most important tool for positioning objects with extreme accuracy – thus their use in actuators. For example: *multiplayer ceramics* use layers thinner than 100 microns, allowing reading of high electric fields with voltage lower than 150V. Multilayer ceramics are used within two kinds of actuators: *direct piezo actuators* and *amplified piezo actuators; loud speaker* voltages are connected to mechanical movement of piezoelectric *polymer film.*

- *Frequency Standard:* Quartz clocks use the piezoelectrical properties to employ a *tuning fork*

---

[208] Scanning Probe Microscopies Technology Include: *Scanning Tunneling Micrscopy* (STM): a powerful technique for viewing surfaces at the atomic level, used in the ultra high vacuum, in air and various other liquid or gas ambients, and at temperatures ranging from zero kelvin to a few hundred degrees Celcius; *Atomic Force Microscope* (AFM) or *Scanning Force Microscope* (SFM): a very high resolution type of scanning probe microscopy, with demonstrated resolution of fractions of nanometer, more than 1000 times than the optical diffraction limit, used for imaging, measuring and manipulating of matter at the nanoscale; *Microthermal Analysis* (MTA): a materials charactarisation technique which combines the thermal analysis principles of Differential Scanning Calorimetry (DSC) with high spatial resolution of scanning probe microscopy; *Near-field Scanning Optical Microscopy* (NSOM/SNOM): a microscopy technique for nanostructure investigation that breaks the far-field resolution limit by exploiting the properties of evanescent waves (a near-field standing wave with an intensity that exhibits exponential decay with distance from the boundary at which the wave was formed.

made from quartz that uses both direct and converse piezoelectricity to generate a regularly timed series of electrical pulses that is used to mark time. The quartz crystal is an elastic material that has a precisely defined natural frequency (caused by its shape and size) at which it prefers to oscillate. The oscillation stabilises the frequency of a periodic voltage applied to the crystal; all radio trasnmitters and receivers and computers usually use a *frequency multiplier* to reach the megahertz and gigahertz ranges.

- *Piezoelectric Motors:* Piezoelectric motors have extremely small distances involved which apply directional force to an axle, to cause the axle to rotate, and is viewed as the stepper motor high precision replacement; piezoelectric elements can be used in *laser mirror alignment,* where their ability to move large masses or *mirror mountain* over microscopic distances, because by precisely controlling the distance between mirrors, the laser electronics can accurately maintain optical conditions inside the laser cavity to optimise the beam output; an *accousto-optic modulator* is a device generated by piezoelectric elements that scatters light off soundwaves in a crystal for fine-tuning a laser's frequency; *Converse piezoelectricy* is used by atomic force- and scanning-microscopes to keep the sensing needle close to the probe; *Inkjet printers* like Epson use piezoelectric crystals to control the flow of ink from the inkjet head to the paper; Common high-performance rail diesel engines use *piezoelectric fuel injectors* instead of the more common *solenoid valve* devices; types of piezoelectric motors, all (except *stick*-slip) use working on *driven by dual orthogonal vibration modes* with a phase shift of 90°, the contact point between two surfaces vibrating in elleptical path: *travelling-wave* used for auto-focus in reflex cameras; *inch worm* for linear motion, *rectangular four-quadrant* with high power density (2.5watt/c³) and speed ranging from 10nm/s to 800 mm/s; *steppin piezo* using stick-slip effect.

Objects that can generate piezoelectricity include:

- rystals or *Crystalline Solids, Ceramics, Bones: Crystals* or *Crystalline Solids* are solid materials whose constituent atoms and molecules or ions are arranged in an orderly repeating pattern extending in all three spatial dimensions. *Ceramics* are inorganic, non-metallic solids prepared by the action of heat and subsequent cooling, and may have crystalline or partly crystalline structure, or may be amorphous, for example, glass. Glass is both crystalline and non-crystalline. *Bones* are rigid organs that form part of the endoskeleten of vertebrates whose function is to move, support and protect the various organs of the body, produce red and white blood cells and store minerals. Bone tissue is a type of dense connective tissue.

- *Polyvinylidene Fluoride* (PVDF)*, Sodium Potassium Niobate* (KNN)*, Bismuth Ferrite* (BiFeO³): The PVDF polymer's intertwined long-chain molecules attract piezoelectricity and repel other when an electric field is applied several times greater than quartz, which is a crystal. *KNN* is a replacement for lead-containing piezoelectric devices because of their toxicity with its properties closer to *lead zirconate titanate* (Pb [ZrnTi]) or PZT. PZT is a ceramic perovskite material that shows a marked piezoelectric effect. BiFeO³ could replace lead-based ceramics with their toxicity.

3.10 Conclusion

UNLIMITED ALTERNATIVE renewable, non-polluting energies other than polluting fossil energies – surround us, coupled with healthy food production and less-polluting transportation methods. The cost-factor to enable us to reorient our food, energy, transportation, and agriculture methods cannot be the main hindrance. Political will is the main hindrance. Solar, wind; thermal and geothermal; biomass; water; hydrogen, fuel cell, nano; and, new transportation and communication technologies, are capable to perform and guarantee us full success.

Chapter 4

FOOD MANIPULATION, BIOTECHNOLOGISATION

4.1 Introduction: What You Eat, Drink, Metabolise

FOOD, CROP, plant and deoxyribonucleic acid (DNA) gene manipulation through biotechnology has become a staple in the 21st century. Changes in food and drink flavor, taste, smell, and texture are tailored artificially or synthetically. It is like searching for a needle in a hay stack trying to determine the nature and content of *what you eat, drink, breathe and metabolise.* The reason for this, is because all food has been manipulated on or tampered with, one way or another before it reaches your table. The standard reasons for tailoring our foods are *"increased yields, increased market share, increased food globalisation, increased food-sufficiency, and of course, increased profits."* Agriculture, transportation, and energy prices, and thus corporate profits, are increased through food manipulation.

Normal-foods-drinks are those produced on natural soil using normal *multicultural* agricultural methods *excluding* any man-made food chemicals, which include: *artificial additives, sweetners, preservatives; synthetic fertilizers, pesticides, herbicides, insecticides; genetically-engineered foods, crops, plants; animal growth promoters or plant growth hormones, and antibiotics. Super-foods-drinks,* and *gene-manipulated-foods-crops,* are those foods and drinks produced using *monocultural* agricultural methosds *including* man-made food chemicals. Food labelling which is decisive to wise- and economical-shopping, and safety to distinguish between *normal* from *artificial* foods and drinks - is mostly not available or accessible.

And selecting and purchasing what foods and drinks to eat and drink has proved a very risky enterprise indeed: many foods are *deliberately* not labelled; others are labelled but *improperly* describe what its real contents are; others are labelled but the labelling *differs* with what actually is contained in the packaging; some manufacturers simply *refuse* to label their foods for fear their foods would not be purchased; some labels are *misleading* or provide *half-truths* or *half-lies*; some manufacturers are *litigating against* those manufacturers who label their food products *truthfully* in attempts to *force labelling not to occur at all*; food labelling regulators are under severe pressure from *aggressive* and *unscrpulous* food lobbyists *against regulating truthfully*, or to *regulate half-truthfully* or *postpone regulating* and *kill* the regulating process with *stretched-out-delays*.

4.2 Super-foods, Artificial-foods, Gene-foods

NORMAL-FOODS and normal-drinks are simply normal-foods and normal-drinks, eaten, drunk or injected by humans, animals or plants, for nutrition or for pleasure. Foods and drinks comprise *carbohydrates, fats, proteins* and *water;* their sources are plants, crops, trees, animals, fungi and bacteria. Some simple diets of normal-foods based on *staple foods* with little variation, are often deficient in certain *nutrients.* Such deficient nutrients are either not present in sufficient amounts in the soils of a region that produces the staple food, or because of the inherent inadeqaucy of the diet. *Food fortification* policies concentrate on adding *micronutrients*[209]or essential trace elements and vitamins, to foodstuffs to ensure that minimum dietary requirements are met.

---

[209] Micronutrients, Macronutrients and Vitamins: Macronutients, Micronutrients and Vitamins are minerals and acids essential for human and animal diets in trace quantities to be complete, healthy and nutritious. *Macronutrients* or *Macrominerals,* include: *carbon, hydrogen, nitrogen, oxygen, phosporous, sulphur, carbohydrates, proteins, fats, water, calcium, salts (sodium+ chloride), magnesium; Micronutrients* or *Microminerals* or *Trace elements,* include: *iron, cobalt, chromium, copper, iodine, manganese, selenium,zinc, molybdenum.; Vitamins* and *DNA Metabolites* (former Vitamins) and their Deficiencies include: *A – Retinoids* (retinol, retinoids, carotenoids) against Night-blindness, Keratomalacia; *B1 – Thiamine,* against Beriberi, Wernicke-Korsakoff syndrome; *B2 – Riboflavin,* against Ariboflavinosis; *B3 – Niacin, Niacinamide,* against Pellagra; *B5 – Pantothenic acid,* against Paresthesis; *B6 – Pyriodoxine, Pyridoxamine, Pyridoxal,* against Anemia peripheral neuropathy; *B7 – Biotin,* against Dermatitis, enteritis; *B9 – Folic acid, folinic acid,* against Deficiency during preganancy associated with birth defects, such as neutral tube

Super-foods include *superfruits, superdrinks, energy drinks, processed-* or *fortified foods* containing high *phytonutrient* or *phytochemicals* which are believed to confer health benefits to consumers. Phytochemicals are chemical compounds such as *beta-carotene,* that occur naturally in plants. Blueberries are considered a super-food fruit because they contain significant amounts of *anthocyanins,*[210] antioxidants, vitamin C, manganese and *dietary fibre. Super-drinks* or *energy-drinks* are generally beverages made up of super-fruit and super-food juices with no sugar added, and contain high antioxidant values. Producers claim that antioxidants contained in super-drinks can protect the body against oxidative damage and can result in stronger immune resistance to flues, viruses and infections; greatly reduce incidence of all cancers; prevent glacoma and muscular degeneration; reduced risk of cholesterol-oxidation, and heart disease and anti-aging cells and overall body.

Energy-drinks (advertised as *soft-drinks)* are supposed to increase a user's mental alertness and physical performance by the addition of *caffeine,*[211] *vitamins,* and *herbal supplements* which may interact to provide a stimulant effect over and above that obtained from caffeine alone. *Super-fruits* refer to fruits which combine exceptional nutrient richness and anti-oxidant quality with appealing taste that can stimulate and retain loyalty for consumer products. Antioxidants are also widely used in *dietary supplements* or *food supplements* or *nutritional supplements* intended to provide nutrients, such as vitamins, minerals, fibre, fatty acids or *amino acids,* that are missing or are not consumed in sufficient quantity in a person's diets. Super-fruits include a growing list of common fruits such as *strawberries, blackberries, blackcurrants, oranges,* which are developed or marketed as *functional foods* or *medicinal foods.* Functional foods claim to have health-promoting or disease-preventing properties beyond the basic functions of supplying nutrients, such as *fortified water. Processed-* or *fortified food* contains harvested crops or slaughtered and butchered animal products, deadly fungi fish - made into attractive, marketable and often long-life food products.

Adverse effects adduced to *energy-drinks* (also advertised as *soft-drinks*) consumption, include:

•
Excessive Caffeine Intake: The consumption of a *single* energy beverage will not lead to excessive caffeine intake, but consumption of two or more a *day* can. Many stimulants, such

defects; *B12 – Cyanocobalamin, Hydroxcobalamin, Methylcobalamin,* against Megaloblastic anemia; *C – Ascorbic acid,* against Scurvy; *D- Ergocalciferol, Cholecalciferol,* against Rickets, Osteomalacia; *E – Tocopherols, Tocotrienols,* against Deficiency very rare; mild hermolytic anemai in newborn infants; *K – Phylloquinone, Menaquinones,* against Bleeding diathesis; *Adenine* or *DNA Metabolite* (former B4); *Adenylic acid* or *DNA Metabolite* (former B8); *Essential fatty acids* (former F) needed in large quantities; *Catechol, Flavin* or *Protein Metabolite* (former J); *Anthranilic acid* or *Protein Metabolite* (former L1); *Adenylthiomethylpentose* or *RNA Metabolite* (former L2); *Carnitine* or *Protein Metabolite* (former O); *S-Methylmethionnine* or *Protein Metabolite* (former U).

[210] Anthocyanins: Are water-soluble vacuolar pigments that occur in all tissues of higher plants, including leaves, stems, roots, flowers, and fruits. Derivatives of anthocyanins include pendant sugars; *Antioxidants* are molecules capable of slowing down or preventingthe oxidation of other molecules. Oxidation is a chemical reaction that transfers electrons from a substance to an oxidizing agent, but can produce free radicals, which start chain reactions that damage cells, which antioxidants terminate.

[211] Caffeine: Is a bitter, psychoactive, white crystalline xanthine alcoloid stimulant, found in, and can be extracted from, *beans, coffee plant* cherries, *tea bush* leaves, *kola nuts, yerba mate, guarana* berries, and *Yaupon Holly.* Caffeine is contained in coffee, tea, soft drinks, and enrgy drinks.

as *ginseng* which are added to caffeine, and other caffeine-containing substances such as *guarana*, concentrate the content of caffeine. Caffeine consumption in amounts greater than 400mg, cause *nervousness, irritability, sleeplessness, increased urination, abnormal heat rhythms,* and *stomach upset.*

- *Concentration of Sugar:* High concentrations of sugar in sports drink or in everyday drinks, beyond the recommended 6-7% carbohydrate which allows maximum absorption and minimise spikes and crashes in blood sugar – will slow fluid absorption into the blood and energy system, increasing the possibility of dehydration. A high level of sugar in the blood stream hinders the body from getting water into the cells that the cells need, because the water is busy trying to dilute concentrations of sugar in the blood stream.

- *Soft Drinks, Carbonated Soft Drinks: Soft Drinks* and *Carbonated Soft Drinks* are beverages that contain no alcohol. Soft Drinks include: *colas, flavored water, sparkling water, iced tea, lemonade, squash, fruit punch,* while *hot chocolate, hot tea, coffee, milk, tap water, juice* and *milkshakes* are not soft drinks; *Carbonated Soft Drinks* include: *soda, soda pop, pop, coke, cola* or *tonic water, cooldrink, colddrink, fizzy drink, soft drink, fizzy drinks, minerals;* many carbonated soft drinks are optionally available in versions sweetened with sugars or with non-caloric sweetners.

- *Abnormal Heart Rhythms, Danger to Expecting Mothers:* Energy drinks in the U.S., have been linked to *nausea, heart rhythms* and *emergency room visits,* as the drinks often cause seizures due to the "crash" following the energy high that occurs after consumption. In the U.S., caffeine dosage is not required to be on the product label for food. The death of an 18-year-old Irish athlete Ross Cooney in France in 2008, let to the banning of the energy drink *Red Bull.* Cooney died as a result of playing a basketball game after consuming four cans of Red Bull, which has excessive amounts of caffeine; Britain, instead of banning the consumption of Red Bull energy drink, issued a warning against its use by pregnant women and children. Denmark also banned the use of Red Bull.

Normal food additives such as *salt, spices, herbs,* and *sulfites* have been used since ancient times to preserve foods and make them more palatable. But *artificial food additives* are any of various chemical substances added to foods, especially processed foods, to produce specific desirable effects. Many modern processed foods such as *low-calorie snack* and *ready-to-eat convenience foods,* are only *possible* foods because of artificial food additives and preservatives.

Artificial food additives include: *additives, sweetners, preservatives, colorings, flavorings, smell flavorants; hydrogenated foods, trans fats, shortenings; soft drinks, non-diet soft drinks, gelatin,* and *natural food dyes.* These artificialities are added to normal-foods and normal-drinks to change their color, taste, smell, appetite, and of course, increase profitability. Encyclopedia Britannica describes *Food Coloring* thus:

- *Food coloring is any of numerous dyes, pigments, or other additives used to enhance the appearance of fresh and processed foods. Coloring ingredients, including natural colors, derived primarily from vegetable sources and sometimes called vegetable dyes; inorganic pigments; combinations of organic and metallic compounds (called lakes); and synthetic coal-tar substances. They are added to orange and potato skins, sausage casings, baked goods, candies, carbonated drinks, gelatin desserts, powdered drink mixes, and many other foods. Many of these additives are also employed as coloring agents in cosmetics, drugs, and products such as toothpastes and mouthwash."*

For example, in brief:

- *Acids, Acidity Regulators, Tracer Gas: Food acids* are added to food to make flavors *sharper*, and also act as *preservatives* and *antioxidants.* Common food acids include *vinegar, citric acid, tartaric acid, malic acid, fumic acid, lactic acid; Acidity Regulators* are used to change or otherwise control acidity and alkalinity of foods; *Tracer Gas* allows for package integrity to prevent foods from being exposed to atmosphere, thus guaranteeing shelf life.

- *Anticaking Agents, Antifoaming Agents, Antioxidents, Preservatives, Humectants, Fermented Foods: Anticaking Agents* keep powders such as milk powder from caking or sticking; *Antifoaming Agents* reduce or prevent foaming in foods; *Antioxidants* such as *vitamin C* act as preservatives by inhibiting the effect of oxygen on food, and can be beneficial to health; *Preservatives* prevent or inhibit spoilage of food due to *fungi, bacteria* and other microorganisms; *Humectants* prevent foods from drying out; *Fermented Foods* are preserved through the fermentation of sugar to alcohol using *yeast,* and the making of *yogurt, pickled* or *soured foods* are fermented, or simply processed with *brine* or *vinegar.*

- *Bulking Agents, Emulsifiers, Stabilizers, Thickeners, Flour Treatment Agents: Bulking Agents* such as *starch,* are additives that increase the bulk of a food without affecting its nutritional value; *Emulsifiers* allow water and oils to remain mixed together in an emulsion, as in *mayonnaise, ice cream* and *homenized milk; Stabilizers, thickeners* and *gelling agents,* like *agar* or *pectin* (used in jam), give foods a firmer texture, and also help to stabilize emulsifiers; *Thickeners* are substances added to mixtures to increase the *viscosity* without substantially modifying its color properties.

- *Food Coloring,, Color Retention Agents, Flour Treatment Agents: Colorings* are added to food to replace colors lost during preparation, or to make food look more attractive; *Coloration Retention Agents* are used to preserve a food's existing color, and the agents are often used for restoring cow's milk to its natural white color as unhealthy cows often bleed into the milk; *Flour Treatment Agents* are added to flour to improve its color or its use in baking.

- *Flavors, Flavor Enhancers, Sweetners, Jellies & Gelatin Desserts: Flavors* are additives that give food a particular taste or smell, and may be derived from natural ingredients or created artificially; *Flavor Enhancers* enhance a food's existing flavors, and may be extracted from natural sources through distillation, solvent extraction, maceration, or created artificially; *Sweetners* are added to foods for flavoring. Sweetners other than *sugar* are added to keep the food energy or *calories* low, or because they have beneficial effects for *diabetes mellitus,* tooth decay and *diarrhoea; Jellies* are the most common culinary use of *gelatin desserts,* in which unprepared gelatin for desserts often marketed as a *flavored powder* or *concentrated gelatinous solid. Prepared gelatin desserts* are marketed in a variety of forms, such as in Kraft Foods, Royal, Hartley's (formerly Roundtree), Aeroplane Jelly.

For example, briefly expanded:

- *Taste, Artificial Sweeteners: Flavorants like* common table salt and sugar enhance salty and sweet tastes. *Artificial Sweeteners* or secondary flavors are *taste flavors* or are flavorants. Artificial sweeteners are mostly sugar substitutes, or food additives that duplicate the effect of sugar in taste, with no food energy. Some sugar substitutes are *natural* and some are *synthetic* or *artificial sweetners,* like *high-intensity sweetners* which are many times sweeter than *sucrose* or *common table sugar.* The sensation of sweetness conveyed by these artificial sweetners often used in complex mixtures, is notably different from sucrose. A bulking agent is often used additionally, if the artificial sweetner has contributed to the texture of the product artificially sweetened., as for example in *soft drinks* labelled as *diet* or *light.* Five intensely-sweet sugar substitutes artificially-synthetised compounds approved for consumption in the U.S., include *saccharin, aspartame, sucralose, neotame, acesulfame pottassium, stevia* (a herbal supplement), *sorbitol, xylitol* (both found in berries, fruits, vegetables and mushrooms), and low-cost *high-fructose corn syrup* (HFCS) sweetner; some *non-sugar artificial sweetners* are *polyols* or *alcohols,* which are generally less sweet than sucrose but have similar bulk properties, and are used in a wide range of foods products.

- *Flavors, Smell Flavorants, Natural Flavoring-, Nature-identical Flavoring-, Artificial Flavoring- Substances: Smell Falvorants* or *Flavorants* are engineered and composed in similar ways as with all industrial fragrances and *fine perfumes,* extracted from the source substance through *solvent extraction, distillation,* or *force* to squeeze it out. Flavor manufacturers first find out the individual naturally-occurring aroma chemicals and mix them appropriately to produce a designed flavor or create a novel non-toxic artificial compound that gives a specific flavor. Flavorists formulate flavors to either imitate or enhance a food product's unique flavor and maintain flavor consistency between different product batches or affect recipe changes. *Flavor substances* give other substances flavor, alter the characteristics of the solute, and cause the solute to become *sweet, sour, scenty, tasty* or *tangy.* Flavorings alter or enhance the flavor of natural foods such as meats and vegetables, or create flavor for food products that do not have the desired flavors such as *candies* and other snacks. The three principal flavoring substances used in foods include: *Natural Flavoring* – obtained from plant or animal raw materials by physical, microbiological or enzymatic processes, and can be used in their natural state or processed for human consumption, but cannot contain any nature-identical or artificial flavoring substances; *Nature-identical Flavorings* – are obtained from chemically-identical flavoring substances naturally present in products intended for human consumption, through synthesis or isolated chemical processes; *Artificial Flavoring* – are not identified in a natural product intended for human consumption, processed or not processed.

- *Food Colouring, Natural Food Dyes:* Fewer people realise that seemingly "natural" foods, such as oranges and salmon are sometimes also dyed to mask natural variations in color effects processing and storage and is commercially advantageous, effectively to: *offset color loss* due to light, air, extremes of temperature, moisture and storage conditions; *mask natural variations* in color; enhance *naturally-occurring colors;* protect flavors and *vitamins* from damage by light; and for *decorative* or *artistic* purposes such as cake icing; *Natural Food Dyes* are produced commercially as a response to consumer concerns on *synthetic dyes,* which include: *Caramel coloring* made from caramelised sugar, used in *cola* products and cosmetics; *Annanto,* a reddish-orange dye made from the seed of the *Achiote* shrub; a *Green dye* made from *Chlorella algae; Cochineal,* a red dye derived from the cochineal insect *Dactylopius coccus; Betanin,* extracted from beets; *Tumeric,* a rhizomateous herbaceous perennial plant of the *ginger* family *Zingiberaceae; Saffron,* a spice derived from the flower of the *Saffron crocus; Paprika,* a spice made from the grinding of dried fruits of *Capsicum annual* (e.g., *bell-* or *chili-peppers); Elderberry, sambacus* or *elder* or *elderberry,* a genus of shrub species or small trees in the *Moschatel* family *Adoxaceae; Brilliant Blue* (BBG) or *Coomasiedyes* or *Coomasie Brilliant Dyes,* are a family of dyes commonly used to stain proteins in *Sodium dodecyl sulfate* and blue native *Polyacrylamide gel electrophoresis* (SDS-PAGE and BV-PAGE, respectively) gels.

- *Natural Food Preservation:* Preservatives are natural or synthetic chemicals added to products, such as foods, pharmaceuticals, paints, biological samples, wood, to prevent decomposition by microbial growth or by undesirable chemical changes. *Natural Food Preservatives,* such as *salt, sugar, vinegar, diatomaceous earth* or *DE, TSS, diatomite, diahydro, kieselguhr* or *kieselgur, Celatom* or *celite* (a naturally-occurring, soft, chalky-like sedimentary rock) are used to preserve foods naturally. Food preservation processes such as *freezing, pickling, smoking* and *salting* are also used. *Citric-* and *ascorbic-acids* from lemon or other citric juice can inhibit the action of the enzyme *phenolase* which turns surfaces of cut apples and potatoes brown. *Boiling* or *heating, oxidation* or use of sulfur dioxide, *toxic inhibition* or smoking or *use of carbon dioxide, vinegar, alcohol;* Drying *or Dehydration;* Osmotic inhibition or *use of syrups, Low temperature in action* or *freezing; ultra high water pressure* or a *fresherised* kind of "cold" or *Pasteurization; vacuum packing; lye; canning; bottling; irradiation; modified atmosphere, burial in the ground; controlled use of microorganism -* kill or denature organism and preserve the foods.

- *Preservatives in Foods:* Preservation in foods are either *antimicrobial,* which inhibit the growth of bacteria and fungi, or *antioxidant* such as *oxygen absorbers,* which inhibit the oidation of food constituents. Common *microbial preservatives* include: *calcium propionate, sodium nitrate, sodium nitrite, sulfies* (sulphur or sulfur dioxide, sodium bisulfite, potassium hydrogen sulfite), and *disodium – Ethylenediaminetetracetic acid –* EDTA *Antioxidants,* BHA – *Butylated hydroxyanisole,* and BHT – *Butylated hydroxytolluene* or *Butylhydroxytoluene); Formaldehyde* (usually in solution); *Glutaraldehyde* (kills insects); *Ethanol* and *Methylchloroisothiazolinone.*

- *Preservatives in Wood:* Preservatives in wood – are added to wood to preserve the growth of fungi, repel insects and termites, and prolong the life of engineered woods, woods, timber or wood structures or increase their durability and resistance; *arsenic, copper, chromium, borate,* and petroleum-based chemical compounds, are commonly applied wood preservatives. *Engineered* or *composite wood* or *man-made wood* include: a range of *derivative wood* products which are manufactured by binding the *strands, particles, fibres* or *veneers* of wood, together with *adhesives,* to form *composite materials* or *composites.* Engineered woods *plywood, hardwoods, softwoods, sawmill scraps, cellusosic* products from *ligni-*containing materials such as *rye strain, wheat straw, rice straw, hemp stalks, kenaf stalks, sugaecane residue,* are used to manufacture lumber. *Adhesives* used in engineered wood, include: *Urea-formaldehyde resins* (UF), which is most common, most cheap and not waterproof; *Phenol-formaldehyde resins* (PF), yellow/brown, and most commonly used for exterior exposure products; *Melamine-formaldehyde resin* (MF), white, heat and water resistant, and often used in exposed surfaces in more costly designs; *Methylene diphenyl diisocyanate* (MDI) or *Polyurethane* (PU) resins, which are expensive, generally waterproof and do not contain formaldehyde.

- *Hydrogenated Foods, Trans Fats, Shortenings: Hydrogenation* is widely applied to the processing of vegetable oils and fats, or convert liquid vegetable oils to solid or semi-solid fats, such as those present in *margarine.* Hydrogen (H^2) is added to organic substances to reduce or saturate them, using catalysts, or *non-catalytic hydrogen* to take place at very high temperatures. *Complete hydrogenation* converts unsaturated fatty acids to saturated ones. *Partially-hydrogenated* vegetable oils or fats are cheaper than *animal source fats,* and are thus available in a wide range of *consistencies,* or have other desirable characteristics, such as *increased oxidative stability* or *longer shelf-life,* or *fat blends* or shortenings; *Shortenings* are semi-solid fat used in food preparation, especially in baked goods, and promotes *short* or *crumbly texture,* such as in short-bread. Shortening is basically just fat or *lard* from an animal or vegetable, used broadly in *butter, lard* and *margarine.* Semi-solid fats are preferred for baking because the way fat mixes with flour produces a more desirable texture in the baked product; *Trans Fats* or *Unsaturated fat* may be *monosaturated* or *polysaturated* but never *saturated. Vaccenic acid* is a class of trans fats which occurs naturally in trace amounts in meat and dairy products from *ruminants.*

- *Fertilizers:* Bryant Walsh[212] summarises the state of the world's food supply, chemical fertilizers, food safety, food-related or food-caused maladies, and irreversible problems confronting us: *"...The U.S., agricultural industry can now produce unlimited quantities of meat and grains at remarkably cheap prices. But it does so at a high cost to the environment, animals and humans. Those hidden prices are the creeping erosion of our fertile farmland,*

---

[212] Bryan Walsh: *"Getting Real About the High Price of Cheap Food."* CNN/TIME, Friday 20 August 2009.

cages for egg-laying chickens so packed that the birds can't even raise their wings and the scary rise of antibiotic-resistant bacteria among farm animals. Add to the price tag the acceleration of global warming – our energy-intensive food system uses 19% of U.S., fossil fuels, more than any other sector of the economy...And perhaps worst of all, our food is increasingly bad for us, even dangerous. A series of recalls involving contaminated foods this year – including an outbreak of salmonella from tainted peanuts that killed at least eight people and sickened 600 – has consumers worried about the safety of their meals - a principal cause of America's obesity epidemic...No one doubts the power of chemical fertilizers to pull more crop from a field. American farmers now produce an astounding 152 bu. of corn per acre, up from 118 as recently as 1990. But the quantity of that fertilizer is flat-out scary: more than 10 million tons for corn alone – and nearly 23 million for all crops. When runoff from the fields of the Midwest [where the corn crop is produced] reaches the Gulf of Mexico, it contributes to what's known as a dead zone, a seasonal, approx. 6,000 sq mi area that has almost no oxygen and therefore almost no sea life. Because of the dead zone, the $2.8 billion Gulf of Mexico fishing industry loses 212,000 metric tons of seafood a year, and around the world, there are nearly 400 similar dead zones. Even as we produce more high-fat, high-calorie foods, we destroy one of our leanest and healthiest sources of protein...".

- *Growth Hormones – Synthetic plant hormones (PGRs) or Plant growth regulators (PGRs) or phytohormones, are chemicals that regulate a plant's growth, such as beta-carotene that occur naturally in plants. Recombinant bovine growth hormone (rBGH)/recombiant bovine somatotropin (rBST), an artificial and synthetic growth hormone for cows, and insulin-like growth factor-1 (IFG-1) that induces cows to be more productive, manufactured by Monsanto,* and found in U.S., milk, cheese and yogurt; growth hormones are either *synthetic steroid* or *naturally-occurring*, used in meat and beef production, for example: *oestradiol, progesterone* (natural female sex hormones), *zeranol, trenbolene, acetate, melengesterol acetate, estrogen* (synthetic or artificial growth promoters), *diethylstibestrol* (DES), a synthetic estrogen drug, and *testerone*.

- *Antibiotics, Antibiotic Resistance:* Antibiotic is a substance or compound that kills or inhibits the growth of bacteria, and belongs to the border group of *antimicrobial* compounds used to treat infections caused by *microorganisms, bacteria, viruses, fungi* and *protozoa*. Most antibiotics are presently semi-synthetic, modified chemically from original compounds in nature, such as *beta-lactams*, including *penicillium, cophalosporins*, and carbapenems; most are produced or isolated from living organisms, such as *aminoglycosides;* others have been created through purely synthetic means, such as *sulfonamides*, and *oxazolidinous*, and are classified into two groups according to their effect on microoorganisms: *bacterial agents* – kill bacteria, for example, *penicillin, erythromycin; bacteriostatic agents* – impair the growth of bacteria; Antibiotics are prescribed orally, intravenously or topically, and are frequently *non-therapeutically* used as growth promoters when fed to cattle to promote growth and fatness; Indiscriminate *over-use* and *misuse* of antibiotics to promote animal or plant growth and cures for diseases, have led to *antibiotic resistance* by the targeted bacteria, germs, and viruses. *Antibiotic resistant* bacteria use the antibiotics themselves to act as a selective pressure which allows their growth within a population to inhibit susceptible bacteria, and survive on *inheritable resistance* or *inherited mutations*. "Inappropriate antibiotic treatment and *overuse* of antibiotics have been a contributing factor to the emergence of resistant bacteria. The problem is further exacerbated by *self-prescribing* of antibiotics by individuals without the guidelines of a clinician and the *non-therapeutic* use of antibiotics as *growth promoters* in agriculture," writes Paul L. Marino.[213] He adds: "Antibiotics are frequently prescribed for individuals in which their use is not warranted, an incorrect or sub-optimal antibiotics is prescribed in some cases for which infections are likely to resolve without treatment."

Gene-foods or **GE-foods** are foods produced from *genetically-engineered* (GE) or *genetically-altered* (GA) or *genetically-manipulated* (GM) crops, ingredients,

---

[213] Paul L. Marino: *The rule of antibiotics is try not to use them, and the second rule is, try not to use too many of them. "* The ICU Book.

animals and organisms (GMOs), which are generally sold in supermarkets without *labelling*. GE-food crops have not evolved in any natural environment and have *never* before been part of the human, animal or plant diet. Such GE-foods and ingredients include *corn, soy, canola, cotton seed oil, lecithin, soy oil,* and *soy proteins* found in foods. GE-fish and GE-cloned animals are part of the GMOs menu.

Everyone would want to know that what he, she or it eats and drinks is safe for health and well-being. This ensures the further and continued purchasing of those foods to reward ourselves, producers, manufacturers, transporters, supermarkets, sales persons, economies, environments and our planet. Everyone would want to know the content of a food before purchasing it; every producer and manufacturer would want to let the consumers know that the former can be trusted and continue to patronize their services, brands and products by *labelling* their products; we also would want to know and be confident that, what appears on a product's label is indeed, what is inside the described content. The best guarantee for maximum corporate or market profits is *best quality, best safety,* and *best transparency* for consumers' products. The consequences for the use of *antibiotics, growth hormones, pesticides, herbicides, insecticides, artificial colors, flavors, sweetners, smells, trans fats, preservatives, additives, agents, emulsifiers, super-, processed-,* and *hydrogenated-foods,* and *GE-foods* and -crops in farming, food production for human, animal and plant consumption, are most disturbing and disquieting, because:

- *Super-foods, Soft-drinks, Antioxidants:* As of 1 July 2007, the marketing of products as *Super-foods* was banned in the European Union (EU), unless accompanied by a specific medical claim, supported by credible scientific research. Companies will have two years to adjust their marketing in line with these rules by EU standards; oxidation reactions are crucial for life, but they can also be damaging, therefore plants and animals maintain complex systems of multiple types of *antioxidants,* such as *catalase, superoxide dismutase* and various *peroxides,* and low levels of antioxidants, or inhibition of the antioxidant enzymes – to avoid *oxidative stress;* oxidative stress may lead to damage or kill cells. *Oxidative stress* might be an important part of many human diseases, and the use of antioxidants in pharmacology is studied for treatments for *stroke* and *neuro-degenerative* diseases; many soft-drinks contain food additives as *food coloring, emulsifiers,* and *preservatives,* while more than 10 teaspoons of added sugars for a 2,000-calorie diet, approved U.S., Food and Drug Administration (FDA's) is *recommended daily allotment* (RDA). Consider replacing *soft-drinks* with other healthier choices in diets, such as *water, milk* and *fruit juice.*

- *Artificial Flavors, Sweetners:* Artificial flavors may be safer to consume than natural flavors due to the standards of purity and mixture consistency that are enforced either by law or by the company. Natural flavors may contain toxins from their sources, while artificial flavors are typically more pure and are required to undergo more testing before being sold for consumption. Flavors from blood products are usually the result of a combination of natural flavors, which set up the basic smell profile of a food product, while artificial flavors modify the smell to accent it; *High fructose corn syrup* (HFCS), is used exclusively as a sweetner because of its low cost; but HFCS is known to have detrimental effects on human health, such as promoting *diabetes, hyperactivity, hypertension,* and a host of other problems.

- *Hydrogenated Foods, Trans Fats:* A side-effect of *incomplete hydrogenation* of fats having implications for human health, is the *isomerization* of the remaining *unsaturated* carbon bonds which cis[214] configuration predominate in the unprocessed fats in most edible fat

---

[214] Cis and Trans: Refer to the arrangement of chains of carbon atoms across the double bonds, where in the cis arrangement, the

sources. Incomplete hydrogenation partially converts these molecules to *trans isomers,* which have been implicated in circulatory diseases including heart disease; *Trans fats* (unlike dietary fats), are not essential, and they do not promote good health. The consumption of trans fats increases one's coronary heart disease by raising levels of "bad" *low-density lipoprotein* (LDL) cholesterol and lowering levels of "good" *high-density lipoprotein* (HDL) cholesterol. Consumption of trans fats, especially LDL, *intermediate-density lipoprotein* (IDL), and *very-low-density lipoprotein* (VLDL) should be reduced or avoided altogether as trans fats from *partially hydrogenated* oils are more harmful than naturally-occurring oils.

- *Food Dyes, Preservatives:* Some modern *synthetic preservatives* have beocme controversial because they have shown to cause *respiratory* or some other health problems. New studies point to *attention-deficit hyperactivity disorders* (ADHD), attributed to *artificial food dyes, synthetic preservatives* and *artificial coloring agents* that aggravate *attention-deficit disorders* (ADD) and ADHD symptoms. Several major studies show academic performance increased and disciplinary problems decreased in large non-ADD student populations when artificial ingredients, and artificial coloring were eliminated from school food programmes. *Allergenic preservatives* in food or medicine can cause *anaphylactic shock* (i.e., acute systemic or multisystem and severe type I *hypersensitivity allergic* reactions in humans and other mammals) in susceptible individuals. This is a condition which is often fatal within minutes without emergency treatment. Norway banned all products containing *coal-tar* and *coal-tar derivatives* in 1978. Many approved U.S., food colorings have been banned by EU regulations: *Tartrazine,* a lemon yellow azo dye, that causes *hives* or *urticaria,* a kind of skin rash notable for dark red, raised, itchy bumps – frequently through allergic and non-allergic reactions; *Erythrosine,* a cherry-pink synthetic fluorone food coloring used in printing inks, biological stain, dental plaque disclosing agent, radiopaque medium, sweets and foods marketed to children such as sweets and candies, popsicles, cake icing and frosting, cake decorating gels – is linked to *thyroid tumors* in rats; *Cochineal* or *carmine,* derived from insects, has been known to cause severe, even life-threatening, allergic reactions in rare cases.

- *Growth Hormones:* Artificial growth hormones or growth promoters fed animals and plants are chemicals that make make cattle, livestocks, sheep, poultry and plants to grow fatter and faster or to produce milk and eggs faster. *Insulin-like growth factor-1* (IFG-1) which is increased in Monsanto's *recombinant bovine growth hormone* (rBGH)/*recombinant bovine somatotropin* (rBST) growth hormones-treated cows, has been found by Canadian and European regulators to survive digestion, and could find its way into the intestines and blood stream of consumers. The studies demonstrate that IFG-1 is an important factor in the growth of *cancers of the breast, prostrate* and *colon.* The U.S., FDA has approved the consumption of rBGH/rBST cow milk trade-named *Posilac* and refused to have such rBGH/rBST-milk to be *labelled,* despite increasing consumer protests.

- *Labelling Foods:* In the U.S., where GE-foods are allowed to be sold and consumed, no labels to identify GE-foods content are allowed. The U.S., Food and Drug Administration (FDA) that approves food for sale and consumption accepts only "voluntary" labelling. The FDA forbids companies that do "voluntary" labelling from labelling GE-foods at all, by making it difficult for companies to eliminate GE-food ingredients to add "NON-GE" labels; but the FDA lets other companies to continue to use GE-food ingredients in secret. Numerous consumer and environmental organisations and groups hope that the new Obama Administration will reverse and roll-back this former Bush Administration monster regulation.

- *GE-Wood Products:* Engineered wood products require more primary energy for their manufacture than solid lumber; GE-wood products-required adhesives may be toxic, such as some resins often seen with *urea-formaldehyde*-bonded products in the finished products releasing *formaldehyde;* cutting and otherwise working with GE-wood products can expose workers to toxic constituents.

---

chains are on the same side of the double bond, resulting in a kinked geometry; in the trans arrangement, the chains are on the opposite sides of the double bonds, and the chain is straight.

4.3 Gene-Manipulated Agriculture, Economy

SYNTHETIC OR artificial hormones, antibiotics, pesticides, herbicides, food additives, flavors, colorings, preservatives, and genetically-manipulated organisms (GMOs), have seemingly possessed the agricultural and food production industries. The agricultural and food-production sectors are on the brink of becoming totally gene-maipulated. The biotechnology industry and its proponents continue to spin out vicious and egregrious myths aimed at manipulating, confusing or conditioning consumers to blindly consume genetically-engineered foods. The myths also cause fears. According to these myths, biotechnology's genetically-engineered crops and foods will better serve the world to end hunger, malnutrition and want, especially in developing economies, and better protect the environment. The *Center for Food Safety* (CFS)[215] reproduces and rebutes some of these preposterous and manipulative "GE-Crop wonder myths" thus:

- *GE-Crops Myth #1:* 'Genetic Engineering is merely an extension of traditional breeding'.

- *REALITY against Myth #1:* GE is a new technology that has been developed to overcome the limitations of traditional breeding. Traditional breeders have never been capable of crossing fish genes with strawberries. But genetically engineered "fishberries" are already in the field. With GE, these types of new organisms can be created and released into the environment. U.S. Food and Drug Administration (FDA) scientists stated that GE is *different* from traditional breeding, and so are the risks. Despite this warning, the FDA continues to assert that GE foods and crops are not different and didn't require special regulations.

- *GE-Crops Myth #2:* 'GE can make foods better, more nutritious, longer-lasting and better-tasting.'

- *REALITY against Myth #2:* The reason for the 70 million acres of GE crops grown in the U.S., today, has nothing to do with nutrition, flavor or any other consumer benefit. There is little benefit aside from the financial gains reaped by the firms producing GE crops. Nearly all of the GE corn, soy, potatoes and cotton grown in the U.S., has been *genetically-altered* so that it can withstand more pesticides or produce its own.

- *GE-Crops Myth #3:* 'GE Crops eliminate pesticides and are necessary for environmentally sustainable farming.'

- *REALITY against Myth #3:* Farmers who grow GE crops actually use more herbicides, not less. For example Monsanto created *Roundup-Ready* (RR) soy, corn and cotton

---

[215] "The True Food Network: *"Myths & Realities of GE Crops."* A Project of the Center for Food Safety, Center for Food Safety (CFS), a national, non-profit, membership organisation in the U.S., dedicated to protecting and promoting human health and organic and other forms of sustainable agriculture against the use of harmful food production technologies; office@centerforfoodsafety.org.

specifically so that farmers would continue to buy Roundup, the company's best-selling chemical weed killer, which is sold with RR seeds. Instead of reducing pesticide use, one study of more than 8,000 university-based field trials, suggested that farmers who plant RR soy use two to five times more herbicide than non-GE farmers who use *integrated* weed-control methods. GE crops may be the greatest threat to sustainable agriculture on the planet. Many organic farmers rely on a natural bacterial spray to control certain crop pests. The advent of GE insect-resistant crops is likely to lead to insects that are immune to this natural pesticide. When this biological pesticide is rendered ineffective, other farmers will turn increasingly to toxic chemicals to deal with the "superbugs" created by GE crops. Meanwhile, organic farmers will be out of options.

- *'GE-Crops Myth #4: The Government ensures that GE is safe for the environment and human health.'*

- *REALITY against Myth #4:* Neither the FDA, the Department of Agriculture (USDA), nor the Environmental Protection Agency (EPA) has done anything long-term on human health or environmental impact studies of GE foods or crops, nor has any mandatory regulation specific to GE food been established. Biotech companies are on the honor system. They have virtually no requirements to show that this new technology is safe. FDA scientists and doctors warned that GE foods could have new and different risks, such as hidden allergies, increased plant-toxin levels and the potential to hasten the spread of *antibiotic-resistant* disease. The FDA has reviewed more than 5,000 applications for experimental GE field trials without denying a single one. FDA officials claimed that they would conduct long-term studies of GE crops, but *have no plans to require any pre-market or pre-release assessment.* Studies conducted after our environment and food supply have been contaminated will be too late.

- *GE-Crops Myth #5 'There is no scientific evidence that GE foods harm people or the environment.'*

- *REALITY against Myth #5:* There is no long-term study showing that GE foods or crops are safe, yet the biotech industry and government (of the U.S.) have allowed our environment and our families to become *guinea pigs* in these experiments. Doctors around the world have warned that GE foods may cause unexpected health consequences that may take years to develop. Laboratory and field evidence shows that GE crops can harm beneficial insects, damage soils and transfer GE genes in the environment, thereby contaminating neighboring crops and potentially creating uncontrollable weeds.

- *GE-Crops Myth #6: 'GE foods are necessary to feed the developing worlds' growing populations.'*

- *REALITY against Myth #6:* In 1998, African scientists at a UN conference strongly objected to Monsanto's promotional GE campaign that used photos of starving African children under the headline *"Let the Harvest Begin."* The scientists, who represented many of the nations affected by poverty and hunger, said gene technologies would undermine the nations' capabilities to feed themselves by destroying established *diversity, local knowledge* and *sustainable agricultural systems.* GE could actually lead to an increase in hunger and starvation. Biotech companies like Monsanto, force growers to sign technical agreements when growing their patented GE crops which stipulates, among other things, that the farmers *cannot save the seeds produced from their GE harvest.* Half the world's farmers rely on saved seeds to produce food that 1.4 billion people rely on for daily nutrition.

The technology that genetically-engineers crops (GMOs) and produces GE foods, crops, and animals in agriculture and other sectors of industry, is based on the science and technology of digging into, and manipulating or altering *deoxyribonucleic acid* (DNA); the DNA is the *basic unit of heredity* in living organisms and all living things. *Genes* in all living things, including humans, animals, and plants hold the information to build and maintain their cells and pass genetic traits to offspring. A *gene* is a portion of the DNA containing the genetic instructions used in the development and functioning of all known living organisms and some viruses.

DNA molecules store information for the long-term or a *set of blueprint* or *recipe* or *codes* containing instructions needed to construct other components of cells, such as *proteins* and *ribonucleic acid* (RNA) molecules. DNA segments that carry this genetic information are called *genes,* locatable in the region of *genomic sequence* corresponding to a unit of inheritance associated with *regulatory regions, transcribed regions* and other *functional sequence regions.* A DNA consists of *nucleotides* or two-way polymers of simple units with their backbones made of sugars and phosphate groups joined by *ester* or chemical bonds that run anti-parallel in opposite directions to each other; each chemical bond has four types of molecules or *bases.* The *sequence* of these four *bases* along the backbone encodes *information* or *genetic codes,* which specify the sequence of amino acids within proteins. The sequence of amino acids are read by copying stretches of DNA into the related *nucleic acid RNA* or *transcription* process.

DNA is organised within cells into long structures or *chromosomes,* which are duplicated before cells divide or DNA *replication; Cells,* the smallest units of an organism and building block of life, are the structural and functional units of all living organisms. Some organisms, such as *bacteria* are *unicellular* or one-celled; other organisms, such as humans, are *multicellular* or multi-celled, with an estimated 100 trillion cells. Animals, plants, fungi and most *protists* or *eukaryotic* organisms store most of their DNA inside the cell *nucleus* or specialised sub-units or *organnles* of their DNA. Chromosomes store *chromatin* proteins, which are complex combinations of DNA, RNA and protein that make up the chromosomes, such as *histones compact* that organise DNA, guide the interactions between the DNA and other proteins, and help control which of the DNA are transcribed.

The *Genome* is a full set of chromosomes or genes in a cell that fuses with another cell or *gamete* during fertilization in organisms that reproduce sexually; the genome are also regular cells or *somatic cells* that form the body of an organism; the *germline cells* or *gametes* are the spermatozoa and ova in mammals; spermatozoa and ova fuse during fertilization to produce a cell or *zygote* from which the entire mammalian embryo develops and contains two full sets of genomes. The *Human Genome Project* (HGP)[216] determines the *sequence*

---

[216] Human Genome Project (HGP): Is an international scientific research project that determines the sequence of chemical base

of chemical bases pairs which make up the Human DNA, or *Homo sapiens,* stored in 23 chromosome pairs, 22 of these being *autosomal* pairs, while the remaining pair is *sex-determining.* An *autosome* is a chromosome that is not a sex chromosome equal in number of copies in males and females. A *diploid cell* has two sets of chromosomes; the *somatic cells* in humans are diploid, and the gametes or germline cells are haploid. The *haploid* has only two-one set of chromosomes, and in human genome it occupies a total of just 3 billion DNA base pairs, with 20,000-25,000 protein-coding genes.

The genetic engineering of organisms' DNAs to produce GE crops, GE plants, GE fish and animal clones, GE foods and foodstuffs by the biotech industry since the 1990s, has greatly affected the issues of the feeding billions of the worlds poor and malnourished. It has affected many conventional and new industries like agriculture, agri-forestry, agri-food, agri-gardening, agri-fishery, agri-business, nano, fuel and transportation. GE technology companies and organisations have gone to unscrupulous lengths to embellish their questionable methods in, and equate them with, traditional breeding methods, in attempts to quieten the alrms raised against their products or to legitimize their products. The differences between GE technology and traditional breeding methods are quite distinct, thus:

- *Traditional Plant Breeding, Animal Breeding, Mutation Breeding, Inbreeding: Plant Breeding* has been practised for thousands of years near the beginning of human civilisation, and is now practised worldwide by gardeners, farmers, government and university institutions and professional plant breeders. The genetics of plants are changed in this process, through selecting plants with *desirable characteristics* for propagation and the use of more complex molecular techniques, such as *cultigen* and *cultivar.* Plant breeding is for the benefit of humankind; *Animal Breeding* involves breeding animals using the genetic value or *best linear unbiased prediction* or *estimated breeding value* (EBV) of superior domestic livestock in growth rate, egg, meat, milk or wool production. Other desirable traits that have revolutionalised agricultural livestock production, such as population genetics, quantitative genetics and statistics are applied.

- *Selective Breeding:* Selective breeding involves plants and animals for particular genetic traits. Strains which are selectively bred are domesticated. Bred animals are *breeds* and bred plants are *varieties, cultigens* or *cultivars;* and crossed animals are *crossbreeds,* crossed plants are *hybrids.* Selective breeding is synonymous with *artificial selection,* i.e., breeding which is intentional for certain traits or combination of traits; *Mutation Breeding* means changes to the *nucleotide sequence* of the genetic material of an organism, which can be caused by copying errors in the genetic material during cell division, exposure to ultraviolet or ionizing radiation, chemical mutagens, viruses, or induced by the organism itself or by cellular processes, such as *hypermutation.* Mutations in multicellular organisms with dedicated reproductive cells are *germline mutation* – which can be passed on to descendants through the reproductive cells, and *somatic mutations* – which involve cells outside the dedicated reproductive group and are not usually transmitted to descendants. Mutations create variations within the gene pool: *less favourable* or *delecterious mutations* can be reduced in frequency in the gene pool by natural selection; *more favourable* or *beneficial* or *advantageous mutations* may accumulate and result in *adaptive evolutional* changes; *Inbreeding* of plants and animals is done between close relatives, but if practised repeatedly, can lead to exposure of recessive, delecterious traits that generally leads to a decreased fitness of a population or *inbreeding depression.* Inbreeding to "fix" desirable characteristics within a population is followed with *culling* to remove animals from a group based on specific criteria, to either reinforce certain *desirable* characteristics or to remove certain *undesirable* characteristics from the group. Culling implies the killing of animals with undesirable

---

pairs which make up DNA and identify and map the approx. 20,000 – 25,000 genes of the human genome or *Homo sapiens.*

characteristics. Inbred plants are used as stocks for creation of *hybrid lines* to make use of the *hetesis effect*. *Self-pollination* is a natural inbreeding process.

- *Genetic Engineering: Genetically Modified Foods* (GMFs) are produced from genetically-engineered organisms (GMOs), whose DNA have been modified, unlike similar organisms, crops or foods developed through *conventional genetic modification* of *Selective Breeding, Plant Breeding, Animal Breeding* or *Mutation Breeding*. GE-modified foods are *transgenic plant* products created in the laboratory: *soybean, corn, canola, cotton seed, fish, animals, ingredients, seeds* and *milk*. Genetic Engineering is divided into four: *Germline Genetic Engineering* – involves changing genes in eggs, sperm, or very early embryos, and the modified genes are inheritable in any children that resulted from the process and in all succeeding generations. Germline GE is controversial because of its potential to create better athletes, and more beautiful people or small people; *Positive Genetic Engineering* - could be used to cure medical conditions, such as aging, death, drastically change people's genomes which could enable people to regrow limbs and other organs, perhaps even extremely complex ones, such as the spine, or making people stronger, faster smarter, or to increase the capacity of their lungs; *Negative Human Genetic Engineering* – involves the treating of genetic disorders caused by abnormalities in genes or chromosomes or by environmental factors. Some types of *recessive gene* disorders confer an advantage in inherited characteristics or similarly of genes for a trait in certain environments. Geneticists diagnose and treat genetic disorders. Gene therapy uses a non-pathogenic virus or other delivery systems to insert a *good copy of the gene* into the cells of the living individual. The modified cells would divide as normal and each division would produce cells that express the desired trait.

- *Genetic Erosion; Genetic, Biological Pollution:* Genetic erosion is a process whereby an already limited gene pool can cause reduced biological fitness and an increased chance of extinction. Individuals in a population identical with regard to a particular *phenotype* trait are *monomorphic,* and *polymorphic,* if the population has several variants of a particular trait; *Genetic,* and *Biological Pollution* is an *undesirable gene flow* into wild populations – from a GE organism (GMO) to a non-GE or non-GMO, or to a *feral,* domestic, non-native or invasive species. Unlike chemicals that are released into the environment, GMOs are *living things* that will reproduce and spread uncontrollably and at will, with little possibility of containment or cleanup; A *feral* organism is one that has escaped from domestication and returned, partly or wholly, to a wild state. The introduction of feral plants and animals to their non-native regions can disrupt ecosystems and may, in some cases, contribute to extinction of indigenous species. But returning *lost species* to their environment can have the opposite effect, bringing damaged ecosystems back into balance.

- *Transgenic Plants, Somatic Genetic Engineering:* Transgenic plants possess genes that have been transferred from a different species, producing plants and animals created in laboratories using *recombinant DNA* technology: plants and animals are designed with specific characteristics by *artificial insertion of genes* from other species or sometimes entirely different kingdoms; *Somatic Genetic Engineering* involves adding genes to cells other than egg or sperm cells. Somatic GE's distinguishing characteristics are that, it is *non-inheritable,* i.e., the new gene would not be passed to the recipients offspring. If a person had a disease caused by a defective gene, a healthy gene could be added to the affected cells to treat the disorder – presently treated through *gene therapy.*

- *Increased Pesticide Use, Superweeds, Threat to Organic Farming:* Most GE crops have been designed to withstand herbicides. Studies show that farmers who grow GE soybeans use 2-5 times more herbicides than farmers who grow natural soy variations; Other studies have shown that GE crops can *cross-pollinate* with related weeds, resulting in "*superweeds*" that become difficult to control. Canadian canola growers have found weeds in their fields resistant to *Roundup* (RR) and *Liberty herbicides,* forcing the growers to use more potent toxic herbicides; GE insect-resistant crops could create "*superbugs*" which will build up a tolerance to a fundamental pest-control toll used by organic farmers; the loss of this tool would be devastating to the safest, most environmentally-friendly food production we have.

- *Population Bottlenecks:* Population Bottlenecks or *Genetic Bottlenecks* occur where a significant percentage of a population or species is killed or prevented from reproducing. Population bottlenecks increase genetic drift as the rate of drift is intersely proportional to the population size, as well as increase inbreeding due to the reduced pool possible mates. If a small group becomes reproductively separated from the main population, a slightly different sort of genetic bottleneck or *founder effect,* can occur. *Founder effect* is the loss of genetic variation that occurs when a new population is established by a very small number of individuals from a larger population.

- *GE-Fish:* Aquatic ecosystems are impacted and threatened by the GE-fish researches. More than 35 species of GE-fish have been developed, often engineered with *growth hormones* to make the fish grow faster enabling them to reach marketable size at an earlier age, according to Greenpeace.[217] Researchers have shown that GE fish with a growth hormone could have a mating advantage due to their increased size. Contrary to biotech industry claims, *sterliisation* of GE fish will not be 100% effective in a commercial situation, and will *not prevent* crossbreeding between GE fish and wild fish. Several governments and inter-governmental organisations have already taken steps to ensure that any GE fish are kept in secure, land-based facilities. However, escaped GE fish will not respect national boundaries. So far, no GE fish or meat products have been approved for consumption by the U.S., FDA under its *15 January 2009 "Final Guidance on Regulation on GE Animals".* But one GE-fish company in the U.S., *Aqua Bounty technologies,* based in Watham, Massachusetts, has requested approval to market *engineered salmon, trout* and *tilapia,* which he has dubbed *"AquaAdvantage Breeds".* These GE fish can grow 10-30 times faster than the normal salmon. The release of GE-fish can cause potentially devastating environmental and human health impacts: a 2001 *National Academy of Sciences* report states that the release of GE-fish into environment may threaten the survival of wild species; a 2004 *Purdue University* research study contains new experimental data that strengthens the plausibility of the *"Trojan gene effect"* first demonstrated in a 1998 study: by incorporating more biological data into a population model, researchers showed that growth-enhanced GE-fish could lead to the extinction or replacement of wild fish populations within only 40 generations; as with the growing of non-native fish, the use and sale of GE-fish in offshore and ornamental aquaculture facilities poses serious threats to the diversity and well-being of native fish. Already the majority of non-native tropical fish invading Florida's waters are from releases of ornamental fish farms and releases by pet owners.

- *Cloning, GE-Animals:* GE-animals include cloned livestock – a new technique with potentially severe risks for food safety. Defects in clones are common, and cloning scientists warn that even small imbalances in clones can lead to hidden food safety problems in a clone's milk or meat. There are few studies on the risks of food from clones, and no long-term food safety studies have been done. Over 90% of cloning attempts fail, and clones that are born have more health problems and higher mortality rates than sexually reproduced animals. Even Ian Wilmut, the leader of the team of scientists that cloned sheep *Dolly,* agrees that determining the health of impacts of food derived from clones must be based on the animal's complete health profile. Numerous opinion polls show that the majority of Americans do not want food from animal clones and are opposed to cloning on moral or ethical grounds. Yet in in January 2008, the U.S., Food, Drug and Alcohol (FDA) administration, told the American public that the milk and meat from cloned livestock are "safe" for human consumption, with as usual, no labelling allowed!

- *Nanotech, DNA:* Nanotechnology uses molecular recognition or *specific interaction* between two or more molecular properties of DNA and other nucleic acids, to create self-assembling branched DNA complexes, and dimentional structures and devices such as *polyhedra-Nanomechanical, algorithmic self-assembly* - to template the arrangement of other molecules such as *gold nanoparticles* and *streptavin proteins. DNA origami* is the nanoscale folding DNA method used to create arbitrary two- and three-dimensional shapes at the nanoscale.

---

[217] Luke Anderson, Greenpeace GE Campaign: *"Genetically Engineered Fish – New Threats to the Environment."* 05 January 2005. A Briefing that examines the development of GE fish, which could soon be produced on commercial scale.

The biotech industry has almost controlled crops, seeds and food production worldwide, in efforts to *supplant* traditional productions of these valuable items of life with genetically-engineered everything. Virtually every plant food on earth, many of these still in laboratories and greenhouses, are being experimented on. Before GE varieties get near commercialisation, the plants are released into the environment first in field tests, where GE companies can observe the crops as they grow and can analyse them when they are harvested. Such tests are rarely or never done to collect data or determine the environmental effects of GE crops. GE companies instead use these tests to promote and propagate their GE crops to farmers for future commercial sales. *Despite consumer pressure against the unlabelled sale of GE foods, the gene giant industries have not stopped their aggressive push to alter the planet's food supply frombeing genetically-manipulated.* It needs only a few more years for the GE food industry to virtually source all of our foods from GE crops, seeds, ingredients, fish, and meat.

GE crops are developed for certain characteristics or traits, usually properties farmers desire. Two types of GE plants make up nearly all the GE crops in the market: crops engineered for *insect-resistance* (IR) or *bacillus thuringiensis* (BT), and crops that can *tolerate* direct spraying of toxic pesticides or *herbicide tolerant* (HT). These two traits, *IR* or BT and *HT,* make up about 70% of all GE crop acreage in the U.S., and include Monsanto's *Roundup Ready* (RR) crops and plants: soy, corn, cotton, and canola. RR is a herbicide that kills natural crops and plants, but farmers can spray RR crops directly and they will survive, while nearby weeds will be killed. BT crops contain a gene from the bacteria bacillus thuringiensis which is toxic to certain insects. IR, BT, HT, RR crops and plants are pesticides that you cannot wash off. Other GE traits biotech companies are pursuing include *viral, fungi* and *bacterial* resistance so crops can survive plant diseases. Apples are being engineered for delayed ripening, coffee for low-caffeine.

4.4 Crops, Seeds, Food Control

THE CROPS, seeds and food production control efforts by the GE Industry are spearheaded by the multinational agricultural, biotechnology Monsanto Company[218] corporation based in the U.S. Monsanto is the world's leading producer of pesticides and herbicides. Monsanto manufactures the special herbicide *glyphosphate* marked as *"Roundup",* GE seed varieties, and holds a 90% market share for various GE crops. Monsanto owns *Agracetus* which exclusively produces *"Roundup Ready"* (RR) soybean seed for the commercial market and *Seminis Inc,* the world's largest conventional seed company. Monsanto developed and markets GE seeds and the synthetic or artificial cow growth hormone *recombinant bovine somatotropin hormone* (rBST) or *recombinant bovine growth hormone* (rBGH)., with the brandname *"POSILAC Brand Dairy Produce".* Monsanto acquired and owns the *Delta & Pine Land Company* which patented the seed *"Terminator"* or *Terminator* genes or *genetic use restriction technology* (GURT).

---

[218] Monsanto Company: Creve Coeur, Missouri, U.S.; Chairman/President/ CEO: Hugh Grant; CF=: Terrell K. Crews; Chief Technology Officer: Rob Fraley; www.monsanto.com

Terminator produces plants, crops and seeds that possess *sterile seeds* that do not flower or grow fruit after initial planting. Clients who buy and plant these GE Terminator plants, crops and seeds are forced to *repurchase* Terminator seed varieties. Monsanto also sponsors Disney's *American Theme Park* in Anaheim, California[219] and the *Svalbard Global Seed Vault* in Norway.[220]

Monsanto's leading brands of GE crops, seeds and products as appear on its website include:

- *Acceleron:* Seed treatment product; *Asgrow Roundup Ready 2Yield; Dekalb winged ear:* e.g., cornhybrids, alfalfa, grain sorghum, spring & winter canola, sunflower.

- *Deltapine plant breeding programmes:* A diverse base of germplasm for "improved" cotton varieties.

- *De Ruiter Seeds breeds:* Produce and sell hybrid vegetable seeds for top vegetable products, including tomato, cucumber, aubergine, pepper, melon and rootstock.

- *Genuity:* A brand that describes the family of traits such as drought tolerance, cold tolerance, nitrogen use efficiency, yield enhancement.

- *Roundup Ready:* Seeds contained in-plant tolerance to Roundup agricultural herbicides with Bollgard & Yieldgard traits against insects.

- *Seminis:* Developer, grower, marketer of vegetable and fruit seeds.

- *Vistive:* New low-linolenic soybeans, soybean oil; *YieldGard+YieldGard VT:* in-plant protection traits against insects in corn.

- *Organisational Brands:* American Seeds, LLC: *in corn and seeds; International Seed Group, Inc* (ISG): an investment holding company that provides specialised, regional vegetable and fruit seed companies with access to capital technology.

- *Licensing:* through Monsant's Holden's/Corn States business, Monsanto

---

[219] Walt Disney World: Walt Disney World and Walt Disney World Resort, are oprated by Walt Disney Parks & Resorts division of The Walt Disney Company, in California, U.S. It is the most visited and largest recreational resort in the world with four theme parks, two water parks, 23 thermal hotels, and numerous shopping, dining, entertainment and recreational avenues.
[220] Svaldbard Global Seed Bank: Is a secure seedbank located on the Norwegian island of Spitsbergen near the town of Longyearbyen in the remote Arctic Svalbard archipelago, established to preserve a wide variety of plant seeds from locations worldwide in an underground cavern. It holds duplicate samples, or "spare" copies of seeds held in greenbanks worldwide, and insurance against the loss of seeds in greenbanks, and as a refuge for seeds in the case of large-scale regional or global crises.

licenses seed germplasm or biotechology traits to more than 250 seed partners in the U.S. Cotton States operates according to a similar model in cotton.

Monsanto practises aggressive sales, marketing, and political lobbying of its GE produce; the company also executes aggressive litigation against any and all who violate its agreements for use of its GE babies. This has made Monsanto controversial around the world and a primary target for the anti-globalisation movement and environmental activists. Monsanto and the biotech industry claim that *"no-one has been harmed by eating GE-foods"*. Sure, but without labelling of GE-foods and ingredients, there is no way to track any harm. Doctors and scientists warn that there is not enough evidence to insure that these GE-foods are safe in the human, animal and plant diet. Medical experts, including over 2000 doctors and health professionals in Europe and the *British Medical Association,* have questioned the safety of GE-foods, because there is ample evidence of risk as follows:

- *Allergies:* By inserting foreign DNA into common foods, without adequate safety testing, the biotech industry is introducing possible new food allergies.

- *Antibiotic Resistance:* The rise of diseases that are resistant to treatment with common antibiotics is already a serious medical concern. Doctors warn that the current use of antibiotic resistance genes in GE crops may add to this risk. In short, GE is an unpredictable technology that, for the sake of corporate profits, puts our environment and health to this risk.

But the various reasons why Monsanto is the Weeping Boy of Environmental Pollution and Foremost Food Manipulator and Contaminant are:

- *Pollution, Convictions: In Indonesia,* in a bid to disguise bribes as "consulting fees" to avoid environmental impact assessments on its GM cotton in Indonesia, Monsdanto bribed high-level Indonesian officials in Indonesia's Ministry of the Environment between 1997-2005 with over USD50,000, for which Monsanto settled later to pay USD1.5m in legal fines;

- *Pollution, Contaminated Sites: In the U.*K: Between 1965-1972, Monsanto paid contractors to illegally dump thousands of highly toxic wastes in south Wales' landfill sites, knowing their chemicals were liable to contaminate wildlife and people. The UK Environmental Agency (EA) said the chemicals were found to be polluting groundwater and the atmosphere 30 years after they were dumped. Brofiscin quarry, near Cardiff, one of the dumpsites, erupted in 2003 spilling fumes over the surrounding area, even as the local community was unaware that the quarry housed toxic wastes. A UK government report showed that 67 chemicals, including *Agent Orange* derivatives, *dioxins* and *polychlorinated biphenyls* (PCBs) exclusively made by Monsanto, were leaking from one unlined porous quarry that was not authorised to take chemical wastes, and groundwater had been polluted since the 1970s. According to the EA, it could cost more than 100 million British pounds to clean up the contaminated sites;

- *Superfund Contaminated Sites, Convictions: In the U.S*: During 1969, Monsanto dumped 45 tons of PCBs – a chemical belonging to the class of *persistent organic polluters* (POPs) into

Snow Creek, a feeder for Chocolocco Creek which supplies much of the area's drinking water. Monsanto also buried millions of pounds of PCBs in open-pit landfills located on hillsides above the plant and surrounding neighborhoods. [The New York Times, 27 Jan., 1969]. Monsanto was also accused of its legacy of environmental damage in Anniston, Alabama, where the local Monsanto factory discharged both *mercury* and *PCB*-laden waste into local creeks for over 40 years. [The Washington Post, 1 Jan., 2002 report]. In the Anniston and other lawsuits brought against Monsanto, Monsanto lost a series of court decisions in USD700 million in damages; although the PCB problem was settled in 1979, Monsanto failed to clear up the levels of PCB already in the natural population, until it was detected by the Federal Soil Detection Service, then partially dredged by Monsanto, but on 22 February 2002, Monsanto was again found guilty of *"negligence, wantonness, suppression of truth, nuisance, tresspass, and outrage."*; The Monsanto-founded town of Sauget, Illinois, for many years employed Sauget's people without providing them tax revenue in return, polluted its environment, but avoided taxation from East Saint Louis even during the city's decline throughout the latter half of the 20th century. The Environmental Protection Agency (EPA) sued Monsanto for contaminating 56 sites or *Superfund sites*, and Monsanto has settled multiple times for damaging the health of its employees or residents near its Superfund sites through pollution and poisoning. The "Kemner vs Monstanto" case (1984-Oct., 1987), where plaintiffs sued Monsanto for poisoning them with *dioxin* in Sturgeon, Missouri. In 2000, GLC sued Monsanto for its USD71 million shortfall in expected sales; *In India:* In 2005, Andhra Pradesh State's Agriculture Minister, following a GE-Approval Committee, the Indian regulatory Authority's release of a fact-finding statement, barred the sale of Monsanto's *bacillus thuringiensis* (Bt) cotton; Andra Pradesh state government later filed several cases against Monsanto and its Mumbai-based licensee *Maharastra Hybrid Seeds,* after Maharastra Hybrid changed the state's order directing the company not to charge a *trait* price of more than Rs. 900 per pack of 450g of *Bt. Cotton* seed.

- *Monsanto Litigates Aggressively:* Since the mid-1990s, Monsanto has sued some 150 U.S., farmers for patent infringement in connection with its GM seeds and technology agreement violations. Monsanto's contracts with farmers prohibit farmers from saving *Terminator* seeds from one season's crop to plant in the next season. Widespread opposition against Monsanto and Terminator technology have been mounted around the globe by environmental organisations and farmer associations: *they are concerned that Terminator seeds increase farmers' dependency on seed suppliers, especially in the third world.* Another *concern is that the "Terminator Effect" will be spread to native vegetation through pollination, rendering all plants unable to reproduce.* But Monsanto's consumers must continue to buy new seed each season or risk lawsuits for violating its patent, which are supported by the U.S., government under the *free-trade considerations.* Monsanto's large market share over the seed buying market only leaves a choice between Monsanto or its top competitor, *DuPont.* In 2004 also, Switzerland's largest agrichemical company *Syngenta,* charged Monsanto with using coercive tactics to monopolise markets, along with suits and counter-suits between the two GM-Giants. On 20 September, Monsanto stood trial in Carcassonne, France, for allegedly illegally importing 100 tonnes of soya seed contaminated with GM varieties – 50 tonnes of which were sold to local farmers – and 50 tonnes returned to the U.S. Monsanto reports that it pursues approx. 500 cases of suspected infringement annually: in 2003, Monsanto sued Oakhurst Dairy in Maine, for advertising that its milk products "did not come from cows treated with bovine growth hormone" by claiming that "such advertisement hurts its business." After Monsanto's patented genes appeared in canola crop on Canadian Percy Schmeister's 40-year old farm, Monsanto sued against patent infringement for "growing GM Roundup-resistant canola," and the Canadian Supreme Court in 2004, ruled in favor of Monsanto Canada, Inc., but awarding no damages to Monsanto. To determine whether Argentina uses black-market-produced "Roundup"soymeal, Monsanto petitioned the Spanish government to "inspect" soymeal seed shipments to Argentina, and change the royalty collection system so that royalties are collected at harvest, rather than upon purchase of the seed.

- *Monsanto Forces Contracts Aggressively:* Monsanto is frequently described by farmers as *"Gestapo"* and *"Mafia"* both because of its frequent resort to lawsuits and the questionable means it uses to collect evidence of patent infringement. The U.S., Center for Food Safety (CFS), a non-profit organisation, has listed 112 lawsuits litigated by Monsanto against farmers for claims of patent violations. Many innocent farmers settle with Monsanto because they cannot afford a time-consuming lawsuit. Monsanto is known for showing up at farmers' houses, making accusations and demanding records. As of May 2008, Monsanto engaged in a campaign to prohibit dairies which do not inject their cows with its synthetic *bovine growth hormone* from advertising this fact on their milk cartons. Monsanto claimed that labelling of hormone-free milk takes advantage of consumers by allowing higher prices for milk suggesting that it is "better" or "safer" than *BTS milk,* when in fact, there is "no difference." Monsanto started lobbying U.S., State lawmakers to implement a similar ban, after the U.S., Federal Trade Commission refused to side with Monsanto on this issue. Pennsylvania State Agriculture Secretary Wolfe attempted to support Monsanto's ban campaign; but Wolfe's attempt to prohibit dairies from using labels that state that "their milk does not contain artificial bovine growth hormone", backfired. In 2002, Monsanto sued Gary Rinehart of Eagleville, Missouri, for claiming Rinehart "violated" their Roundup Ready (RR) soybean patent, although Rinehart was no farmer or seed dealer. Monsanto sued the Pilot Grove Cooperative Elevator in Piulot Grove, Missouri, claiming that Pilot Grove's offering services to farmers was tantamount to inducing farmers to pirate Monsanto's seeds, even though Pilot Grove had been cleaning seeds for decades before companies such as Monsanto could patent organisms. In 1997, two U.S., Fox News TV broadcaster reporters were fired for "cooperating with Monsanto in suppressing an investigative report" on the health risks associated with Monsanto's bovine growth hormone produce Posilac.

- *Toxicity of GE Produce: Toxicity, Human, Animal,* and *Plant Health; Saftey,* and *Ecosystems' Contamination* and *Pollution by GE-genes* are the foremost issues at the centre of the controversy that surrounds GE-food products. The risk and effect of *horizontal gene transfer* (HGT) or *lateral gene transfer* (LGT) are great concerns, with the possibility that genes might spread from modified crops to wild relatives; Greenpeace and WWF consider the available data on GMOs and GE food products not sufficient enough to prove that these laboratory-chemically-produced foods and products do not pose risks to health. Both call for additional and more rigorous testing before marketing of these foods. Pest- and herbicide-resistant GMO crops could reduce the numbers of pest insects in farmlands and impact biodiversity, or decrease the use of insecticides; GE-fish could escape aquaculture facilities into waterways and destroy wild species. The European Union (EU) has adamantly resisted authorising the marketing or production of GE-foods and products in the EU-Zone, in contrast to the U.S., where GE-foods and products are allowed amidst strong protests. The 2003-2004 EU legislation against GMOs and GE-foods and products called for *strict rules on labelling, traceability and risk assessments by all biotech companies, GMOs and all products produced from GMOs, mandatory for GMO ingredients.*

- *Child Labor, Seeds of Suicides:* In India, a subsidiary of Monsanto employs child labor in the manufacture of cotton-seeds: kids handle polsonous pesticides such as *endosulfan,* an organochlorine compound used as an insecticide or *acaricide* or *miticide,* for less than Rs. 20 (half a dollar) per day wages; Indian farmers have been lured to GM seeds promoted by Cargill and Monsanto with the promise of greater profits. Monsanto's GM seeds are subcontracted to a joint venture *Mahyco Monsanto Biotech* (MMB) India Ltd, which markets its BT Cotton seeds under various names. The result is farmers' accumulation of mounting debts from GE seed poor yields, increased need for pesticides, and higher cost of Bt cotton seed. This has led some farmers into the equivalent of indentured servitude and to more than 450 farmer-suicides. In July 2004, India's PM Manmohan Singh promised to set aside federal-budget funds to aid struggling farmers and families of suicide victims: each struggling farming household was to receive Rs. 150,000 (USD3,400), and families mourning a relative who committed suicide was to receive Rs. 50,000 (USD1.136). At the time PM Singh made his promise to help, more than 3,000 farmers had committed suicide.

Thousands of products, particularly in U.S., supermarket shelves, contain GE ingredients. Anti-GE food campaigns have convinced some prominent supermarkets to now engage the initiative to refuse to stock GE-foods and to purge off their shelves those GE-foods which have already been stocked. Many national supermarket chains which stocked GE-foods and even adamantly refused to discriminate against GE-foods and stocked them, have now agreed to cooperate. Supermarkets such as *Whole Foods, Wild Oats, Trader Joe's, HEB, Albert's,* etc., have now agreed to clean their stocks free of GE-foods.

According to the Center of Food Safety (CFS), supermarkets are strategically important for market intervention against GE-foods because:

- *Supermarkets* control a large amount of food production through their store brand products. In the U.S., store brands typically account for 25-40% of supermarket sales. Store brands are also the supermarkets' way of gaining customer loyalty. Supermarkets are the part of the food industry that has the most exposure and immediate accountability to the public, and also have the *least to gain* from GE-foods.

- *Supermarkets* must answer to their customers and the public, and they must maintain a good public profile and trust. The supermarket industry is highly competitive and the profit margins are relatively slim, so any threat of losing customers, sales or image, is a serious worry.

- *Supermarkets* are great places to engage people: most people do the majority of their shopping during the weekends when they have a little more time. This allows you the time to talk with them about GE-food; people are already thinking about food and making decisions about their purchases so what you say has immediate relevance; you are at the magical "point". Food companies and advertisers pay a lot of money to be where you are – use it wisely; supermarkets are public, high-volume, mainstream and accessible – even better than a street!

- *Why Supermarkets Campaigns are Effective:* Supermarkets watch their bottom-line and the movement of their sales very closely. Again, this is a highly competitive industry with narrow profit margins – every dent will be felt; Supermarkets protect their brands – they cannot afford to lose their trust. If their store brands are thought of as lower in quality, so too, will be their store; Supermarket managers pride themselves on responding to the needs of their immediate community of customers. Most store managers also receive bonuses based on their store sales. Supermarkets strive to be seen as the consumers's friend and they want to build up a relationship with their customers (hence loyalty cards, special points, store coupons on receipts, etc.); Supermarkets watch their competitors closely: if one moves on an issue and does well, they will follow suit.

4.5 Conclusion

THE BATTLE to control food production by the biotech industry and flood the world with genetically engineered foodstuffs is raging and may rage for generations to come. Coupled with climate change and global warming, the trendy-slide toward being fed and stuffed with genetically-manipulated food is forced on human beings, animals, plants and environment. Not health or safety but our corporate profits matter more. Our transportation, agriculture and fuel industries, principal players in the latest and catastrophic *food and oil crisis,* are playing partnership with the gene food industry.

4.6 General Conclusion

OUR PLANET's economies, livelihoods, forests, ecosystems' diversity and waterways are gradually but surely being squeezed to death and extinction by deliberate but *preventable acts of human profligacy* through:

- *roduction and use of Fossil fuels, petroleum products and derivatives and their* pollution and contamination of air, water and soil; *Food additives, preservatives, colorings'* poisoning; *Synthetic plant and animal growth hormons'* contamination and poisoning; *Antibiotic-resistant bacteria to medication;* Sponsoring and encouraging *irreversible climate change and global warming* catastrophes; Funding and coercing the *intrusive gene manipulated plants, crops and food Pandora-Box dilemma;* and funding and coercing the *escalating global malnutrition, hunger, poverty, and unemployment* staples.

- *We stand to lose everything, if nothing is done.*

- *We stand to lose everything, if half-hearted, lip-serviced measures are applied.*

- *The comfort is that, not all is lost yet.*

- *The discomfort is that, we are losing more faster due to the inadequate, inappropriate* and *ineffective* measures applied to combat the total disintegration of the Earth.

The battle to *reverse motion to prevent, stop and stabilise the Earth from total disintegration,* must be prosecuted *stock, barrel and bullet,* through:

- *Revitalised and regenerative methods of agriculture and afforestation;*

- *Increasing use of renewable energies;*

- *Improving methods of intelligent transportation and communication;*

- *Increasing action to produce and use less and lesser fossil fuels, petroleum products and derivatives.*

- *Monsanto* and the *GE-biomonsters* must be fought tooth-and-nail to preserve biodiversity, ecosystems, safe and healthy food, and cleanup the heavily polluted environments.

4.7 Sources & Acknowledgments

II. Introduction: Apocalyptic Happenings

1.
European Commission: *"Mitigating the effect of rising global food prices."* 18/07/2008 – Global Price Rise: Commision proposes Special Financing facility worth €1 billion to help developing country farmers. EC Agriculture and Development/Food Prices...

2.
U Biofuel Directive 2003.

3.
Intergovernmental Panel on Climate Change (IPCC): *"Climate Change 2007: Synthesis Report"*. http://www.ipc.ch/assessment-report/art4/syrart4-syr.pdf; IPCC-Sec@wmo.int.

4.
International Assessment of Agricultural Science & Technology for Development (IAASTD): *"IAASTD Agriculture Questions & Answers"*. IAASTD http://www.agassessment.org/

5.
artin Ahola: *"Giant multinationals squeezing out millions of small-scale producers. The market is just too narrow and we need to open it up more fairly to small-scale producers."* ActionAid/South Centre, Accra, Ghana.

Chapter 1

RENEWABLE ENERGIES AND FOOD PRICES

6.
Hydrocarbons.

7.
known Existing 118 Elements in Nature, their Atomic Numbers, Symbols and Atomic Weights.

8.
The 48 Metal elements and their Symbols.

9.
Non-metals.

10.
Sunlight.

11.
Wind.

12.
Akira Yanagisawa: *"Institute of Energy Economics"*. 3/088, http://eneken.ieej.or.jp/en/data/pdf/421, p. 13.

13.

Michelle Foss: *"Oxford Institute for Energy Studies Working Paper",* 2/07.http://www.oxfordenergy.org/p.34.

14.

Larry Chorn, Business Week, 5/13/08.

15.

U.S., Senate Permanent Subcommittees on Investigations: Staff Report, 606. http://levin.senate.gov/newsroom/s

16.

Clarence Cazalot, Marathon Oil: CNNMoney, 11/1207, http://money.cnn.com/2007/11/12/markets/oil_hundred? Postversion=2007111216.

17.

Stephen Simon: Home testimony, 14/1/08.http://global.house.gov/tools/assets/files/0453.pdf. .

18.

John Hofmeister: Financial Post, 5/22/08. http://www.financialpost.com/reports/oil-watch/story.html?=532747.

19.

Gerry Ramm: Senate Testimony. http://www.commerce.senate.gov/public/-files/RammSenateCommerce 06308Testimony.pdf.

20.

Centre for Research on Globalisation. Editor: Prof. Michel Chossudovsky. Globalresearch.ca; crgeditor@yahoo.com.

21.

U.S. Government's Commodity Futures Trading Commission (CFTC), established 1974.

22.

Ibid.,U.S., Permanent Subcommittee on Investigations: Staff Report 27 June 2006; and House committee on Energy & Commerce's Hearing, December 2007.

23.

U.S., Senate Report: *"The Role of Market Speculation in Rising Oil and Gas Prices."* June 2006.

24.

World Petroleum Congress 19th (WPC) Meeting, Madrid, Spain, June 2008: *"A World in Transition: Delivering Energy for Sustainable Growth."* http://www.19wpc.com/find-newsletter.pdf; info@19wpc.com.

25.

WPC 19th Congress Technical Programme.

26.

Harry Tchilingguirian, Senior Oil Market Analyst, BNP Pariba – Commodity Derivatives: *"Oil Prices down $100: Funds or Fundamentals?"*

27.

The Organisation of Petroleum Exporting Countries (OPEC), founded in 1950.

28.

U.S., Sub-prime Crisis.

29.

H.E. Abdalla Salem El-Badri, OPEC Secretary General: *"Meeting the Challenges in the international oil market."* An OPEC/IEA Luncheon Address to the 19th WPC, Spain, 29 June –3 July 2008.

30.

International Energy Agency / Information Energy Agency (IEA), established in 1974.

31.

Jean-Yves Garnier, head of Energy Statistics Division, IEA: *"The Joint Oil Data Initiative: The 5th JODI Conference."* Paris, 18 Nov., 2004.

32.

U.S., Democratic Presidential Campaign for the White House 2008: *"Obama Release on 'Enron Loophole' : Obama Announces Plan to Fully Close the Enron Loophole,Crack Down on Excessive Energy Speculation."* The Page-Politics upto the Minute by Mark Halperin, Sunday 22 June 2008. TIME/CNN.

33.

U.N. Food and Agricultural Organisation (FAO): *"High-Level Conference on World Food Security 2000: The Challenge of Climate Change and Bioenergy."*

34.

Colin J. Campbell: *"OIL CRISIS."* http://healthandenergy.com/oil.crisis.htm.

35.

International Panel on Climate Change (IPCC): *"Climate Change 2007. Synthesis Report."* http://www.ipcc.ch/assessment_report/art4/syr.pdf;IPCC_Sec@wmo.int.

36.

Oil & Petroleum Companies.

37.
U.N. Framework Convention on Climate Change (UNFCCC): *"Objectives & Principles."* UN 1992, GE.05-62220 (E) 2007.GHGdata@unfccc.int; secretariat@unfccc.int.

38.
Clean Technology Funds (CTF).

39.
Global Environmental Facility (GEF), established 1991. Spokesperson: Ms Maureen Shields Lorenzetti, mlorenzetti@thegef.org; www.gefweb.org.

40.
Persistent Organic Pollutants (POPs).

41.
Group of G8 Nations.

42.
"Environment: World Leaders avoid setting greenhouse gas target." TIME/CNN, Business/07//09/gas.summit/index.html.

43.
U.N. Global Renewable Energy Forum (GREF):*"PROPOSAL FOR AN IPCC SPECIAL REPORT ON RENEWABLE ENEERGY."* Brazil 18-21 may 2008.

44.
"WORLD OIL OUTLOOK 2008." OPEC's Executive Summary, ISBN-938-3-200-01253-0 OPEC Secretariat, Obere Donaustrasse 93, A-1020 Vienna, Austria; www.opec.org.

45.
Precious Metals.

46.
Gold Standard.

47.
Bretton Woods [Monetary] System.

48.
Report Buyer: *"Analysing the Major Gold Mines Worldwide."* Aug. 2007, Avurian Research, 54 Maltings Place, London, UK; service@reportbuyer.com.

49.
AME Mineral Economics – Gold: *"List of Campanies with Ownerships in Gold Operations."* AME House, 342 Kent ST, Sydney NSW 200, GPO Box 362, Sydney NSW 200, Australia. http://www.ame.com.au/companies/Au/campanies. ame@ame.com.an.

50.
Dan Neil, Pulitzer Prize-winning automative critic and syndicated columnist for the Los Angeles Times and TIME, taking a look at the greatest lesson of the automotive industry on the 5oth anniversary of the Ford Edsel: *"The 50 Worst Cars of All Time – 1899-2007."* Monday 05 Nov., 2007, TIMR/CNN.

51.
TDM Encyclopedia: *Transportation Costs & Benefits; Resources for Measuring Transportation Costs and Benefits."* 22 July 2008, Victoria Transport Policy Institute (VTPI), Canada.

52.
Automobile Associations.

53.
International Energy Agency & International Information Agency (IEA) & Co.: *"The Value of Travel Time: A Review of British Evidence."* Journal of Transport Economics & Policy, vol. 32, No. 3, Sept 1998, pp. 285-316.

54.
Wardman, 1998: *"The Value of Travel Time: A Review of British Evidence."* Journal of Transport Economics & Policy, vol. 32, No. 3, Sept., 1998, pp. 285-316.

55.
TRB, 1997: *"Quantifying Congestion."* TRB (www.trb.org), NCHRP Project 7-13.

56.
Bureau of Transportation Statistics (www.bts.gov) & Co.

57.
USEPA Transportation Air Quality Center (www.epa.gov/otag), & Co.

58.
AGU, 1998: *"Climate Chamnge and Greenhouse Gases."* American Union (www.agu.org).

59.
USEPA, 1999: *"Indicators of EnvironmentalImpacts of Transportation."* Office of Policy & Planning, USEPA (www.itre.ncsu.edu/cte) & Co.

60.
Green/Tischishya, 2000: *"Costs of Oil Dependence: A 2000 Update."* Oak Ridge National Laboratory, ORNL-TM-2000/152 (www.osti.gov/bridge/

61.
Robert Burchelle, et al., 1998 : *"The Costs of Sprawl-Revisited."* TCRP Report 39, Transportation Research Board (www.trb.org).

62.
Richard Untemann/Arne Vernez Moudon, 1989: *"Street Design Reassessing the Safety, Sociability and Economics of Streets."* University of Washington at Seattle, U.S.

63.
Equity.

64.
Thomas W. Sanchez/Marc Brenman, 2007: *"The Right To Transportation:Moving To Equity."* Planners Press (www.planning.org).

65.
Elmer Johnson, 1993: *"Avoiding the Collision of Cities and cars."* American Academy of Arts & Scineces (Chicago, U.S.); Institute for Science & Technology Policy, Murdoch University (http://wwwistp:murdoch.edu.au).

66.
Tasso Adamopoulos: *"Transportation Costs, Agricultural Productivity and Cross-Country Income Differences."* Abstract and Introduction, Nov., 2005, York University, 4700 Keele ST, Toronto, Ontario, Canada, M3JIP3 (www.york.ca).

67.
Haber-Bosch-Process or *Nitrogen Fixation Reaction*.

68.
Ammonia (NH^3) gas occurrence in nature.

69.
International Fertilizer Industry Association (IFA), 28, rue Marbeuf, 75008 Paris, France; http://www.fertilizer.org.

70.
International Fertilizer Industry Association (IFA): *"What are some common fertilizer products and intermediates?"*

71.
Fertilizer and Pesticide Organisations and their functions.

72.
Natural Resources Defense Council (NRDC): *"ISSUES: WATER: Sewage Pollution Threatens Public Health: Ageing sewer systems and rollbacks of the environmental law are compounding the problem."* webmaster@nrdc.org.

73.
Original TAP Databases Form 1995: *"Aquatic Plant Extracts"* from NOSB National Database.

74.
K. R. Baldwin, 2001: *"Soil fertility for organic farming."* http://www.ncsu.edu/organic_farming_systems/news/.

75.
B. Hall/P. Sullivan, 2001: *"Alternative Soil Amendments: Appropriate Technology Transfer for Rural Areas."* http://attra.ncat.org/attra-pub/PDF/altsoil.pdf.

76.
Organic Trade Association, 2002: *"Comarative Analysis of U.S., National Organic Programme (7CFR 205) and the EU Organic Legislation (EEC2092/9) & Amendments."* http://www.ota.com/pics/documents/NoPEUunidiedreport.

77.
Eric Henry, 2005: *"Report of Alkaline Extraction of Aquatic Plants."* http://www.omri.org/AdvisoryCouncil/Aquatic.

78.
Nitrogen.

79.
CropLife International (CI).

80.
Types of Pesticides.

81.
International Fertilzer Industry Association (IFA) Key sources.

82.
Enzyme Growth Inhibition.

83.
Nitrous Oxide or *'Laughing Gas'*.

84.
American Medical Association (AMA): *"Pesticide-Related Illnesses Associated with the Use of a Plant Growth Regulator- Italy, 2001."* JAMA, 2001, 286 pp., etc.

85.
Pesticide Poisoning.

86.
Eutrophication.

87.
Vegan Peace *"Farm Sanctuary"*.

88.
Wikipedia, the free encyclopedia: *"Agribusiness."*

89.
CropLife International (CI): *"What is 'agbiotechnology'"?*

90.
Michael Grunwald: *Obama's Agenda: Get America Back On Track: First Things First."* TIME's Commemorative Issue: President-Elect Barack Obama, 17 Nov., TIME/CNN. http://www.time.com.

91.
U.S. Invasion of Iraq, 20 March 2003.

92.
Afghanistan War, 7 Oct., 2001.

93.
The Wall Street, New York.

94.
Weak Dollar.

95.
Dollar Index (USDX).

96.
AP Writer Jean Averson on U.S. Unemployment: *"Jobless Ranks Hit 10 million, 25-year High."* TIME/CNN, 07 Nov., 2008.

97.
Mark Halperin: *"More from the November 6-7 CNN Poll."* CNN//Opinion Research Corp Data. The Page – Politics up to the Minute, Tues., 11 Nov., 2008.

98.
Emergency Economic Stabilisation Act 2008: *"Bailout of the U.S., Financial System."*

99.
World Governments' Currencies.

100.
Market Economy & Deregulation.

101.
Recession.

102.
Joseph Stiglitz: *"Nobel Laurete: How to Get Out of the Financial Crisis."* TIME/CNN, Saturday 18 Oct., 2008, 3 pp.

103.
The White House, Office of the Secretary, G-20 Saturday Statement After World Summit: *DECLARATION ON FINANCIAL MARKETS AND THE WORLD ECONOMY."* Preamble of the Initial Meeting of the Group of 20 held in Washington on 15 Nov., 2008.

104.
Michael Grunwald: *"Obama's Agenda: Get America Back on Track: First Things First."* 17 Nov., 2008, TIME/CNN Commemorative Issue: President-Elect Barack Obama.

105.
Eliot Spitzer: *"How to Ground the Street."* Washington Post, 16 Nov., 2008, p. B01. http://www.washingtonepost.com/wp-dyn/content/article/2008/11.

106.
Monetary Inflation.

107.
Steve H. Hanke, Prof. Applied Economics, The John Hopkin's University/Senior Fellow, The Cato Institute: *"New Hyperinflation Index (HHIZ) Puts Zimbabwe Inflation at 516 Quintillion Percent."* Cato Institute, 100 Massachusetts Ave., N.W. Washington D.C., 2001-5403; http://www.cato.org/zimbabwe; MeriNews, India's First Journalism Portal: *"Inflation in Zimbabwe Breaks all records."* & Co.

108.
Lesley Wroughton: *"IMF Warns of Deepening Recession in Rich Countries."* Reuters, 8 Nov., 2008. http://www.reuters.com/businessnews/news.

109.
IMF World Outlook Growth Projections, Oct., 2008.

110.
International Labour Organisation (ILO): *"Global Financial crisis To Increase Unemployment by 20 Million."* ILO News, ILO/08/45/Press release, 20 Oct., 2008.

111.

Marcy Gordon: *"Feds Take Over Two Failed Banks."* TIME/CNN, Business/Tech, 22 Nov., 2008.

112.

Aaron Smith, CNNMoney.com staff writer: *"Job cuts mount as year-end nears. AT&T, DuPont, Viacom, Credit Suisse and others take another 22,000-plus jobs out of the workforce."* TIME/CNN Business/Tech 04 Dec., 2008.

113.

Adam Smith: *"The Credit Crisis Spreads to Europe."* TIME/CNN, Business/Tech, 11 Aug., 2008.

114.

Unemployment Welfare.

115.

Unemployment Classified.

116.

Laura McInnis, Reuters: *"Global Unemployment Rate to Climb in 2008."* Reuters, 24 Jan., 2008.

117.

The U.S., Central Intelligence Agency (CIS): "World Fact Book: *Rank, Order-Unemployment Rate."* Covering most of the countries of the world, updated 06 Nov., 2008.

118.

Diarmid Campbell-Lendrum/Carlos Corvalon: *"Climate Change and Developing-Country Cities: Implications For Environmental Health and Equity."* Department of Public Health & Environment, WHO, Geneva, 27, Switzerland; campbellendrumd@who.int.

119.

Lourdes Salvatore: *"Environmental Pollution Costs Billions in Illnesses Each Year."* American Chronicle, 28 Nov., 2008.

120.

Volatile Organic Compounds (VOCs).

121.

U.S. Department of Commerce/National Oceanic & Atmospheric Administration (NOAA): Research, Earth System Research Laboratory, Global Monitoring Division (GMD): *"Annual GHGs by Sectors."* & Co.

122.

U.S. Environmental Protection Agency (EPA): *"EPA Greenhouse Gas Emissions: 2008 Inventory of Greenhouse Gas Emissions & Sinks Report."* Presented annually since 1990 through 2005.

123.

EPA Voluntary Programme.

124.

Bryan Walsh: *"The World's Most Polluted Places."* 12 Sept., 2007. TIME/CNN.

125.

Margie Mason, Associated Press (AP) Reporter: *"World's Highest Drug Level Found in India Stream."* Sunday, 25 Jan., 2009, quoted by TIME/CNN: World.

126.

Regenerative Agriculture.

127.

Abu Dhabi: *"The Masdar Initiative."* The Government of Abu Dhabi, United Arab Emirate (UAE).

128.

Coco Masters: *"A Japanese Town That Kicked the Oil habit."* 22 Dec., 2008, TIME/CNN.

129.

Alternative Currency.

130.

Primitive Technology.

131.

David Westcott: *"Primitive Technology."* Gibbs-SmithPublishers. Amazon.com.

132.

David Stiles: *"Rustic retreats: A Build-It-YourselfGuide."* Storey Publishing, Amazone.com; etc..

133.

Rob Roy: *"The Complete Book of Underground Houses: How To Build a Low-Cost House."* Sterling Publication Co., Inc., New York. Amazon .com.

134.

Elaine Aubert: *"Keeping Food Fresh: Old World techniques & recipes."* Amazon.com.

135.

Altruism.

136.

Deborah Madison: *"Preparing Food Without Greasing or Canning: Traditional TechniquesUsing Salt, Oil, Sugar, Alcohol,Vinegar,Cold Storage & Lactic Fermentation."* Gardeners and Farmers of Centre Terre Vivante.Amazon.com.

137.

Eliot Coleman/Barbara Darnoush: *"For Season harvest Organic Vegetables from Your Home Garden All Year Long."* Amazon.com

138.

Suzanne Ashworth: *"Seed to See: Saving & Growing Techniques."* Amazon.com.

139.

Gail Damerow (ed.): *"Banyard in Your Backyard: A Beginner's Guide to Raising Chickens, Ducks, Rabbits, Goats, Sheep, and Cows."* Amazon.com.

140.

Steve Brill: *"Identifying and Harvesting Edible and Medicinal Plants In Wild* (And Not So Wild) *Place."* Amazon.com.

141.

Carbon Dioxide Information Analysis Center (CDIAC): *"CDIAC 2004 Data Collected for the UN on CO² Emissions by Sovereign States."*

Chapter 2

RENEWABLE ENERGIES AND FOOD PRODUCTION

142.

Hormones & Plant Growth Agents.

143.

Dr. George Washington Carver: *"How the Farmer can save his Sweet Potatoes and ways of Preparing them for the Table."*4th ed., Tuskegee Institute, U.S.

144.

Field Extension Practitioners.

145.

Dr. Booker T. Whatley: *"Handbook on how to make $100,000 Farming 25 acres: With Special Plans for Prospering on 10 to 200 acres."* Ed. George De Vault. Emmaus, P.A., Regenerative Agriculture Association, U.S.

146.

Dr. Christine E. Jones *"Regenerative Land Managment : a whole lot of landscape approach to the regenerationof water balance and water quality."* Australia.

147.

International Assessment of Agricultursl Knowledge, Science and Technology (IAASTD): *"Agricultural Questions & Answers."*

148.

Dan Sullivan, (IAASTD): *"Global Report."* Johannesburg, South Africa,7 April 2008.

149.

Rodale Institute: *"Leaders in Organic Solutions since 1947."*

150.

International Organic Agricultural Movement (IFOAM): *"Building Sustainable Organic Sectors Report."* Bonn,Germany.

151.

International Cooperative Alliance (ICA): *"ICA Global Ranking by Turnover USD (2000 Data)."*

152.

Larry Harrington, Natural Resources Group – Cento Internacional de Mejoramiento de Maizy Trigo (CIMMYT): *"Diversity by Design."* Consultative Group on International Agricultural Research (CGIAR) News, vol. Number 3, June 1997.

153.

Dr. Romen Corsini, Idealizer of the Integrated Mini-Sugar Plant (GERIPA) Concept, sponsored by FAPESA, & CO.

154.

Bagasse.

155.

Judith D. Schwartz: *"Alternative Currencies Grow in Popularity."* 14 Dec., 2008, TIME/CNN.

156.
Community Exchange System (CES).

157.
U.N. International Local Employment Trading System (UNILETS).

158.
BBX:BBX, phone 02 9622 9000; Fax 02 9622 9011 Australia; chris@bbxaustralia.com.au

159.
Barter Card or Bartercard: Founded in 1979, a non-profit IRTA member.

160.
Stephen DeMeulenaere: *"2007 Micro Grant Fund for Complementary Currency Systems in Asia, Africa and Latin America."*

161.
Access Foundation: *"First International Complementary currency Summit: Expanding the Impact of new monetary innovations worldwide."* 31 July-5 Aug., 2005.

162.
Betina Corke: *"Global Resource Bank. (GRB)."*

163.
Robin Upton/Clive Beresford: *"Modern Money is Zero-Sum; Re-establishing altruism as a variable social norm Altruism Community Working for Love, Not Money."* Altruism International (AI).

164.
Altruism.

165.
GNU/Linus Project.

166.
Global Renewable Energy Forum Bulletin.

167.
William Moomaw: *"Renewable Energy and Climate Change: An Overview."* The Fletcher School, Tufts University, U.S.

168.
Samuel Asfaho, South Centre Geneva: *Suggestions to Tackle the Commodities Problems."* Executive Summary. International Alliance Against Hunger (IAAH) launched by the U.N. 16 Oct., 2003.

169.
Michael W. Masters: *"Energy Speculation Causes Fuel Inflation."* 22 June 2008, Grand Rapids, MI, U.S.

170.
Bill Mollison & David Holmgren.

171.
David Holmgren: *"Permaculture: Principles and Pathways Beyond Sustainability."* Holmgren Design Services, 16 Fourteenth ST, Hepburn Victoria, 3461, Australia; info@holmgren.com.au; http://www.holmgren.com.au.

172.
World Bank: *"World Bank 2007 Global Economic Prospects: Managing the Next Wave of Globalisation."* Press Release No. 2007/159/DEC., 13 Dec., 2006.

173.
The World Bank: *"Poverty At A Glance: New Estimates."* 1990-2009. The World Bank, Washington, DC, U.S.

174.
Micro-loans.

175.
Prof. Muhammad Yunis, Nobel Prize Laureate, 2006, Founder of Grameen Bank, Oct., 1983.

176.
U.N. International Fund for Agricultural Development (IFAD): *"IFAD's Strategic Framework for 2007-2010: Enabling the Rural Poor to Overcome Poverty."*

177.
Guardian.co.uk.: *"UN to buy surplus crops from small-scale farmers."* Xan Rice, Nairobi, Kenya. 24 Sept., 2008.

178.
Julian Borger/Juliette Jowitt : *"Nearly a billion people worldwide are starving, UN agency warns: Rising prices mean 14% now undernourished; Urgency over food crisis lost amid credit crunch."* The Guardian, London, 10/12/08.

179.
La Via Campesina, International Peasant Movement: *Small-scale sustainable farmers are cooling down the earth."* International Operative Secretariat (IOS) JS Mampang Prapat, Indonesia.

180.
Foods To Avoid.
181.
Earth Negotiations Bulletin.
182.
Greenpeace International: *"Greenpeace Blacklist: Official Blacklist & Official Blacklist of Companies."*
183.
2008 G-20 Washington Summit/APEC Peru 2008/2009 G-20 London Summit: *Financial Crisis of 2007: Late 2008 Recession."*
184.
Associated Press (AP): *Natural disasters 'killed over 220,000' in 2008."* 29 Dec., 2008.

Chapter 3:

RENEWABLE ENERGIES AND FOOD PRICES

185.
L ist of Worldwide Largest and Smaller Selected Photovoltaic (PV) Power Stations/Plants by Country.
186.
Photovoltaics
187.
List of Worldwide Solar Thermal Power stations.
188.
Convection.
189.
Cristina L. Archer/Mark Z. Jacobson: *"Evidence of global windpower."* Abstract. Journal of Geophysical Research-Atmospheres in 2005; stanford.edu; lozeji@stanford.edu.
190.
Worldwide Installed Windpower Capacity & List of Turbine Manufacturers
191.
List of Offshore Wind Farms that are Operating or Under Construction Worldwide.
192.
Tectonic Plate Boundaries.
193.
Installed Geothermal Electric Capacities, 2007.
194.
Carbon Cycle.
195.
Anthropogenic Effects.
196.
Lignocellulosic Ethanol.
197.
Anaerobic Digestion (AD).
198.
UN Global Bioenergy Partnership (GBEP).
199.
Water or Hydrological Cycle.
200.
Worldwide Major Hydroelectric Schemes Under Construction with 2000 MW & Proposed Ones.
201.
Types of Fuel Cells.
202.
Battery Cells.
203.
Carbon Allotropes.
204.
Quantum Dots.
205.
Early Computers.
206.
Autonomous or Adaptive Cruise Control (ACC) Vehicles Samples.
207.
Scanning Probe Microscopies Technology.

208.
   Micronutrients, Macronutrients and Vitamins.

Chapter 4

FOOD MANIPULATION, BIOTECHNOLOGISATION

209.
   Anthocyanins.
210.
   Caffeine.
211.
   Bryant Walsh : *"Getting Real About the High Price of Cheap Food."* Friday, 20 Aug., 2009, TIME/CNN.
212.
   Paul L. Marino: *The rule of anitbiotics is, try not to use them, and the second rule is, try not to use too many of them."* The ICU Book.
213.
   Cis and Trans.
214.
   The True Food Network: *"Myths & Realities of GE Crops."* A Project of the Center for Food Safety. Center for Food Safety (CFS). office@centerforfoodsafety.org.
215.
   Human Genome Project (HGP).
216.
   Luke Anderson, Greenpeace GE Campaign: *"Genetically-Engineered Fish – New Threats to the Environment."* 05 Jan., 2005.
217.
   Monsanto Company. www.monsanto.com.
218.
   Walt Disney World.
219.
   Svaldbard Global Seed Bank, Norway.

www.ingramcontent.com/pod-product-compliance
Lightning Source LLC
Chambersburg PA
CBHW081055170526
45166CB00006B/2074